SPECIAL FUNCTIONS for Engineers and Applied Mathematicians

LARRY C. ANDREWS
University of Central Florida

MACMILLAN PUBLISHING COMPANY
A Division of Macmillan, Inc.
NEW YORK

Collier Macmillan Publishers
LONDON

For Louise

Copyright © 1985 by Macmillan Publishing Company
A division of Macmillan, Inc.

All rights reserved. No part of this book may be reproduced or transmitted in any form or by any means, electronic or mechanical, including photocopying, recording, or by any information storage and retrieval system, without permission in writing from the Publisher.

Macmillan Publishing Company
866 Third Avenue, New York, NY 10022

Collier Macmillan Canada, Inc.

Printed in the United States of America

printing number
1 2 3 4 5 6 7 8 9 10

Library of Congress Cataloging in Publication Data

Andrews, Larry C.
  Special functions for engineers and applied mathematicians.

  Bibliography: p.
  Includes index.
  1. Functions, Special.   I. Title.
QA351.A75   1984   515.9   84-15435
ISBN 0-02-948650-5

Contents

	Preface	vii
1	**INFINITE SERIES, IMPROPER INTEGRALS, AND INFINITE PRODUCTS**	**1**
1.1	Introduction	1
1.2	Infinite Series of Constants	2
1.3	Infinite Series of Functions	15
1.4	Asymptotic Series	26
1.5	Fourier Trigonometric Series	32
1.6	Improper Integrals	38
1.7	Infinite Products	45
2	**THE GAMMA FUNCTION AND RELATED FUNCTIONS**	**50**
2.1	Introduction	50
2.2	Gamma Function	51
2.3	Beta Function	66
2.4	Incomplete Gamma Function	71
2.5	Digamma and Polygamma Functions	74
3	**OTHER FUNCTIONS DEFINED BY INTEGRALS**	**92**
3.1	Introduction	92
3.2	The Error Function and Related Functions	93

3.3	The Exponential Integral and Related Functions	103
3.4	Elliptic Integrals	108

4 LEGENDRE POLYNOMIALS AND RELATED FUNCTIONS — 116

4.1	Introduction	116
4.2	The Generating Function	117
4.3	Other Representations of the Legendre Polynomials	132
4.4	Legendre Series	137
4.5	Convergence of the Series	147
4.6	Legendre Functions of the Second Kind	155
4.7	Associated Legendre Functions	160

5 OTHER ORTHOGONAL POLYNOMIALS — 166

5.1	Introduction	166
5.2	Hermite Polynomials	167
5.3	Laguerre Polynomials	176
5.4	Generalized Polynomial Sets	184

6 BESSEL FUNCTIONS — 195

6.1	Introduction	195
6.2	Bessel Functions of the First Kind	196
6.3	Integral Representations and Integrals of Bessel Functions	205
6.4	Bessel Series	215
6.5	Bessel Functions of the Second and Third Kinds	220
6.6	Differential Equations Related to Bessel's Equation	228
6.7	Modified Bessel Functions	232
6.8	Other Bessel Functions	241
6.9	Asymptotic Formulas	248

7 BOUNDARY-VALUE PROBLEMS — 252

7.1	Introduction	252
7.2	Spherical Domains: Legendre Functions	253
7.3	Circular and Cylindrical Domains: Bessel Functions	264

8 THE HYPERGEOMETRIC FUNCTION — 272

8.1	Introduction	272
8.2	The Pochhammer Symbol	273
8.3	The Function $F(a, b; c; x)$	276

8.4	Relation to Other Functions	285
8.5	Summing Series	292
9	**THE CONFLUENT HYPERGEOMETRIC FUNCTION**	**298**
9.1	Introduction	298
9.2	The Functions $M(a;c;x)$ and $U(a;c;x)$	299
9.3	Relation to Other Functions	308
9.4	Whittaker Functions	315
10	**GENERALIZED HYPERGEOMETRIC FUNCTIONS**	**321**
10.1	Introduction	321
10.2	The Set of Functions $_pF_q$	322
10.3	Other Generalizations	327

BIBLIOGRAPHY — 333

APPENDIX: A LIST OF SPECIAL-FUNCTION FORMULAS — 335

SELECTED ANSWERS TO EXERCISES — 348

INDEX — 351

Preface

MODERN ENGINEERING AND PHYSICS APPLICATIONS demand a more thorough knowledge of applied mathematics than ever before. In particular, it is important to have a good understanding of the basic properties of *special functions*. These functions commonly arise in such areas of application as heat conduction, communication systems, electro-optics, nonlinear wave propagation, electromagnetic theory, quantum mechanics, approximation theory, probability theory, and electric circuit theory, among others. Special functions are sometimes discussed in certain engineering and physics courses, and math courses like partial differential equations, but the treatment of special functions in such courses is usually too brief to focus upon many of the important aspects such as the interconnecting relations between various special functions and elementary functions. This book is an attempt to present, at the elementary level, a more comprehensive treatment of special functions than can ordinarily be done within the context of another course. It provides a systematic introduction to most of the important special functions that commonly arise in practice and explores many of their salient properties. I have tried to present the special functions in a broader sense than is often done by not introducing them as simply solutions of certain differential equations. Many special functions are introduced by the generating function method, and the governing differential equation is then obtained as one of the important properties associated with the particular function.

In addition to discussing special functions, I have injected throughout the text by way of examples and exercises some of the techniques of applied analysis that are useful in the evaluation of nonelementary integrals, summing series, and so on. All too often in practice a problem is labeled "intractable" simply because the practitioner has not been exposed to the "bag of tricks" that helps the applied analyst deal with formidable-looking mathematical expressions.

During the last ten years or so at the University of Central Florida we have offered an introductory course in Special Functions to a mix of advanced undergraduates and first-year graduate students in mathematics, engineering, and physics. A set of lecture notes developed for that course has finally led to this textbook. The prerequisites for our course are the basic calculus sequence and a first course in differential equations. Although complex-variable theory is often utilized in studying special functions, knowledge of complex variables beyond some simple algebra and Euler's formulas is not required here. By not developing special functions in the language of complex variables, the text should be accessible to a wider audience. Naturally, some of the beauty of the subject is lost by this omission.

The text is not intended to be an exhaustive treatment of special functions. It concentrates heavily on a few functions, using them as illustrative examples, rather than attempting to give equal treatment to all. For instance, an entire chapter is devoted to the Legendre polynomials (and related functions), while the other orthogonal polynomial sets, including Hermite, Laguerre, Chebyshev, Gegenbauer, and Jacobi polynomials, are all lumped together in a single separate chapter. However, once the student is familiar with Legendre polynomials (which are perhaps the simplest set) and their properties, it is easy to extend these properties to other polynomial sets. Some applications occur throughout the text, often in the exercises, and Chapter 7 is devoted entirely to applications involving boundary-value problems. Other interesting applications which lead to special functions have been omitted, since they generally presuppose knowledge beyond the stated prerequisites.

Because of the close association of infinite series and improper integrals with the special functions, a brief review of these important topics is presented in the first chapter. In addition to reviewing some familiar concepts from calculus, this first chapter also contains material that is probably new to the student, such as the Cauchy product, index manipulation, asymptotic series, Fourier trigonometric series, and infinite products. Of course, our discussion of such topics is necessarily brief.

I owe a debt of gratitude to the many students who took my course on Special Functions over the years while this manuscript was being developed. Their patience, understanding, and helpful suggestions are greatly appreciated. I want to thank my colleague and friend, Patrick J. O'Hara, who graciously agreed on several occasions to teach from the lecture notes in their early rough form, and who made several helpful suggestions for improving the final version of the manuscript. Finally, I wish to express my appreciation to Ken Werner, Senior Editor of Scientific and Technical Books Department, for his continued faith in this project and efforts in getting it published.

1

Infinite Series, Improper Integrals, and Infinite Products

1.1 Introduction

Because of the close relation of *infinite series* and *improper integrals* to the special functions, it is useful to review some basic concepts of series and integrals. *Infinite products*, which are generally less well known, are introduced here mostly for the sake of completeness. Infinite series are important, of course, in almost all areas of pure and applied mathematics. In addition to numerous other uses, they are used to define functions and to calculate accurate numerical values of transcendental functions. In beginning courses dealing with infinite series the primary problem is deciding whether a given series converges or diverges. In practice, however, the more crucial problem may actually be summing the series. If a convergent series converges too slowly, the series may be worthless for computational purposes. On the other hand, the first few terms of a divergent series in some instances may give excellent results. Improper integrals and infinite products are used in much the same fashion as infinite series, and in fact, their basic theory closely parallels that of infinite series.

In the application of mathematics it frequently happens that two or more limiting processes have to be performed successively. For example, we often perform the derivative (or integral) of an infinite sum of functions by taking the sum of derivatives (or integrals) of the individual terms of the series. However, in many cases of interest, performing two limit operations in one order may yield an answer different from that obtained from the other order. That is, the order in which the limiting processes are carried out is critical. Therefore, it is of utmost importance to know the conditions under which such interchanges are permissible, and that is one of the considerations of this chapter.

Because our coverage of topics here is primarily a review, the treatment is intentionally cursory. In this regard we will state only the most relevant theorems, and then usually without proof. For a deeper discussion of the subject matter, the reader is advised to consult one of the standard texts on advanced calculus.

1.2 *Infinite Series of Constants*

If to each positive integer n we can associate a number S_n, then the ordered arrangement

$$S_1, S_2, \ldots, S_n, \ldots \qquad (1.1)$$

is called an *infinite sequence*, and we call S_n the *general term* of the sequence. Should it happen that

$$\lim_{n \to \infty} S_n = S \qquad (1.2)$$

where S is finite, the sequence (1.1) is said to *converge* to S, and is said to *diverge* otherwise.

An *infinite series* results when an infinite sequence of numbers $u_1, u_2, \ldots, u_k, \ldots$ is summed, i.e.,

$$u_1 + u_2 + \cdots + u_k + \cdots = \sum_{k=1}^{\infty} u_k \qquad (1.3)$$

In this case the number u_k is called the *general term* of the series. Closely associated with the infinite series (1.3) is a particular sequence

$$\begin{aligned}
S_1 &= u_1 \\
S_2 &= u_1 + u_2 \\
&\vdots \\
S_n &= u_1 + u_2 + \cdots + u_n = \sum_{k=1}^{n} u_k \\
&\vdots
\end{aligned} \qquad (1.4)$$

called the *sequence of partial sums*. If the partial sums (1.4) converge to a finite limit S, we say the infinite series (1.3) *converges*, or sums, to the value S. The series (1.3) *diverges* when the limit of partial sums fails to exist, i.e., fails to approach a unique finite value.

Example 1: Determine whether the following series converges or diverges:
$$\sum_{k=1}^{\infty} \left(\frac{1}{k} - \frac{1}{k+1} \right)$$

Solution: To show convergence or divergence we need to find the sum of the first n terms and examine its limit. Here we see that
$$\sum_{k=1}^{n} \left(\frac{1}{k} - \frac{1}{k+1} \right) = \left(1 - \frac{1}{2}\right) + \left(\frac{1}{2} - \frac{1}{3}\right) + \cdots + \left(\frac{1}{n} - \frac{1}{n+1}\right)$$
$$= 1 - \frac{1}{n+1}$$
where only the first and last term do not cancel. Thus,
$$\lim_{n \to \infty} S_n = \lim_{n \to \infty} \left(1 - \frac{1}{n+1}\right) = 1$$
and we conclude that the series *converges*, and in particular, converges to the value unity.

When a series diverges, it may do so for different reasons. For example, making use of the well-known formula
$$S_n = \sum_{k=1}^{n} k = \tfrac{1}{2} n(n+1) \qquad (1.5)$$
it is clear that $S_n \to \infty$ as $n \to \infty$, and therefore the infinite series Σk diverges.* In other instances the partial sums may not approach any particular limit, as for the series
$$\sum_{k=1}^{\infty} (-1)^{k-1} = 1 - 1 + 1 - 1 + \cdots + (-1)^{k-1} + \cdots \qquad (1.6)$$
The partial sums are $S_1 = 1$, $S_2 = 0$, $S_3 = 1, \ldots$, so that in general $S_n = 1$ for odd n and $S_n = 0$ for even n. Hence, we say that (1.6) diverges, since a unique limit of S_n does not exist.

1.2.1 The Geometric Series

The special series
$$1 + r + r^2 + \cdots + r^k + \cdots = \sum_{k=0}^{\infty} r^k \qquad (1.7)$$

---
*We will occasionally find it convenient to use the symbol Σu_k to denote $\Sigma_{k=1}^{\infty} u_k$.

is called a *geometric series*. The value r is called the *common ratio* since it is the ratio between the $(k + 1)$th term and the kth term. This series is important because it has a wide variety of applications, and because it can be summed exactly in those cases for which it converges.

From elementary algebra we know that the sum of the first n terms of (1.7) is given by (see problem 1)

$$S_n = \sum_{k=0}^{n-1} r^k = \frac{1 - r^n}{1 - r} \qquad (1.8)$$

where we stop the summation at $n - 1$, since the series begins at $k = 0$. Taking the limit of (1.8) as n tends to infinity leads to

$$\lim_{n \to \infty} S_n = \begin{cases} \dfrac{1}{1 - r}, & |r| < 1 \\ \text{no finite limit}, & |r| \geq 1 \end{cases} \qquad (1.9)$$

where we are using the fact that $r^n \to 0$ for increasing n when $|r| < 1$. Hence, we have derived the important result

$$\sum_{k=0}^{\infty} r^k = \frac{1}{1 - r}, \qquad |r| < 1 \qquad (1.10)$$

which not only establishes the values of r for which the series converges, but also provides the actual sum of the series.

Example 2: Test the series $3 - 2 + \frac{4}{3} - \frac{8}{9} + \cdots + 3(-\frac{2}{3})^k + \cdots$ for convergence.

Solution: By writing the series in the form

$$3 - 2 + \tfrac{4}{3} - \tfrac{8}{9} + \cdots = 3\left(1 - \tfrac{2}{3} + \tfrac{4}{9} - \tfrac{8}{27} + \cdots\right)$$

we recognize it as a geometric series (multiplied by 3) with $r = -\frac{2}{3}$. Since r is less than unity in absolute value, we deduce that the series converges, and moreover, converges to the value

$$3 \sum_{k=0}^{\infty} \left(-\tfrac{2}{3}\right)^k = \frac{3}{1 - \left(-\tfrac{2}{3}\right)} = \frac{9}{5}$$

1.2.2 Summary of Convergence Tests

Generally speaking, the only series that are useful in practice are those that converge. For that reason we attach a great deal of importance to the task of deciding whether a particular series converges or not. In the case of the geometric series we were able to get the nth partial sum S_n into "closed form" and examine its limit directly as $n \to \infty$. By so doing, we not only answered the question of convergence or divergence, but actually summed the series. Unfortunately, the geometric series is one of the rare examples for which we are able to get S_n into a form from which we can evaluate its limit for large n. What is required then is a handful of *tests* that can be applied to the series in question, from which its convergence or divergence can be established. A great many such convergence tests have been developed over the years, some simple to apply and others quite sophisticated.

The development of various convergence tests is taken up in courses on calculus (both elementary and advanced). Our intention here is to simply recall some of the elementary tests for reference purposes.

We first observe that if $\Sigma u_k = S$, where S is finite, then $S_n \to S$ and $S_{n-1} \to S$ as $n \to \infty$; hence, necessarily,

$$\lim_{n \to \infty} (S_n - S_{n-1}) = S - S = 0$$

But, since $S_n - S_{n-1} = u_n$, we find that a *necessary condition* (but not a sufficient condition) for the series Σu_n to converge is that*

$$\lim_{n \to \infty} u_n = 0 \tag{1.11}$$

Remark: The general term of a series can be denoted by u_k or u_n, or any other dummy index can be used. We will switch back and forth between indices for convenience.

A series is called *positive* if the terms of the series are either all positive or all negative. In other cases the terms of the series will vary in sign, some terms positive and some negative. If the consecutive terms have opposite signs, we call the series an *alternating series*. Series containing terms both positive and negative converge more rapidly (when they converge) than do positive series, due to the partial cancellation of the negative terms with the positive terms. Because of these distinctions, we introduce notions of different kinds of convergence.

Definition 1.1. The series Σu_n is said to *converge absolutely* if the associated series of positive terms $\Sigma |u_n|$ converges.

Definition 1.2. If the series Σu_n converges but the related series $\Sigma |u_n|$ diverges, the original series is said to *converge conditionally*.

For alternating series, we have the following important theorem.

---

*If $\lim_{n \to \infty} u_n \neq 0$, then of course the series Σu_n diverges.

Theorem 1.1 (*Alternating-series test*). If after a certain point the absolute values of the terms of an alternating series decrease monotonically to zero, the series converges (conditionally at least). Also, the sum of a convergent alternating series always lies between the partial sums S_n and S_{n+1} for each n.

If an alternating series converges by the alternating-series test, we must further investigate its convergence to determine whether it converges absolutely. This we do by testing the related series of positive terms, for which we have the following convergence tests.

Remark: If a positive series converges, it necessarily converges absolutely. (Why?) Hence, the term conditional convergence applies only to series that vary in sign, such as alternating series.

Theorem 1.2 (*Comparison test*). A positive series Σu_n converges absolutely if each term (after a finite number) is less than or equal to the corresponding term of a known convergent positive series Σa_n, i.e.,

$$u_n \leq a_n, \qquad n > N$$

The positive series Σu_n *diverges* if each term (after a finite number) is greater than or equal to the corresponding term of a known divergent positive series Σb_n, i.e.,

$$u_n \geq b_n, \qquad n > N$$

Theorem 1.3 (*Comparison test*). If Σu_n and Σa_n are positive series and

$$\lim_{n \to \infty} \frac{u_n}{a_n} = c \neq 0,$$

then Σu_n and Σa_n converge or diverge together.

Remark: Theorem 1.3 can be extended to the case where c is either zero or infinity. That is, if $c = 0$ and Σa_n converges, then Σu_n also converges; if $c \to \infty$ and Σa_n diverges, then Σu_n diverges.

The following theorem is probably the most widely used test of convergence.

Theorem 1.4 (*Ratio test*). Let Σu_n denote any series for which

$$\lim_{n \to \infty} \left| \frac{u_{n+1}}{u_n} \right| = L.$$

(1) If $L < 1$, the series Σu_n converges absolutely.
(2) If $L > 1$, the series Σu_n diverges.
(3) If $L = 1$, the test fails (no conclusion).

The ratio test is a particularly useful test for those series involving factorials or exponentials, but fails in those case where the general term is a rational function of n.

Theorem 1.5 (*Integral test*). Let $f(n)$ denote the general term of the series Σu_n. If the function $f(x)$ is positive, continuous, and nonincreasing for $x \geq a$, then the positive series Σu_n converges or diverges according to the convergence or divergence of the improper integral $\int_a^\infty f(x)\,dx$.*

An important series for comparison purposes is the *p-series*†

$$\sum_{n=1}^{\infty} \frac{1}{n^p} = 1 + \frac{1}{2^p} + \frac{1}{3^p} + \cdots + \frac{1}{n^p} + \cdots \qquad (1.12)$$

To find the values of p for which this series converges, we can take $f(x) = 1/x^p$ and use the integral test. Thus, for $a > 0$,‡

$$\int_a^\infty x^{-p}\,dx = \begin{cases} \left.\dfrac{x^{1-p}}{1-p}\right|_a^\infty, & p \neq 1 \\ \left.\log x\right|_a^\infty, & p = 1 \end{cases}$$

from which we deduce that the series converges for $p > 1$ and diverges for $p \leq 1$. The special value $p = 1$ leads to

$$\sum_{n=1}^{\infty} \frac{1}{n} = 1 + \frac{1}{2} + \frac{1}{3} + \cdots + \frac{1}{n} + \cdots \qquad (1.13)$$

called the *harmonic series*. It plays an important role in the use of comparison tests. Although the series diverges, it does so at a very slow rate. For example, the first million terms add up to a number only slightly larger than 14.

1.2.3 Operations with Series

In applications the need arises to combine various series by such operations as addition and multiplication. To perform these operations it is usually

---

*The convergence of improper integrals is discussed in Section 1.6.
†For $p > 1$, the p-series is also called the *Riemann zeta function* (see Section 2.5.4).
‡We use $\log x$ to denote the natural logarithm, also commonly denoted by $\ln x$.

important to establish the absolute convergence of all series involved in the process, since such operations can then be performed by the familiar rules of algebra or arithmetic. Specifically, we have that:

1. The sum of an absolutely convergent series is independent of the order in which terms are added.
2. Two absolutely convergent series may be added termwise, and the resulting series will converge absolutely.
3. Two absolutely convergent series may be multiplied (Cauchy product), and the resulting series will also converge absolutely.

The significance of property 1 above can best be realized by considering what can happen if the series we wish to sum is not absolutely convergent. The classic example of a series converging conditionally, but not absolutely, is the *alternating harmonic series* $\sum_{n=1}^{\infty}(-1)^{n-1}/n$. Generally we associate the sum of this series with the value $\log 2$ (see Section 1.3), i.e., we write

$$\sum_{n=1}^{\infty} \frac{(-1)^{n-1}}{n} = 1 - \tfrac{1}{2} + \tfrac{1}{3} - \tfrac{1}{4} + \tfrac{1}{5} - \tfrac{1}{6} + \cdots = \log 2 \qquad (1.14)$$

However, if we rearrange the terms of the series according to

$$\begin{aligned}
1 - \tfrac{1}{2} + \tfrac{1}{3} - \tfrac{1}{4} + \cdots &= \left(1 - \tfrac{1}{2}\right) - \tfrac{1}{4} + \left(\tfrac{1}{3} - \tfrac{1}{6}\right) - \tfrac{1}{8} + \left(\tfrac{1}{5} - \tfrac{1}{10}\right) \\
&\quad - \tfrac{1}{12} + \left(\tfrac{1}{7} - \tfrac{1}{14}\right) - \tfrac{1}{16} + \cdots \\
&= \tfrac{1}{2} - \tfrac{1}{4} + \tfrac{1}{6} - \tfrac{1}{8} + \tfrac{1}{10} - \tfrac{1}{12} + \tfrac{1}{14} - \tfrac{1}{16} + \cdots \\
&= \tfrac{1}{2}\left(1 - \tfrac{1}{2} + \tfrac{1}{3} - \tfrac{1}{4} + \tfrac{1}{5} - \tfrac{1}{6} + \cdots\right)
\end{aligned}$$

we may conclude that the sum of the series is $\tfrac{1}{2}\log 2$. We arrive at this conclusion not because we have cleverly omitted some terms of the series. Indeed, each term of the series (1.14) does eventually appear exactly once, but in the final arrangement the whole series appears multiplied by the factor $\tfrac{1}{2}$.

What is being illustrated here is that by rearranging the terms of a conditionally convergent series, that series may be made to converge to any desired numerical value, or can even be made to diverge. Thus it is clear that conditionally convergent series must be handled very carefully.

Also, if the series is *not positive and diverges*, we can sometimes produce a convergent series from it by rearranging or regrouping the terms. For example, if we write

$$\sum_{n=1}^{\infty}(-1)^{n-1} = (1-1) + (1-1) + \cdots + (1-1) + \cdots$$

the partial sums (of terms in parentheses) are all zero, and hence we may deduce that the series convergences to the value zero. If a series is positive and diverges, no rearrangement of terms can make the series converge.

If two series are absolutely convergent, no rearrangement of their terms will alter the sum or difference of the two series. But again, if both of the series forming the sum or difference are divergent, it is not clear what will happen. For instance, by writing

$$\sum_{n=1}^{\infty} \left(\frac{1}{n} - \frac{1}{n+1} \right) = \sum_{n=1}^{\infty} \frac{1}{n} - \sum_{n=1}^{\infty} \frac{1}{n+1}$$

we can treat the series on the left as the difference of two divergent series, as shown on the right. Although we might be tempted to say that the series on the left diverges because of its relation to the divergent series on the right, we have actually shown (in Example 1) that the series on the left converges, and in fact converges absolutely.

In forming the product of two series, we are led to *double infinite series* of the form*

$$\sum_{m=0}^{\infty} u_m \sum_{k=0}^{\infty} v_k = \sum_{m=0}^{\infty} \sum_{k=0}^{\infty} A_{m,k}$$

where the summand $A_{m,k} = u_m v_k$ can be treated as a function of two variables. We find that by making a change of index the above double sum can often be simplified, or even partially summed. For example, suppose we introduce the change of index $m = n - k$, or equivalently, $n = m + k$. Now, since $m \geq 0$, the index k must satisfy the condition $n - k \geq 0$, or $k \leq n$. Hence, we deduce that (for absolutely convergent series)

$$\sum_{m=0}^{\infty} \sum_{k=0}^{\infty} A_{m,k} = \sum_{n=0}^{\infty} \sum_{k=0}^{n} A_{n-k,k} \qquad (1.15)$$

Equation (1.15) illustrates that all absolutely convergent double infinite sums can be replaced by a single infinite series of finite sums. This property is particularly useful in numerical computations. If the two series forming the product are each absolutely convergent, we can also interchange the order of the infinite sums and then apply (1.15). In fact, all possibilities of this kind should be explored when trying to simplify double infinite series.

On rare occasions we find it necessary to make a different change of indices in our double infinite sums than illustrated above. For example, if we set $m = n - 2k$, it follows that $k \leq n/2$. But since $n/2$ is not always an integer, it is conventional to introduce the bracket notation

$$[n/2] = \begin{cases} n/2, & n \text{ even} \\ (n-1)/2, & n \text{ odd} \end{cases} \qquad (1.16)$$

---

*Many of the infinite series that we encounter will start with the index value zero.

Hence, with this index change we deduce that (for absolutely convergent series)

$$\sum_{m=0}^{\infty}\sum_{k=0}^{\infty} A_{m,k} = \sum_{n=0}^{\infty}\sum_{k=0}^{[n/2]} A_{n-2k,k} \qquad (1.17)$$

and upon combining (1.15) and (1.17), it also follows that

$$\sum_{n=0}^{\infty}\sum_{k=0}^{n} A_{n-k,k} = \sum_{n=0}^{\infty}\sum_{k=0}^{[n/2]} A_{n-2k,k} \qquad (1.18)$$

Theorem 1.6 (*Cauchy product*). If $\sum_{n=0}^{\infty} a_n$ and $\sum_{n=0}^{\infty} b_n$ are both absolutely convergent series, then so is their Cauchy product defined by

$$\sum_{n=0}^{\infty} a_n \cdot \sum_{n=0}^{\infty} b_n = \sum_{n=0}^{\infty} c_n$$

where

$$c_n = \sum_{k=0}^{n} a_k b_{n-k}$$

Other theorems on the product of two infinite series have been developed. For example, it has been shown that if both Σa_n and Σb_n converge, and if one of them converges absolutely, then Σc_n converges. Also, it is possible for both Σa_n and Σb_n to converge while the product series Σc_n diverges.

1.2.4 Factorials and Binomial Coefficients

In simplifying products of infinite series, as well as numerous other applications, we frequently encounter series involving *binomial coefficients*. Perhaps the simplest way of introducing these coefficients is by considering the expanded product of $(a + b)^n$. For example,

$$(a + b)^1 = a + b$$
$$(a + b)^2 = a^2 + 2ab + b^2$$
$$(a + b)^3 = a^3 + 3a^2b + 3ab^2 + b^3$$

and in general,

$$(a + b)^n = a^n + na^{n-1}b + \frac{n(n-1)}{2!} a^{n-2}b^2 + \cdots$$
$$+ \frac{n(n-1)\cdots(n-k+1)}{k!} a^{n-k}b^k + \cdots + b^n \qquad (1.19)$$

The coefficient of the general term in (1.19) can be expressed more simply in terms of factorials by writing

$$\frac{n(n-1)\cdots(n-k+1)}{k!} = \frac{n!}{k!(n-k)!}$$

for which we also introduce the notation*

$$\binom{n}{k} = \frac{n!}{k!(n-k)!}, \qquad n = 0, 1, 2, \ldots, \quad k = 0, 1, \ldots, n \quad (1.20)$$

Adopting this notation, we can now write (1.19) more compactly as

$$(a+b)^n = \sum_{k=0}^{n} \binom{n}{k} a^{n-k} b^k \qquad (1.21)$$

The symbol $\binom{n}{k}$ is what we call a *binomial coefficient*. Besides its connection in (1.21) with the expansion of $(a+b)^n$, the binomial coefficient also occurs in combinatoric problems, probability theory, and algorithm development. In these other applications the upper index is often not an integer, or even a positive number. For such situations we cannot use (1.20) to define the binomial coefficient, but rather we resort to

$$\binom{r}{0} = 1, \quad \binom{r}{k} = \frac{r(r-1)\cdots(r-k+1)}{k!}, \quad k = 1, 2, 3, \ldots \qquad (1.22)$$

As simple consequences of the definition of binomial coefficient and properties of factorials, we have the following useful relations:

$$\binom{n}{0} = \binom{n}{n} = 1 \qquad (1.23)$$

$$\binom{n}{1} = \binom{n}{n-1} = n \qquad (1.24)$$

$$\binom{n}{k} = \binom{n}{n-k} \qquad (1.25)$$

$$\binom{n+1}{k+1} = \binom{n}{k+1} + \binom{n}{k}, \quad 0 \le k \le n-1 \qquad (1.26)$$

$$\binom{-r}{k} = (-1)^k \binom{r+k-1}{k} \qquad (1.27)$$

Example 3: Show that

$$\binom{-r}{k} = (-1)^k \binom{r+k-1}{k}$$

---

*$0! = 1$.

Solution: From (1.22), we have

$$\binom{-r}{k} = \frac{-r(-r-1)\cdots(-r-k+1)}{k!}$$

$$= (-1)^k \frac{r(r+1)\cdots(r+k-1)}{k!}$$

$$= (-1)^k \frac{(r+k-1)(r+k-2)\cdots(r+1)r}{k!}$$

$$= (-1)^k \binom{r+k-1}{k}$$

where the last step again follows from (1.22).

There are literally thousands of identities involving binomial coefficients that have been discovered over the years. Fortunately, only a few of these are required in most applications. In addition to (1.23)–(1.27) above, the following summation formulas are also very important:

$$\sum_{k=0}^{n} \binom{n}{k} = 2^n \qquad (1.28)$$

$$\sum_{k=0}^{n} \binom{r+k}{k} = \binom{r+n+1}{n} \qquad (1.29)$$

$$\sum_{k=0}^{n} \binom{k}{m} = \binom{n+1}{m+1}, \qquad m = 0, 1, 2, \ldots \qquad (1.30)$$

$$\sum_{k=0}^{n} \binom{r}{k}\binom{s}{n-k} = \binom{r+s}{n}, \qquad n = 0, 1, 2, \ldots \qquad (1.31)$$

$$\sum_{k=0}^{n} \binom{n}{k}\binom{s+k}{m}(-1)^k = (-1)^n \binom{s}{m-n} \qquad (1.32)$$

Equation (1.28) follows directly from (1.21) with $a = b = 1$. To prove (1.29) requires repeated application of (1.26), whereas (1.30) follows from (1.29) with two applications of (1.25). Equation (1.31) is verified below (Example 4), and (1.32) is left to the exercises.

Remark: In Section 8.5 we will present another method of summing certain series of binomial coefficients by use of the *hypergeometric function*.

Example 4: Verify Equation (1.31) above.

Solution: Starting with the obvious identity

$$(1+x)^r (1+x)^s = (1+x)^{r+s}$$

and replacing each binomial with its series, we find*

$$\sum_{n=0}^{\infty} \binom{r}{n} x^n \cdot \sum_{n=0}^{\infty} \binom{s}{n} x^n = \sum_{n=0}^{\infty} \binom{r+s}{n} x^n$$

The left-hand side can be simplified by use of the Cauchy product (Theorem 1.6), which leads to

$$\sum_{n=0}^{\infty} \sum_{k=0}^{n} \binom{r}{k} \binom{s}{n-k} x^n = \sum_{n=0}^{\infty} \binom{r+s}{n} x^n$$

and now, by comparing coefficients of like terms of x^n, we obtain the result†

$$\sum_{k=0}^{n} \binom{r}{k} \binom{s}{n-k} = \binom{r+s}{n}, \qquad n = 0, 1, 2, \ldots$$

EXERCISES 1.2

1. Show that the nth partial sum of a geometric series satisfies

$$1 + r + r^2 + \cdots + r^{n-1} = \frac{1 - r^n}{1 - r}, \qquad r \neq 1$$

Hint: Observe that

$$S_n = 1 + r + r^2 + \cdots + r^{n-1}$$
$$rS_n = \phantom{1 + {}} r + r^2 + \cdots + r^{n-1} + r^n$$

and subtract termwise.

In problems 2–5, find the sum of the geometric series.

2. $\sum_{k=0}^{10} 2^k$.

3. $\sum_{k=0}^{100} (-1)^k$.

4. $\sum_{k=2}^{10} (\tfrac{1}{2})^k$.

5. $\sum_{n=0}^{\infty} \sin^{2n} x$, $|x| < \pi/2$.

In problems 6 and 7, use geometric series to express the repeating decimal as a rational number.

6. $3.42121212\ldots$

7. $2.123123123\ldots$

---

*When r and s are not integers the binomial series becomes an infinite series, and in this case x is restricted to the interval $|x| < 1$ (see Section 1.3.2).
†We are actually using Theorem 1.13 in Section 1.3.2.

14 • Special Functions for Engineers and Applied Mathematicians

In problems 8–14, determine whether the series *converges absolutely, converges conditionally*, or *diverges*.

8. $\sum_{n=2}^{\infty} \dfrac{1}{n \log n}$.

9. $\sum_{n=1}^{\infty} (n+2)^{-3/2}$.

10. $\sum_{n=0}^{\infty} \dfrac{(2n)!}{(n!)^2}$.

11. $\sum_{n=1}^{\infty} (-1)^n \left(1 + \dfrac{1}{n^4}\right)$.

12. $\sum_{n=1}^{\infty} (-1)^n \dfrac{1}{2n-1}$.

13. $\sum_{n=1}^{\infty} \dfrac{\log n}{n^p}$.

14. $\sum_{n=3}^{\infty} \dfrac{(-1)^n}{\sqrt{n} \log(\log n)}$.

15. By using the Cauchy product, verify the identity

$$e^a e^b = e^{a+b}$$

Hint: Recall that $e^a = \sum_{n=0}^{\infty} \dfrac{a^n}{n!}$.

16. Show that

(a) $\dbinom{n}{k} = \dbinom{n}{n-k}$.

(b) $\dbinom{n+1}{k+1} = \dbinom{n}{k+1} + \dbinom{n}{k}$, $0 \le k \le n-1$.

17. Show that

(a) $\dbinom{-\frac{1}{2}}{n} = \dfrac{(-1)^n (2n)!}{2^{2n}(n!)^2}$.

(b) $\dbinom{-2k-1}{n} = (-1)^n \dfrac{(n+2k)!}{(2k)! n!}$, $k = 0, 1, 2, \ldots$.

In problems 18–20, verify the given formula.

18. $\sum_{k=0}^{n} \dbinom{r+k}{k} = \dbinom{r+n+1}{n}$.

Hint: Use Equation (1.26).

19. $\sum_{k=0}^{n} \dbinom{k}{m} = \dbinom{n+1}{m+1}$, $m = 0, 1, 2, \ldots$.

Hint: Use problem 18 and Equation (1.25).

20. $\sum_{k=0}^{n} \dbinom{n}{k}\dbinom{s+k}{m}(-1)^k = (-1)^n \dbinom{s}{m-n}$, $m = 0, 1, 2, \ldots$, $m \ge n$.

1.3 Infinite Series of Functions

Of special importance to us are those series that result when the general term u_n is a function of x, i.e., $u_n = u_n(x)$. The nth partial sum then defines the function

$$S_n(x) = \sum_{k=1}^{n} u_k(x) \tag{1.33}$$

and similarly, the sum of the series becomes

$$\lim_{n \to \infty} S_n(x) = f(x) \tag{1.34}$$

The question of concern here is whether there exists any values of x for which (1.34) is meaningful. If the value of x is fixed and the resulting series sums to $f(x)$, we say the series *converges pointwise* to $f(x)$. All such values of x for which the series converges pointwise constitute the domain of the function f.

Example 5: Test the series $\sum_{n=0}^{\infty} x^n$ for convergence.

Solution: Applying the ratio test, we find

$$\lim_{n \to \infty} \left| \frac{x^{n+1}}{x^n} \right| = \lim_{n \to \infty} |x| = |x|$$

and deduce that the series converges pointwise for $|x| < 1$ and diverges for $|x| > 1$. The cases $|x| = 1$ must be treated separately, but it can easily be established that the series diverges for both $x = 1$ and $x = -1$. Our conclusion here is consistent with previous results, since the series in question is just the geometric series once again with $f(x) = 1/(1 - x)$.

In some applications it is important to establish a different kind of convergence of the series, for which we have the following definition.

Definition 1.3. If, given some $\varepsilon > 0$, there exists a number $N = N(\varepsilon)$ independent of x, and if

$$|f(x) - S_n(x)| < \varepsilon$$

for all x in $a \le x \le b$ and all $n > N$, we say that $S_n(x)$ *converges uniformly* to $f(x)$ in $a \le x \le b$ as $n \to \infty$.

Uniform convergence is illustrated in Fig. 1.1. It is clearly a stronger requirement than is pointwise convergence, which treats convergence at

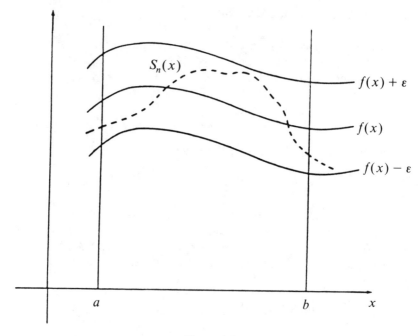

Figure 1.1

individual points, but it is also more difficult to establish in practice. The key to uniform convergence is *continuity* of the function f (see Theorem 1.8).

The most commonly used test for establishing uniform convergence of an infinite series of functions is the famous *Weierstrass M-test*.

Theorem 1.7 (*Weierstrass M-test*). If ΣM_n is a convergent series of positive constants such that $|u_n(x)| \leq M_n$ ($n = 1, 2, 3, \ldots$) for all x in $a \leq x \leq b$, then the series $\Sigma u_n(x)$ is uniformly (and absolutely) convergent over the interval $a \leq x \leq b$.

Normally, if a series converges uniformly it converges absolutely, but not always. That is, neither type of convergence necessarily implies the other. For example, the series $\sum_{n=1}^{\infty}(-1)^{n-1}x^n/n = \log(1 + x)$ converges uniformly for $0 \leq x \leq 1$, but not absolutely. (Why?) Also, the series

$$\sum_{n=0}^{\infty}(1 - x)x^n = \begin{cases} 1, & 0 \leq x < 1 \\ 0, & x = 1 \end{cases}$$

converges absolutely but not uniformly in the interval $0 \leq x \leq 1$. Thus the Weierstrass *M*-test is not suitable for series that converge uniformly but not absolutely.

1.3.1 Properties of Uniformly Convergent series

Establishing that a given series converges uniformly in an interval is useful for performing certain operations on the series termwise.

Theorem 1.8. If each term $u_n(x)$ is continuous in $a \leq x \leq b$ and the series

$$f(x) = \sum_{n=1}^{\infty} u_n(x)$$

converges uniformly in $a \leq x \leq b$, then f is a continuous function in this same interval.

Note that Theorem 1.8 requires uniform convergence to conclude that f is continuous. To show that pointwise convergence is not sufficient, consider the series

$$f(x) = x + \sum_{n=2}^{\infty} (x^n - x^{n-1}), \qquad 0 \leq x \leq 1 \tag{1.35}$$

Clearly each term of the series is a continuous function. Also, the sum of the first n terms is $S_n(x) = x^n$, which converges to zero in the interval $0 \leq x < 1$ and to unity when $x = 1$. Hence, the sum of the series is

$$f(x) = \begin{cases} 0, & 0 \leq x < 1 \\ 1, & x = 1 \end{cases} \tag{1.36}$$

which is clearly not a continuous function in the closed interval $0 \leq x \leq 1$.

Theorem 1.9. If each term $u_n(x)$ is continuous in $a \leq x \leq b$ and the infinite series $\Sigma u_n(x)$ converges uniformly in $a \leq x \leq b$, then termwise integration of the series is permitted, i.e.,

$$\int_a^b \left(\sum_{n=1}^{\infty} u_n(x) \right) dx = \sum_{n=1}^{\infty} \int_a^b u_n(x)\, dx$$

. Theorem 1.9 is particularly important in applied mathematics, since the integral of an infinite series arises frequently there. The difficulty in many situations, however, is that we may not be able to show that the given series converges uniformly prior to performing termwise integration. In such situations we tacitly assume the conditions of Theorem 1.9 and formally carry out all computations. It is essential in these situations to justify the derived result by some independent means.

In order to illustrate the use of Theorem 1.9, consider the infinite series

$$\frac{1}{1+x} = \sum_{n=0}^{\infty} (-1)^n x^n, \qquad -1 < x < 1 \tag{1.37}$$

It can be shown that this series converges uniformly in any closed interval contained within the indicated open interval (see Section 1.3.2). Hence, termwise integration of (1.37) leads to

$$\int_0^x \frac{dt}{1+t} = \sum_{n=0}^{\infty} (-1)^n \int_0^x t^n \, dt, \quad -1 < x < 1$$

where we introduce the dummy variable t to avoid confusion. Completing the integration, we obtain

$$\log(1+x) = \sum_{n=0}^{\infty} (-1)^n \frac{x^{n+1}}{n+1}$$

and by making the change of index $n \to n-1$, we get the more familiar form

$$\log(1+x) = \sum_{n=1}^{\infty} (-1)^{n-1} \frac{x^n}{n}, \quad -1 < x < 1 \quad (1.38)$$

Notice that setting $x = 1$ in (1.38) leads to

$$\log 2 = \sum_{n=1}^{\infty} \frac{(-1)^{n-1}}{n} \quad (1.39)$$

where the right-hand side is the alternating harmonic series (see Section 1.2.3). It is interesting to observe that the result (1.39) is valid even though the value $x = 1$ is outside the original interval of (pointwise) convergence. This example illustrates that the process of integration of an infinite series can sometimes extend the interval of convergence of the integrated series beyond that of the original series.

The conditions stated in Theorem 1.9 are satisfied for many of the series that commonly arise in practice, and for this reason we find that most of the time termwise integration of the series is permitted. The same is not true, however, for termwise differentiation of a series, even under the same conditions. That is, uniform convergence of the series does not validate its differentiation.

Theorem 1.10. If $u_n(x)$ and $u_n'(x)$ are continuous functions in the interval $a \le x \le b$ for each n, and if

$$f(x) = \sum_{n=1}^{\infty} u_n(x)$$

converges in $a \le x \le b$ and the series $\Sigma u_n'(x)$ converges uniformly in $a \le x \le b$, then

$$f'(x) = \sum_{n=1}^{\infty} u_n'(x)$$

Basically, the requirement for termwise differentiation of a series is the uniform convergence of the differentiated series. For example, the series

$$f(x) = \sum_{n=1}^{\infty} \frac{\sin n^2 x}{n^2} \qquad (1.40)$$

converges uniformly in every finite interval (by the Weierstrass M-test), whereas the series

$$f'(x) = \sum_{n=1}^{\infty} \cos n^2 x \qquad (1.41)$$

diverges for all x. Clearly, termwise differentiation of an infinite series must be handled with caution.

1.3.2 Power Series

By a *power series*, we mean an expression of the form

$$c_0 + c_1(x-a) + \cdots + c_n(x-a)^n + \cdots = \sum_{n=0}^{\infty} c_n(x-a)^n \qquad (1.42)$$

where the c's are constants and a is some fixed value.

Theorem 1.11. Every power series has a radius of convergence ρ such that the series converges absolutely when $|x - a| < \rho$ and diverges when $|x - a| > \rho$.*

If $\rho > 0$, then for every ρ_1 such that $0 \le \rho_1 < \rho$, the power series converges uniformly for $|x - a| \le \rho_1$. The question of convergence of a power series when $|x - a| = \rho$ can be answered separately by one of our previous convergence tests, since (1.42) is just a series of constants in this case.

Theorem 1.12 (Abel's theorem). If the radius of convergence of a power series is ρ, then the sum

$$f(x) = \sum_{n=0}^{\infty} c_n(x-a)^n$$

is a continuous function for $|x - a| < \rho$.

Proof: Since the series converges uniformly for $|x - a| \le \rho_1$, $0 \le \rho_1 < \rho$, it follows (from Theorem 1.8) that f is a continuous function on this

---
*In some cases $\rho = 0$, and hence the series converges only for $x = a$.

interval. But since this is true for every ρ_1 between 0 and ρ, we conclude that f is continuous for all x in the interval $|x - a| < \rho$. ∎

As simple consequences of Theorem 1.9 and 1.10, it follows that convergent power series can always be differentiated and integrated termwise. If the power series converges uniformly for all x such that $|x - a| \le \rho$, the *integrated* series will also converge uniformly for $|x - a| \le \rho$. On the other hand, the *differentiated* series may not converge at the endpoints. For example, the series

$$f(x) = \sum_{n=2}^{\infty} \frac{x^n}{n(n-1)}, \quad -1 \le x \le 1$$

converges uniformly for all $|x| \le 1$, but the related series

$$f'(x) = \sum_{n=1}^{\infty} \frac{x^n}{n}, \quad -1 \le x < 1$$

does not converges at $x = 1$. Moreover, the series

$$f''(x) = \sum_{n=0}^{\infty} x^n, \quad -1 < x < 1$$

does not converge at either endpoint.

One way of generating a power series for a given function f is illustrated in the following discussion.

If f is continuous and differentiable, then

$$\int_a^x f'(t)\, dt = f(x) - f(a) \tag{1.43}$$

which we can rearrange as

$$f(x) = f(a) + \int_a^x f'(t)\, dt \tag{1.44}$$

If f also has a second derivative f'', we can then replace f in (1.44) by the function f' to obtain

$$f'(x) = f'(a) + \int_a^x f''(t)\, dt$$

The substitution of this last expression for f' in (1.44) leads to

$$f(x) = f(a) + \int_a^x \left[f'(a) + \int_a^t f''(t_1)\, dt_1 \right] dt$$

$$= f(a) + f'(a)(x - a) + \int_a^x \int_a^t f''(t)(dt)^2 \tag{1.45}$$

where we are using the notation

$$\int_a^x \int_a^t f''(t)(dt)^2 \equiv \int_a^x \left[\int_a^t f''(t_1)\, dt_1 \right] dt \tag{1.46}$$

Now assuming the function f has n derivatives, we can repeat this process over and over until we obtain

$$f(x) = f(a) + f'(a)(x-a) + \frac{f''(a)}{2!}(x-a)^2 + \cdots$$
$$+ \frac{f^{(n-1)}(a)}{(n-1)!}(x-a)^{n-1} + R_n \tag{1.47}$$

where

$$R_n = \int_a^x \cdots \int_a^t f^{(n)}(t)(dt)^n \tag{1.48}$$

Equation (1.47) is known as *Taylor's formula with remainder*.

If the function $f^{(n)}(t)$ satisfies the inequality

$$m \leq f^{(n)}(t) \leq M, \qquad a \leq t \leq x \tag{1.49}$$

where m and M are constants, then it can be shown that

$$m \int_a^x \cdots \int_a^t (dt)^n \leq R_n \leq M \int_a^x \cdots \int_a^t (dt)^n$$

which reduces to

$$\frac{m(x-a)^n}{n!} \leq R_n \leq \frac{M(x-a)^n}{n!} \tag{1.50}$$

Hence, if $f^{(n)}(t)$ is also continuous over $a \leq t \leq x$, then there exists some value ξ such that

$$R_n = \frac{f^{(n)}(\xi)}{n!}(x-a)^n, \qquad a < \xi < x \tag{1.51}$$

known as the *Lagrangian form of the remainder* after n terms.

Finally, if the function f has the property that, for $|x-a| < \rho$, $R_n \to 0$ as $n \to \infty$, then it follows that

$$f(x) = \sum_{n=0}^{\infty} \frac{f^{(n)}(a)}{n!}(x-a)^n, \qquad |x-a| < \rho \tag{1.52}$$

We refer to (1.52) as the *Taylor series* for the function f. The special case of (1.52) that occurs when $a = 0$ is known as *Maclaurin's series*, i.e.,

$$f(x) = \sum_{n=0}^{\infty} \frac{f^{(n)}(0)}{n!} x^n, \qquad |x| < \rho \tag{1.53}$$

22 • Special Functions for Engineers and Applied Mathematicians

Most of the elementary functions that arise in the calculus can be represented by a Taylor series, where the interval of convergence is determined by the ratio test. Many of the special functions that we will encounter in subsequent chapters can also be represented by a Taylor series (or a Maclaurin series).

Example 6: Expand $f(x) = (1 + x)^a$ in a Maclaurin series, where a is a parameter not restricted to integer values.

Solution: Repeated differentiation of the function reveals that

$$f'(x) = a(1 + x)^{a-1}$$
$$f''(x) = a(a - 1)(1 + x)^{a-2}$$
$$\vdots$$
$$f^{(n)}(x) = a(a - 1) \cdots (a - n + 1)(1 + x)^{a-n}$$

Hence, by setting $x = 0$ in f and all its derivatives, we find that the series (1.53) leads to

$$(1 + x)^a = 1 + ax + \frac{a(a - 1)}{2!} x^2 + \cdots$$
$$+ \frac{a(a - 1) \cdots (a - n + 1)}{n!} x^n + \cdots$$

which we can express more compactly in the form

$$(1 + x)^a = \sum_{n=0}^{\infty} \binom{a}{n} x^n$$

where $\binom{a}{n}$ denotes the binomial coefficient (see Section 1.2.4).

By applying the ratio test to the above series, known as the *binomial series*, we find that it converges for $|x| < 1$.

The binomial series in Example 6 is important in much of our work to follow, and we will have many occasions to refer back to this result.

The following theorem assures us of the uniqueness of representation of a function by a Taylor series for a fixed value of a.

Theorem 1.13 (*Uniqueness*). If $f(x) = \Sigma c_n(x - a)^n$ and $g(x) = \Sigma b_n(x - a)^n$ both have nonzero radii of convergence, and $f(x) = g(x)$ wherever the two series converge, then $c_n = b_n$, $n = 0, 1, 2, \ldots$.*

---

*If $\Sigma c_n(x - a)^n = 0$, then necessarily $c_n = 0$ for all n.

1.3.3 Operations with Power Series

If
$$f(x) = \sum_{n=0}^{\infty} a_n x^n \qquad (1.54)$$

and
$$g(x) = \sum_{n=0}^{\infty} b_n x^n \qquad (1.55)$$

have a common interval of convergence, then the series of their sum and product, i.e.,

$$f(x) + g(x) = \sum_{n=0}^{\infty} (a_n + b_n) x^n \qquad (1.56)$$

and

$$f(x)g(x) = \sum_{n=0}^{\infty} \left(\sum_{k=0}^{n} a_k b_{n-k} \right) x^n \qquad (1.57)$$

also converge on this common interval of convergence. We recognize (1.57) as simply the Cauchy product introduced in Theorem 1.6. Finally, since power series are merely a special type of infinite series, the theorems in Section 1.3.1 concerning integration and differentiation of infinite series apply directly to convergent power series.

Example 7: Find the Maclaurin series for $e^x \sin x$.

Solution: By using well-known results, we have

$$e^x = \sum_{n=0}^{\infty} \frac{x^n}{n!} \quad \text{and} \quad \sin x = \sum_{n=0}^{\infty} \frac{(-1)^n x^{2n+1}}{(2n+1)!}$$

However, we cannot directly apply (1.57), since the series for $\sin x$ involves only odd powers of x. To remedy this situation, we rewrite the sine series in the form

$$\sin x = \sum_{n=0}^{\infty} \cos\left[(n-1)\frac{\pi}{2}\right] \frac{x^n}{n!}$$

where

$$\cos\left[(n-1)\frac{\pi}{2}\right] = \begin{cases} 0, & n \text{ even} \\ (-1)^{(n-1)/2}, & n \text{ odd} \end{cases}$$

Thus, the Cauchy product now leads to

$$e^x \sin x = \sum_{n=0}^{\infty} c_n x^n$$

where

$$c_n = \sum_{k=0}^{n} \frac{\cos[(k-1)\pi/2]}{k!(n-k)!}$$

Although it often happens that the expression for c_n cannot be simplified, here we find that we can actually evaluate the finite sum. By using the Euler formula $\cos x = \frac{1}{2}(e^{ix} + e^{-ix})$, together with properties of the binomial series, we find

$$c_n = \frac{1}{2n!} \sum_{k=0}^{n} \binom{n}{k} [e^{i(k-1)\pi/2} + e^{-i(k-1)\pi/2}]$$

$$= \frac{e^{-i\pi/2}}{2n!} \sum_{k=0}^{n} \binom{n}{k} e^{ik\pi/2} + \frac{e^{i\pi/2}}{2n!} \sum_{k=0}^{n} \binom{n}{k} e^{-ik\pi/2}$$

$$= \frac{e^{-i\pi/2}}{2n!} (1 + e^{i\pi/2})^n + \frac{e^{i\pi/2}}{2n!} (1 + e^{-i\pi/2})^n$$

Now writing

$$(1 + e^{i\pi/2})^n = e^{in\pi/4}(e^{i\pi/4} + e^{-i\pi/4})^n = 2^n e^{in\pi/4} \cos^n(\pi/4)$$
$$(1 + e^{-i\pi/2})^n = 2^n e^{-in\pi/4} \cos^n(\pi/4)$$

we deduce that

$$c_n = \frac{1}{2n!} [e^{i(n-2)\pi/4} + e^{-i(n-2)\pi/4}] 2^n \cos^n(\pi/4)$$

$$= \frac{2^{n/2}}{n!} \cos[(n-2)\pi/4]$$

and hence, we obtain our result*

$$e^x \sin x = \sum_{n=0}^{\infty} \frac{2^{n/2}}{n!} \cos[(n-2)\pi/4] \, x^n$$

which converges for all x.

---

*Notice that the terms corresponding to $n = 0, 4, 8, \ldots$ are all zero.

Closely related to the Cauchy product defined by Equation (1.57) is the power formula

$$\left(\sum_{k=0}^{\infty} a_k x^k\right)^n = \sum_{k=0}^{\infty} c_k x^k \qquad (1.58)$$

where

$$c_0 = a_0^n, \qquad c_k = \frac{1}{k a_0} \sum_{m=1}^{k} (mn - k + m) a_m c_{k-m}, \qquad k = 1, 2, 3, \ldots \qquad (1.59)$$

The reader should verify that this power formula is equivalent to the Cauchy product for $n = 2$. By repeated application of the Cauchy product for $n = 3, 4, 5, \ldots$, the above result can readily be obtained.

EXERCISES 1.3

In problems 1–4, use the ratio test to determine the interval of convergence. Check the (finite) endpoints of the interval for convergence by a separate test.

1. $\sum_{n=1}^{\infty} \dfrac{x^n}{n}$.

2. $\sum_{n=1}^{\infty} \dfrac{1 \times 3 \times 5 \times \cdots \times (2n-1)}{2 \times 5 \times 8 \times \cdots \times (3n-1)} (x-1)^n$.

3. $\sum_{n=0}^{\infty} \dfrac{n!}{2^n} x^n$.

4. $\sum_{n=0}^{\infty} n x^n$.

In problems 5–8, test the series for uniform convergence on the indicated interval.

5. $\sum_{n=1}^{\infty} \dfrac{\cos nx}{n^2}$, $-10 \le x \le 10$.

6. $\sum_{n=2}^{\infty} \dfrac{x^n}{n(\log n)^2}$, $-1 \le x \le 1$.

7. $\sum_{n=0}^{\infty} \left(\dfrac{1}{nx+2} - \dfrac{1}{nx+x+2}\right)$, $0 \le x \le 1$.

8. $\sum_{n=1}^{\infty} x(1+x)^{-n}$, $0 \le x \le 1$.

9. Using termwise integration, show that

$$\int_0^1 \left(\sum_{n=0}^{\infty} \frac{x^n}{n!}\right) dx = e - 1$$

In problems 10–13, indicate those series that can be differentiated termwise in the indicated interval.

10. $\sum_{n=0}^{\infty} \frac{x^n}{\sqrt{n}}$, $-1 \le x \le 0$.

11. $\sum_{n=0}^{\infty} \frac{e^{-nx}}{n(n+1)^2}$, $0 \le x \le 10$.

12. $\sum_{n=1}^{\infty} \left(\frac{x}{x-1}\right)^n$, $-4 \le x \le -3$.

13. $\sum_{n=1}^{\infty} \left(\frac{x^n}{n} - \frac{x^{n+1}}{n+1}\right)$, $0 \le x \le 1$.

14. Starting with the geometric series

$$\frac{1}{1-x} = \sum_{n=0}^{\infty} x^n, \quad -1 < x < 1$$

(a) make a change of variable to derive the Maclaurin series for

$$f(x) = \frac{1}{1+x^2}$$

(b) Use the answer in (a) to determine the Maclaurin series for arctan x. Give the interval of convergence.

15. Starting with the binomial series

$$(1+x)^a = \sum_{n=0}^{\infty} \binom{a}{n} x^n, \quad -1 < x < 1$$

(a) find the Maclaurin series for $f(x) = (1-x^2)^{-1/2}$.
(b) Use the answer in (a) to determine the Maclaurin series for arcsin x. Give the interval of convergence.

In problems 16–19, use the Cauchy product to find the Maclaurin series representation for the given function.

16. $f(x) = (1-x)^{-2}$.

17. $f(x) = \cos^2 x$.

18. $f(x) = \sin^2 x$.

19. $f(x) = e^x \cos x$.

20. Use (1.58) and (1.59) to determine the first *four* terms of $\cos^3 x$.

1.4 Asymptotic Series

In computational analysis we often seek to represent a given function $f(x)$ by some simpler function, say $g(x)$, that accurately describes the numerical values of $f(x)$ in the vicinity of a particular point $x = a$. Thus we write $f(x) \sim g(x)$ as $x \to a$ to mean

$$\lim_{x \to a} \frac{f(x)}{g(x)} = 1$$

Generally we confine our attention to either the case $x \to 0$ or $x \to \infty$, although we could also choose any other value of x.

For the case x near zero, we seek a representation of the form

$$f(x) \sim \sum_{n=0}^{\infty} c_n x^n, \qquad x \to 0, \tag{1.60}$$

from which we can deduce the simple *asymptotic formulas* $f(x) \sim c_0$ or $f(x) \sim c_0 + c_1 x$, and so on. Ordinarily we might obtain the representation (1.60) from the Maclaurin series expansion of $f(x)$. In such cases the series *converges* for all values of x such that $|x| < \rho$. That is, if

$$S_n(x) = \sum_{k=0}^{n} c_k x^k$$

then

$$\lim_{n \to \infty} |f(x) - S_n(x)| = 0 \tag{1.61}$$

for each fixed x in the region $|x| < \rho$. By taking a sufficient number of terms of the series, our calculations for $f(x)$ can be as accurate as desired. However, the representation (1.60) does not have to be a Maclaurin series, nor is there any requirement that the series converge in order to be useful for computations. That is, we define (1.60) to be an *asymptotic power series* for $f(x)$ as $x \to 0$ if and only if

$$\lim_{x \to 0} \frac{|f(x) - S_n(x)|}{|x|^n} = 0 \tag{1.62}$$

for each fixed n. By this condition we are requiring the sum of the terms of (1.60) out to the term $c_n x^n$ to approximate the function $f(x)$ more closely than $|x|^n$ approximates zero, by choosing x sufficiently close to zero. Hence, if the series (1.60) diverges, we find that the accuracy of computation is closely tied to the actual value of x and number of terms n. This means that after a certain number of terms the accuracy of computation will actually get *worse* instead of better—a sharp contrast as compared with convergent series.

1.4.1 Large Arguments

For large values of x we seek representations of the form

$$f(x) \sim \sum_{n=0}^{\infty} \frac{a_n}{x^n}, \quad x \to \infty \tag{1.63}$$

We call (1.63) an *asymptotic series* for large arguments; it is also commonly called a *semiconvergent series*. A precise definition of asymptotic series was first provided by J. H. Poincaré (1854–1912) in 1886; he stated that (1.63) is an asymptotic series if and only if, for each fixed n,

$$\lim_{x \to \infty} x^n |f(x) - S_n(x)| = 0 \tag{1.64}$$

where

$$S_n(x) = \sum_{k=0}^{n} \frac{a_k}{x^k}$$

Asymptotic series like (1.63) are intriguing in that they usually *diverge* for all values of x, but are still useful for computational purposes. In such cases, once again, too many terms of the series can lead to gross errors in computations, and therefore it is important to know just how many terms to retain for a particular computation. The error incurred in most cases turns out to be less than the first term omitted in the approximation.

Not all functions have an asymptotic series of the form (1.63). For example, neither e^x nor $\sin x$ has such an asymptotic expansion. If the function $f(x)$ itself has no asymptotic series, it may happen that there exists a suitable function $h(x)$ such that the quotient $f(x)/h(x)$ has an asymptotic series. In this case we write

$$f(x) \sim h(x) \sum_{n=0}^{\infty} \frac{a_n}{x^n}, \qquad x \to \infty \tag{1.65}$$

Necessary and sufficient conditions for $f(x)$ to possess an asymptotic series have been developed, but we will not discuss them.*

If $f(x)$ has an asymptotic series, it may turn out that other functions have the same asymptotic series. That is to say, an asymptotic series does not uniquely determine the function from which it was generated. However, if a function has an asymptotic series, it has only one such series.

There are several ways in which asymptotic series can be derived. For our first example, we wish to consider the case where the function is defined by an integral of the form

$$F(x) = \int_x^{\infty} f(t)\, dt \tag{1.66}$$

A simple and often effective way of developing the series in such cases consists of repeated integration by parts. Each new integration yields the next term in the expansion, and the error committed in stopping after n

---

*See F.W.J. Olver, *Asymptotics and Special Functions*, New York: Academic, 1974.

terms can be expressed by the remaining integral, for which error bounds can often be deduced.

Example 8: Find an asymptotic series for the function defined by

$$F(x) = \int_x^\infty \frac{e^{-t}}{t} \, dt$$

Solution: Using integration by parts with

$$u = \frac{1}{t}, \qquad dv = e^{-t} \, dt$$

$$du = -\frac{dt}{t^2}, \qquad v = -e^{-t}$$

we find

$$F(x) = -\left.\frac{e^{-t}}{t}\right|_x^\infty - \int_x^\infty \frac{e^{-t}}{t^2} \, dt$$

$$= \frac{e^{-x}}{x} - \int_x^\infty \frac{e^{-t}}{t^2} \, dt$$

Continued integration by parts leads to

$$F(x) = e^{-x}\left[\frac{1}{x} - \frac{1}{x^2} + \frac{1 \times 2}{x^3} - \cdots \right.$$

$$\left. + (-1)^{n-1}\frac{1 \times 2 \times \cdots \times (n-1)}{x^n}\right]$$

$$+ (-1)^n 1 \times 2 \times \cdots \times n \int_x^\infty \frac{e^{-t}}{t^{n+1}} \, dt$$

from which we deduce

$$F(x) \sim \frac{e^{-x}}{x} \sum_{n=0}^\infty \frac{(-1)^n n!}{x^n}, \qquad x \to \infty$$

Applying the ratio test, it can be shown that the asymptotic series in Example 8 diverges for all x. Yet it can also be shown that the error $E_n(x)$ committed in approximating $F(x)$ by the first n terms of the series is bounded by*

$$|E_n(x)| \leq \frac{n!}{x^n}, \qquad x \gg 1$$

---

*See N.N. Lebedev, *Special Functions and Their Applications*, New York: Dover, 1972, p. 33.

For large enough x and small n, the error $E_n(x)$ can be made quite small. On the other hand, if n is too large, the use of the asymptotic series to compute $F(x)$ can lead to extremely large errors.

Another way in which the asymptotic series is sometimes derived is illustrated by the following example.

Example 9: Find an asymptotic series for the function defined by

$$F(x) = \int_0^\infty e^{-xt}(1+t^2)^{-1} dt, \qquad x > 0$$

Solution: Here we find it convenient to start by making the change of variable $s = xt$, which leads to the expression

$$F(x) = \frac{1}{x}\int_0^\infty e^{-s}\left(1 + \frac{s^2}{x^2}\right)^{-1} ds$$

Then, by expanding $(1+s^2/x^2)^{-1}$ in a binomial series (see Example 6) and integrating the result termwise, we obtain

$$F(x) \sim \frac{1}{x}\sum_{n=0}^\infty \binom{-1}{n}\frac{1}{x^{2n}}\int_0^\infty e^{-s}s^{2n} ds, \qquad x \to \infty$$

This last integral can be evaluated by repeated integration by parts.* Upon so doing, we finally deduce that

$$F(x) \sim \sum_{n=0}^\infty \frac{(-1)^n (2n)!}{x^{2n+1}}, \qquad x \to \infty$$

where we have made use of the identity

$$\binom{-1}{n} = (-1)^n \binom{n}{n} = (-1)^n$$

The technique used in Example 9 is nonrigorous, and even somewhat incorrect in that the particular binomial series in the example converges only for $s < x$ and we are allowing s to be arbitrarily large. Moreover, as is usual, it has led to a series that diverges for all values of x.

In spite of the fact that they usually diverge, asymptotic series behave very much like convergent power series. For example, the asymptotic expansions of two functions can be added to form the asymptotic series of the sum of two functions. These same asymptotic series can be multiplied to form an asymptotic series of the product of the two functions. Also, the asymptotic series of a function can be integrated termwise (as if it were a

---

*$\int_0^\infty e^{-s}s^{2n} ds = (2n)!$, $n = 0, 1, 2, \ldots$.

uniformly convergent series of continuous functions), and the result will be an asymptotic expansion of the integral of the original function. Under more stringent conditions, the asymptotic series may even be differentiated termwise to produce the asymptotic series of the derivative of the original function.

EXERCISES 1.4

In problems 1–7, derive the given asymptotic series. Check convergence.

1. $\int_0^\infty e^{-xt}\cos t\, dt \sim \sum_{n=0}^\infty \dfrac{(-1)^n}{x^{2n+1}}$, $x \to \infty$.

2. $\int_0^\infty e^{-xt}\sin t\, dt \sim \sum_{n=1}^\infty \dfrac{(-1)^{n-1}}{x^{2n}}$, $x \to \infty$.

3. $\int_{-\infty}^x \dfrac{e^t}{t}\, dt \sim \dfrac{e^x}{x} \sum_{n=0}^\infty \dfrac{n!}{x^n}$, $x \to \infty$.

4. $\int_0^x \dfrac{dt}{\log t} \sim \dfrac{x}{\log x} \sum_{n=0}^\infty \dfrac{n!}{(\log x)^n}$, $x \to \infty$.

    Hint: Let $u = \log t$.

5. $\int_0^\infty \dfrac{e^{-xt}}{1+t}\, dt \sim \sum_{n=0}^\infty \dfrac{(-1)^n n!}{x^{n+1}}$, $x \to \infty$.

6. $\int_x^\infty e^{-t} t^{a-1}\, dt \sim x^{a-1} e^{-x}\left[1 + \sum_{n=1}^\infty \dfrac{(a-1)(a-2)\cdots(a-n)}{x^n}\right]$,
    $x \to \infty$, $a > 0$.

7. $\int_x^\infty e^{-t^2}\, dt \sim \dfrac{e^{-x^2}}{2x}\left[1 + \sum_{n=1}^\infty (-1)^n \dfrac{1 \times 3 \times \cdots \times (2n-1)}{(2x^2)^n}\right]$, $x \to \infty$.

8. Given
$$F(x) = \int_0^\infty \dfrac{e^{-t}}{1+xt}\, dt, \qquad x \geq 0$$
   show that
$$F(x) \sim \sum_{n=0}^\infty n!(-1)^n x^n, \qquad x \to 0^+$$
   Hint: Verify that Equation (1.62) is satisfied by first establishing
$$S_n(x) = \sum_{k=0}^n k!(-1)^k x^k = \int_0^\infty e^{-t} \sum_{k=0}^n (-1)^k (xt)^k\, dt$$

1.5 Fourier Trigonometric Series

The expansion of a function f in a power series requires (at least) that f be infinitely differentiable. However, many functions of practical interest do not satisfy such strong differentiability requirements, due to discontinuities, lack of smoothness, etc., and therefore cannot be represented in a power series. For such cases there are other types of series representations.

A particular type of series having a wide range of applications is the *Fourier trigonometric series* (or simply *Fourier series*)*

$$f(x) = \tfrac{1}{2}a_0 + \sum_{n=1}^{\infty}\left(a_n \cos\frac{n\pi x}{p} + b_n \sin\frac{n\pi x}{p}\right) \qquad (1.67)$$

where the constants a_0, a_n, and b_n are called the *Fourier coefficients* of the series. If the series representation is to be valid for all values of x, then clearly f must be a periodic function with period $2p$, since the right-hand side of (1.67) has this property. In other cases, the series (1.67) is useful for representing the function f only in the interval $-p \le x \le p$, so that the periodicity is of no concern.

Formally identifying the Fourier coefficients depends upon the evaluation of the integrals

$$\int_{-p}^{p}\cos\frac{n\pi x}{p}\,dx = \int_{-p}^{p}\sin\frac{n\pi x}{p}\,dx = \int_{-p}^{p}\sin\frac{n\pi x}{p}\cos\frac{k\pi x}{p}\,dx = 0 \qquad (1.68a)$$

and

$$\int_{-p}^{p}\cos\frac{n\pi x}{p}\cos\frac{k\pi x}{p}\,dx = \int_{-p}^{p}\sin\frac{n\pi x}{p}\sin\frac{k\pi x}{p}\,dx = \begin{cases} 0, & k \ne n \\ p, & k = n \end{cases} \qquad (1.68b)$$

where n and k both assume positive integer values. The details of verifying these integral relations are left to the exercises.

Assuming that termwise integration of (1.67) is permitted, we find

$$\int_{-p}^{p} f(x)\,dx = \tfrac{1}{2}a_0 \int_{-p}^{p} dx + \sum_{n=1}^{\infty}\left(a_n \int_{-p}^{p}\cos\frac{n\pi x}{p}\,dx + b_n \int_{-p}^{p}\sin\frac{n\pi x}{p}\,dx\right)$$

from which we deduce

$$a_0 = \frac{1}{p}\int_{-p}^{p} f(x)\,dx \qquad (1.69)$$

---
*It is customary to write the constant term in (1.67) as $\tfrac{1}{2}a_0$.

If we now multiply (1.67) by $\cos(k\pi x/p)$ and integrate once again, we have

$$\int_{-p}^{p} f(x)\cos\frac{k\pi x}{p}\,dx = \tfrac{1}{2}a_0 \overbrace{\int_{-p}^{p} \cos\frac{k\pi x}{p}\,dx}^{0}$$
$$+ \sum_{n=1}^{\infty}\left(a_n \underbrace{\int_{-p}^{p} \cos\frac{n\pi x}{p}\cos\frac{k\pi x}{p}\,dx}_{0\ (n\neq k)} + b_n \underbrace{\int_{-p}^{p} \sin\frac{n\pi x}{p}\cos\frac{k\pi x}{p}\,dx}_{0}\right)$$

This time all terms on the right go to zero except for the coefficient of a_n corresponding to $n = k$, and here we find

$$\int_{-p}^{p} f(x)\cos\frac{k\pi x}{p}\,dx = a_k \int_{-p}^{p} \cos^2\!\left(\frac{k\pi x}{p}\right)dx$$
$$= pa_k$$

or

$$a_k = \frac{1}{p}\int_{-p}^{p} f(x)\cos\frac{k\pi x}{p}\,dx, \quad k = 1, 2, 3, \ldots \tag{1.70a}$$

By a similar process, the multiplication of (1.67) by $\sin(k\pi x/p)$ and subsequent integration provides the final formula

$$b_k = \frac{1}{p}\int_{-p}^{p} f(x)\sin\frac{k\pi x}{p}\,dx, \quad k = 1, 2, 3, \ldots \tag{1.70b}$$

In summary, we have formally shown that if f has the representation

$$f(x) = \tfrac{1}{2}a_0 + \sum_{n=1}^{\infty}\left(a_n\cos\frac{n\pi x}{p} + b_n\sin\frac{n\pi x}{p}\right) \tag{1.71}$$

then the Fourier coefficients are given by [changing the index back to n and combining (1.69) and (1.70a)]

$$a_n = \frac{1}{p}\int_{-p}^{p} f(x)\cos\frac{n\pi x}{p}\,dx, \quad n = 0, 1, 2, \ldots \tag{1.72}$$

and

$$b_n = \frac{1}{p}\int_{-p}^{p} f(x)\sin\frac{n\pi x}{p}\,dx, \quad n = 1, 2, 3, \ldots \tag{1.73}$$

Example 10: Find the Fourier trigonometric series for the periodic function

$$f(x) = \begin{cases} 0, & -\pi < x < 0 \\ x, & 0 < x < \pi \end{cases}, \quad f(x + 2\pi) = f(x)$$

Solution: The Fourier coefficients computed from (1.72) and (1.73) with $p = \pi$ lead to

$$a_0 = \frac{1}{\pi}\int_{-\pi}^{\pi} f(x)\,dx = \frac{1}{\pi}\int_0^\pi x\,dx = \frac{\pi}{2}$$

$$a_n = \frac{1}{\pi}\int_0^\pi x\cos nx\,dx = \begin{cases} 0, & n = 2, 4, 6, \ldots \\ -\dfrac{2}{\pi n^2}, & n = 1, 3, 5, \ldots \end{cases}$$

and

$$b_n = \frac{1}{\pi}\int_0^\pi x\sin nx\,dx = \frac{(-1)^{n+1}}{n}, \qquad n = 1, 2, 3, \ldots$$

Substituting these results into (1.71), we obtain

$$f(x) = \frac{\pi}{4} - \frac{2}{\pi}\left(\cos x + \frac{\cos 3x}{3^2} + \frac{\cos 5x}{5^2} + \cdots\right)$$
$$+ \left(\sin x - \frac{\sin 2x}{2} + \frac{\sin 3x}{3} - \cdots\right)$$

or more compactly,

$$f(x) = \frac{\pi}{4} - \frac{2}{\pi}\sum_{n=1}^{\infty}\left[\frac{\cos(2n-1)x}{(2n-1)^2} + \frac{(-1)^{n+1}}{n}\sin nx\right]$$

We might observe that the function f in Example 10 is not differentiable at $x = 0$ and multiples of π. Thus, while it surely doesn't have a power-series expansion over any interval containing these points, its Fourier series converges for all x, even at the points of discontinuity (see Theorem 1.14 below).

Theorem 1.14 (*Pointwise convergence*). If $f(x + 2p) = f(x)$ for some p, and if f and f' are at least piecewise continuous in $-p \le x \le p$, then the Fourier series of f converges pointwise to $f(x)$ at all points of continuity of f. At points of discontinuity of f, the series converges to the average value $\frac{1}{2}[f(x^+) + f(x^-)]$.*

Remark: A function f is said to be *piecewise continuous* in an interval if it has only a finite number of discontinuities, and further, if all discontinuities are finite. This class of functions is discussed in more detail in

---

*For a proof of Theorem 1.14, see H.S. Carslaw, *Introduction to the Theory of Fourier's Series and Integrals*, New York: Dover, 1950.

Section 4.5.1. Also, $f(x^+)$ and $f(x^-)$ denote the limits of f at x from the right and left, respectively.

Theorem 1.14 is also valid for nonperiodic functions which satisfy the other stated conditions in some interval $c \le x \le c + 2p$, where c is any real number. In such cases the convergence at the endpoints of the interval will lead to the value $\frac{1}{2}[f(c^+) + f(c + 2p^-)]$. The Fourier coefficients are then computed by performing the integrations over the interval $c \le x \le c + 2p$. Finally, we remark that if we add to Theorem 1.14 the condition that f is also continuous, the Fourier series will then converge *uniformly*.

1.5.1 Cosine and Sine Series

If $f(-x) = f(x)$, we say that f is an *even function*, whereas if $f(-x) = -f(x)$, we say that f is an *odd function*. If the function f falls into one of these two classifications, certain simplifications in handling Fourier series takes place. Such simplifications are primarily consequences of the following result (see problems 8 and 9):

$$\int_{-p}^{p} f(x)\,dx = \begin{cases} 2\int_{0}^{p} f(x)\,dx & \text{if } f(x) \text{ is even} \\ 0 & \text{if } f(x) \text{ is odd} \end{cases} \quad (1.74)$$

If f is an even function, the product $f(x)\cos(n\pi x/p)$ is an even function while the product $f(x)\sin(n\pi x/p)$ is an odd function. (Why?) In this case, using (1.74), we see that the Fourier coefficients satisfy

$$a_n = \frac{1}{p}\int_{-p}^{p} f(x)\cos\frac{n\pi x}{p}\,dx$$

$$= \frac{2}{p}\int_{0}^{p} f(x)\cos\frac{n\pi x}{p}\,dx, \qquad n = 0, 1, 2, \ldots \quad (1.75)$$

and

$$b_n = \frac{1}{p}\int_{-p}^{p} f(x)\sin\frac{n\pi x}{p}\,dx = 0, \qquad n = 1, 2, 3, \ldots \quad (1.76)$$

Hence, for an even function f the Fourier series reduces to

$$f(x) = \tfrac{1}{2}a_0 + \sum_{n=1}^{\infty} a_n \cos\frac{n\pi x}{p} \quad (1.77)$$

where a_n ($n = 0, 1, 2, \ldots$) is defined by (1.75). We call such a series a *cosine series*.

Using a similar argument, when f is an odd function, the Fourier series reduces to the *sine series*

$$f(x) = \sum_{n=1}^{\infty} b_n \sin \frac{n\pi x}{p} \qquad (1.78)$$

where $a_n = 0$ $(n = 0, 1, 2, \ldots)$ and

$$b_n = \frac{2}{p}\int_0^p f(x)\sin \frac{n\pi x}{p}\, dx, \qquad n = 1, 2, 3, \ldots \qquad (1.79)$$

EXERCISES 1.5

In problems 1–6, determine the Fourier trigonometric series of each function.

1. $f(x) = \begin{cases} 1, & -p < x \leq 0, \\ \frac{1}{2}, & 0 < x \leq p. \end{cases}$

2. $f(x) = x, \; -\pi < x < \pi,$

3. $f(x) = |x|, \; -\pi \leq x \leq \pi.$

4. $f(x) = x^2, \; -1 < x \leq 1.$

5. $f(x) = \begin{cases} x, & -2 < x < 0, \\ 2 - x, & 0 < x < 2. \end{cases}$

6. $f(x) = \begin{cases} x + \pi, & -\pi < x < 0, \\ x - \pi, & 0 < x < \pi. \end{cases}$

7. Verify the integral relations (1.68a) and (1.68b).

   Hint: Use the trigonometric identities
   $$\sin A \sin B = \tfrac{1}{2}[\cos(A - B) - \cos(A + B)]$$
   $$\cos A \cos B = \tfrac{1}{2}[\cos(A - B) + \cos(A + B)]$$
   $$\sin A \cos B = \tfrac{1}{2}[\sin(A - B) + \sin(A + B)]$$

8. Prove that if f is an even function,
   $$\int_{-p}^{p} f(x)\, dx = 2\int_0^p f(x)\, dx$$

9. Prove that if f is an odd function,
   $$\int_{-p}^{p} f(x)\, dx = 0$$

10. Prove that
    (a) the product of two odd functions is even,
    (b) the product of two even functions is even,
    (c) the product of an even and an odd function is odd.

11. To what numerical value will the Fourier series in problem 1 converge at

(a) $x = 0$?
(b) $x = -p$?
(c) $x = p$?

12. A sinusoidal voltage $E \sin t$ is passed through a *half-wave rectifier*, which clips the negative portion of the wave. Find the Fourier series of the resulting waveform: $f(t) = 0$, $-\pi < t < 0$; $f(t) = E \sin t$, $0 < t < \pi$; and $f(t + 2\pi) = f(t)$.

13. Find the Fourier series of the periodic function resulting from passing the voltage $v(t) = E \cos 100\pi t$ through a half-wave rectifier (see problem 12).

14. A certain type of *full-wave rectifier* converts the input voltage $v(t)$ to its absolute value at the output, i.e., $|v(t)|$. Assuming the input voltage is given by $v(t) = E \sin \omega t$, determine the Fourier series of the periodic output voltage.

15. From the Fourier series developed in Example 10, show that

(a) $1 + \dfrac{1}{3^2} + \dfrac{1}{5^2} + \dfrac{1}{7^2} + \cdots = \dfrac{\pi^2}{8}$,

(b) $1 - \dfrac{1}{3} + \dfrac{1}{5} - \dfrac{1}{7} + \cdots = \dfrac{\pi}{4}$.

16. Starting with the Fourier series representation

$$\frac{x}{2} = \sum_{n=1}^{\infty} \frac{(-1)^{n-1}}{n} \sin nx, \qquad -\pi < x < \pi$$

obtain a Fourier series for x^2, $-\pi < x < \pi$, by integrating termwise.

17. If $f(x + 2p) = f(x)$, show that for any constant c

$$\int_{c-p}^{c+p} f(x)\, dx = \int_{-p}^{p} f(x)\, dx$$

Hint: Write

$$\int_{c-p}^{c+p} f(x)\, dx = \int_{c-p}^{-p} f(x)\, dx + \int_{-p}^{c+p} f(x)\, dx$$

and let $x = t + 2p$ in the first integral on the right-hand side.

1.6 Improper Integrals

Integrals which have an infinite limit of integration or an infinite discontinuity in the integrand between the limits of integration are called *improper integrals*. If a certain amount of care is not exercised in the evaluation of such integrals, we may derive results like

$$\int_{-1}^{1/2} \frac{dx}{x^2} = -\frac{1}{x}\bigg|_{-1}^{1/2} = -3$$

This is clearly an absurd result, since the integrand is always positive and therefore cannot lead to a negative value for the integral.

Our treatment of improper integrals here will be brief, since the theory so closely parallels that of infinite series.

1.6.1 Types of Improper Integrals

We say the function f is *bounded* on the interval $a \le t \le b$ provided there is some constant B such that

$$|f(t)| \le B, \qquad a \le t \le b$$

If this is not true, we say that f is *unbounded* on $a \le t \le b$. For example, the function $f(t) = e^{-t}$ is bounded for all $t \ge 0$, since $|e^{-t}| \le 1$, $t \ge 0$, whereas $g(t) = 1/t$ is unbounded on any interval containing $t = 0$.

If f is unbounded on $a \le t \le b$, then its integral over this interval is by definition improper. If f has only one infinite discontinuity and it occurs at $t = c$, then we write*

$$\int_a^b f(t)\, dt = \lim_{\varepsilon \to 0^+} \int_a^{c-\varepsilon} f(t)\, dt + \lim_{\varepsilon \to 0^+} \int_{c+\varepsilon}^b f(t)\, dt \qquad (1.80)$$

If both limits on the right exist, we say the integral *converges* to the sum of the limits; otherwise, it *diverges*.

Another type of improper integral arises when one or both limits of integration are infinite. In such cases we write

$$\int_a^\infty f(t)\, dt = \lim_{b \to \infty} \int_a^b f(t)\, dt \qquad (1.81a)$$

$$\int_{-\infty}^b f(t)\, dt = \lim_{a \to -\infty} \int_a^b f(t)\, dt \qquad (1.81b)$$

and

$$\int_{-\infty}^\infty f(t)\, dt = \lim_{a \to -\infty} \int_a^c f(t)\, dt + \lim_{b \to \infty} \int_c^b f(t)\, dt \qquad (1.82)$$

---
*If f has several infinite discontinuities, the interval can be decomposed and each limit evaluated separately.

Once again, the integrals are said to *converge* when the limits on the right exist, and *diverge* otherwise.

In some cases an integral may be classified improper for more than one reason. For example, the integral

$$\int_0^\infty e^{-t} t^{-1/2} \, dt$$

is improper because of the infinite limit of integration, but also because the integrand has an infinite discontinuity at $t = 0$.

As we did for infinite series, we distinguish between conditional and absolute convergence of improper integrals. For example, if the integral $\int_a^\infty |f(t)| \, dt$ converges, we say that $\int_a^\infty f(t) \, dt$ *converges absolutely*. However, if $\int_a^\infty f(t) \, dt$ converges but $\int_a^\infty |f(t)| \, dt$ diverges, we say the first integral *converges conditionally*.

1.6.2 Convergence Tests

Thus far our discussion of convergence and divergence of improper integrals has been based upon direct evaluation of the integral and appropriate limits. For many integrals this is not possible. For example, the integral

$$I = \int_0^\infty \frac{e^{-t}}{t^2 + 1} \, dt$$

cannot be evaluated by any direct method of integration from the calculus, and yet we may still wish to arrive at a conclusion regarding its convergence. For instance, it would be a waste of time (and money) to attempt to evaluate this integral numerically if it could be shown that the integral in fact diverges. For this reason, various tests of convergence or divergence have been developed which answer the question without directly evaluating the integral and taking appropriate limits.

Improper integrals involving either $\cos t$ or $\sin t$ are quite prevalent in practice. Such integrals are analogous to alternating series which contain the factor $(-1)^n$. Without proof, we state the following important theorem concerning the convergence of these integrals.

Theorem 1.15. If f is continuous and decreasing for all $t \geq a$, and furthermore, if $\lim_{t \to \infty} f(t) = 0$, then the integrals

$$\int_a^\infty f(t) \cos t \, dt \quad \text{and} \quad \int_a^\infty f(t) \sin t \, dt$$

both converge (at least conditionally).

If $\int_a^\infty f(t)\,dt$ converges, then the integrals in Theorem 1.15 both converge *absolutely*. The convergence is only *conditional*, however, if $\int_a^\infty f(t)\,dt$ diverges.

The following two limit tests are quite useful in proving either absolute convergence or divergence of certain improper integrals.

Theorem 1.16. If f is continuous for all $t \geq a$, and if
$$\lim_{t \to \infty} t^p f(t) = A, \qquad p > 1$$
where A is finite, then $\int_a^\infty f(t)\,dt$ converges absolutely.

Theorem 1.17. If f is continuous for all $t \geq a$, and if
$$\lim_{t \to \infty} tf(t) = A \neq 0$$
where A can be finite or infinite, then $\int_a^\infty f(t)\,dt$ diverges. If $A = 0$, the test fails.

We have stated the above theorems for improper integrals of a particular type; similar theorems have been developed for other types. Also, there are numerous other convergence tests that have been devised over the years, but we will not discuss them.

Example 11: Show that $\int_0^\infty e^{-t^2}\,dt$ converges absolutely.

Solution: By taking $p = 2$ and applying the hypothesis of Theorem 1.16, we see that
$$\lim_{t \to \infty} t^2 e^{-t^2} = 0$$
and thus we conclude that the integral converges absolutely.

Example 12: Show that $\int_2^\infty 1/(\log t)\,dt$ does not converge.

Solution: Here we find that
$$\lim_{t \to \infty} \frac{t}{\log t} = \infty$$
and thus by Theorem 1.17 the integral diverges.

1.6.3 Pointwise and Uniform Convergence

It frequently happens that the integral of interest is of the form
$$F(x) = \int_a^\infty f(x,t)\,dt \qquad (1.83)$$

where x is a parameter that can assume various values. Such integrals may converge for certain values of x and diverge for other values. Hence, if for certain fixed values of x the integral sums to $F(x)$, we say the integral *converges pointwise* to $F(x)$. The collection of all such points constitutes the domain of the function F.

Remark: Integrals of the type (1.83) are similar to the series of functions discussed in Section 1.3.

For many purposes it is important to establish *uniform convergence* of integrals like (1.83). The notion of uniform convergence of improper integrals can be introduced by analogy with infinite series. Here we find it convenient to define the "partial integral"

$$S_R(x) = \int_a^R f(x,t)\,dt \qquad (1.84)$$

Definition 1.4. If, given some $\varepsilon > 0$, there exists a number Q, independent of x in the interval $c \leq x \leq d$, such that

$$|F(x) - S_R(x)| < \varepsilon$$

whenever $R > Q$, then the integral (1.83) is said to *converge uniformly* to $F(x)$ in the interval $c \leq x \leq d$.

Analogous to Theorem 1.7 is the following *Weierstrass M-test* for improper integrals.

Theorem 1.18 (*Weierstrass M-test*). Let $f(x,t)$ be a continuous function of x and t, for all $t \geq a$ and all x in the interval $c \leq x \leq d$, for which $|f(x,t)| \leq M(t)$ when $t \geq t_0 > a$, where t_0 is some fixed value. Then, if the improper integral $\int_{t_0}^\infty M(t)\,dt$ converges, it follows that $\int_a^\infty f(x,t)\,dt$ converges uniformly in $c \leq x \leq d$.

Example 13: Show that $\int_1^\infty e^{-t} t^{x-1}\,dt$ converges uniformly in $1 \leq x \leq 2$.

Solution: If we select $M(t) = t^2 e^{-t}$, then clearly

$$|e^{-t} t^{x-1}| \leq t^2 e^{-t}, \qquad 1 \leq x \leq 2, \quad t \geq 1$$

Also,

$$\lim_{t \to \infty} t^2 M(t) = 0$$

so by virtue of Theorem 1.16 (with $p = 2$) the integral $\int_1^\infty M(t)\,dt$ converges. It now follows from the Weierstrass M-test that the given integral $\int_1^\infty e^{-t} t^{x-1}\,dt$ converges uniformly in $1 \leq x \leq 2$.

42 • Special Functions for Engineers and Applied Mathematicians

The following three theorems on uniform convergence are important in much of our work in later chapters.

Theorem 1.19. If $f(x,t)$ is continuous in $c \le x \le d$, $t \ge a$, and $\int_a^\infty f(x,t)\,dt$ converges uniformly to $F(x)$ in $c \le x \le d$, then $F(x)$ is continuous in $c \le x \le d$.

Theorem 1.20. If $f(x,t)$ is continuous in $c \le x \le d$, $t \ge a$, and $\int_a^\infty f(x,t)\,dt$ converges uniformly to $F(x)$ in $c \le x \le d$, then

$$\int_c^d F(x)\,dx = \int_a^\infty \int_c^d f(x,t)\,dx\,dt$$

Theorem 1.21. If

$$f(x,t) \quad \text{and} \quad \frac{\partial f}{\partial x}(x,t)$$

are continuous in $c \le x \le d$, $t \ge a$, the integral $\int_a^\infty f(x,t)\,dt$ converges to $F(x)$ in $c \le x \le d$, and if

$$\int_a^\infty \frac{\partial f}{\partial x}(x,t)\,dt$$

converges uniformly in the interval $c \le x \le d$, then

$$F'(x) = \int_a^\infty \frac{\partial f}{\partial x}(x,t)\,dt, \qquad c \le x \le d$$

Notice that the conditions required to justify differentiation under the integral sign are much more stringent than those to justify integration under the integral sign. Analogously to infinite series, we see that the basic requirement for differentiation under the integral sign is uniform convergence of the integral of the derivative of f.

Example 14: Derive the integral formula

$$\int_0^\infty \frac{e^{-pt} - e^{-qt}}{t}\,dt = \log\left(\frac{q}{p}\right), \qquad 0 < p < q$$

Solution: Because the integral is not an elementary integral, we cannot derive the result directly. However, an indirect approach can be used which relies on the notion of uniform convergence.

To begin, we observe that the integral

$$\frac{1}{x} = \int_0^\infty e^{-xt}\,dt$$

converges uniformly for $0 < p \leq x \leq q$. Hence, by application of Theorem 1.20 we have that

$$\int_p^q \frac{dx}{x} = \int_0^\infty \int_p^q e^{-xt}\, dx\, dt$$

from which we deduce our result

$$\log\left(\frac{q}{p}\right) = \int_0^\infty \frac{e^{-pt} - e^{-qt}}{t}\, dt$$

EXERCISES 1.6

In problems 1–6, determine if the integrals exist, and if so, evaluate them by an appropriate method.

1. $\int_0^1 t^{-1/2}\, dt.$

2. $\int_0^1 \frac{dt}{t}.$

3. $\int_2^\infty t(t^2 + 1)^{-3}\, dt.$

4. $\int_{-\infty}^\infty \frac{dt}{t^2 + 1}.$

5. $\int_0^2 (4 - t^2)^{-1/2}\, dt.$

6. $\int_0^\infty \sin t\, dt.$

In problems 7–10, use the limit tests (Theorems 1.16 and 1.17) to prove absolute convergence or divergence of the integrals.

7. $\int_0^\infty \frac{\cos t}{\sqrt{1 + t^3}}\, dt.$

8. $\int_0^\infty \frac{5e^{-t} - 3}{(1 + 2t^2)^{1/3}}\, dt.$

9. $\int_0^\infty e^{-t} t^{-1/2}\, dt.$

10. $\int_1^\infty t^{-2}(1 + t) e^t\, dt.$

In problems 11–15, use Theorem 1.18 to prove that the integral converges uniformly in the indicated interval.

11. $\int_1^\infty \frac{x\, dt}{3x^2 + t^2},\ 1 \leq x \leq 2.$

12. $\int_0^\infty \frac{\sin xt}{t}\, dt,\ 1 \leq x \leq 10.$

13. $\int_0^1 e^{-t} t^{x-1}\, dt,\ 0.1 \leq x \leq 1.$

14. $\int_0^\infty e^{-x^2 t^2}\, dt,\ 1 \leq x \leq 10.$

15. $\int_0^\infty \frac{\cos xt}{t^2 + 4}\, dt,\ -10 \leq x \leq 10.$

16. Use the integral relation

$$\frac{1}{1+x^2} = \int_0^\infty e^{-t}\cos xt\, dt$$

to deduce the function $F(x)$ represented by

$$F(x) = \int_0^\infty e^{-t}\frac{\sin xt}{t}\, dt$$

17. Use the integral relation

$$\frac{x}{x^2+1} = \int_0^\infty e^{-xt}\cos t\, dt, \qquad x > 0$$

to deduce the value of the integral

$$I = \int_0^\infty te^{-2t}\cos t\, dt$$

18. Use the integral relation

$$(b^2 - x^2)^{-1/2} = \int_0^\infty \cos(xt)\, J_0(bt)\, dt, \qquad b > x \geq 0$$

where $J_0(x)$ is a *Bessel function* (see Chapter 6), to deduce the relation

$$\sin^{-1}\!\left(\frac{1}{b}\right) = \int_0^\infty \frac{\sin t}{t} J_0(bt)\, dt, \qquad b > 1$$

19. Given that

$$\frac{\pi}{2} x^{-1/2} = \int_0^\infty \frac{dt}{t^2 + x}, \qquad x > 0$$

show that (for $n = 1, 2, 3, \ldots$)

$$\int_0^\infty \frac{dt}{(t^2 + x)^{n+1}} = \frac{\pi(2n)!}{2^{2n+1}(n!)^2} x^{-n-1/2}$$

20. Given that (for $n = 0, 1, 2, \ldots$)

$$P_n(x) = \frac{(-1)^n}{n!\sqrt{\pi}} \int_{-\infty}^\infty e^{-(1-x^2)t^2} \frac{d^n}{dx^n}\left(e^{-x^2 t^2}\right) dt$$

show that

$$xP_n'(x) - P_{n-1}'(x) = nP_n(x)$$

[$P_n(x)$ is the nth Legendre polynomial. See Chapter 4.]

1.7 Infinite Products

Given the infinite sequence of positive numbers $u_1, u_2, \ldots, u_n, \ldots$, we can express their product by the notation

$$u_1 \times u_2 \times u_3 \cdots u_n \cdots = \prod_{n=1}^{\infty} u_n \qquad (1.85)$$

By analogy with infinite series, we define the *partial product*

$$P_n = \prod_{k=1}^{n} u_k \qquad (1.86)$$

and investigate the limit

$$\lim_{n \to \infty} P_n = P \qquad (1.87)$$

If P is finite (but not zero) we say the infinite product (1.85) *converges* to P; otherwise, that it *diverges*. The product (1.85) may diverge because the limit (1.87) fails to exist, but also because $P = 0$, in which case we say the infinite product *diverges to zero*. We will not discuss infinite products that diverge to zero.

Because the infinite product will become infinite if $\lim_{n \to \infty} u_n > 1$ or diverge to zero if $0 < \lim_{n \to \infty} u_n < 1$, we find it convenient to write $u_n = 1 + a_n$ and then discuss infinite products of the form $\prod_{n=1}^{\infty} (1 + a_n)$. Based upon the above remarks, it is clear that a necessary (but not sufficient) condition for the infinite product $\prod_{n=1}^{\infty} (1 + a_n)$ to converge is that (see problem 1)

$$\lim_{n \to \infty} a_n = 0 \qquad (1.88)$$

Remark: Our original assumption was that the sequence $u_1, u_2, \ldots, u_n, \ldots$ was composed of positive numbers. Hence it follows that $a_n > -1$ for all n. However, should m of the original numbers be negative, we can replace their product by $(-1)^m$ times the product of their absolute values.

Example 15: Find the value of the infinite product $\prod_{n=2}^{\infty} (1 - 1/n^2)$.

Solution: We first make the observation that

$$\lim_{n \to \infty} a_n = -\lim_{n \to \infty} \frac{1}{n^2} = 0$$

which is required for convergence. To find the value of the product, we try to obtain an expression for the partial product P_n and take its limit.

The product of the first n terms leads to

$$\begin{aligned}P_n &= \prod_{k=2}^{n}\left(1 - \frac{1}{k^2}\right) \\ &= \prod_{k=2}^{n} \frac{(k-1)(k+1)}{k^2} \\ &= \frac{1 \times 2 \times 3 \times \cdots \times (n-1) \times 3 \times 4 \times 5 \times \cdots \times (n+1)}{2 \times 2 \times 3 \times 3 \times 4 \times 4 \times \cdots \times n \times n} \\ &= \frac{n+1}{2n}\end{aligned}$$

where in the last step we have canceled all common factors. Thus, by taking the limit

$$\lim_{n \to \infty} P_n = \lim_{n \to \infty} \frac{n+1}{2n} = \frac{1}{2}$$

we conclude that

$$\prod_{n=2}^{\infty}\left(1 - \frac{1}{n^2}\right) = \frac{1}{2}$$

1.7.1 Associated Infinite Series

In many cases of interest we are unable to find an explicit expression for the partial product P_n and examine its limit as we did in Example 15. When this is the case, it is useful to have tests of convergence as we did in studying infinite series. Although we could devise convergence tests based directly on the product, there are related infinite series whose convergence or divergence will settle the question in regards to the infinite product.

For example, closely associated with all infinite products is the infinite series of logarithms derived from

$$\log \prod_{n=1}^{\infty}(1 + a_n) = \sum_{n=1}^{\infty} \log(1 + a_n) \tag{1.89}$$

where it is assumed that no $a_n = -1$. If we denote the partial product and partial sum, respectively, by

$$P_n = \prod_{k=1}^{n}(1 + a_k), \qquad S_n = \sum_{k=1}^{n} \log(1 + a_k)$$

then clearly

$$\lim_{n \to \infty} P_n = \lim_{n \to \infty} \exp(S_n) = \exp\left(\lim_{n \to \infty} S_n\right)$$

through properties of limits. Therefore we see that P_n approaches a limit value P ($P \neq 0$) if and only if S_n has a limit value S.

A result more useful than considering the associated series of logarithms is contained in the following theorem.

Theorem 1.22. If $0 \leq a_n < 1$ for all $n > N$, the infinite products $\prod_{n=1}^{\infty}(1 + a_n)$ and $\prod_{n=1}^{\infty}(1 - a_n)$ converge or diverge according to whether the infinite series $\sum_{n=1}^{\infty} a_n$ converges or diverges.

1.7.2 Products of Functions

When the general term of the product is a function of x, we are led to infinite products of the form

$$f(x) = \prod_{n=1}^{\infty} [1 + a_n(x)] \tag{1.90}$$

If, for a fixed value of x, the product (1.90) equals $f(x)$, we say the product *converges pointwise* to $f(x)$. The general theory of representing functions by infinite products of the form (1.90) goes beyond the intended scope of this text, and thus we will treat only some special cases.

Remark: The notion of *uniform convergence* of infinite products plays an important role in the theory of infinite products, much as it does in the theory of infinite series and improper integrals. The usual way in which uniform convergence is established for infinite products is by (another) *Weierstrass M-test*. The interested reader should consult E.D. Rainville, *Special Functions*, New York: Chelsea, 1960, p. 6.

Recall from algebra that an nth-degree polynomial $p_n(x)$ with n real roots (zeros) can be expressed in the product form*

$$p_n(x) = (x - x_1)(x - x_2) \cdots (x - x_n) = \prod_{k=1}^{n} (x - x_k) \tag{1.91}$$

We might well wonder if functions with an infinite number of zeros have similar product representations. It turns out this is sometimes indeed the case. For example, the zeros of $\sin \pi x$ occur at $x = \pm n$ ($n = 0, 1, 2, \ldots$), and it can be shown that (see problem 8)

$$\sin \pi x = \pi x \prod_{n=1}^{\infty} \left(1 - \frac{x}{n}\right)\left(1 + \frac{x}{n}\right)$$

---

*For simplicity of notation we are assuming $p_n(x) = x^n + \cdots$, where the leading coefficient is unity.

or
$$\sin \pi x = \pi x \prod_{n=1}^{\infty}\left(1 - \frac{x^2}{n^2}\right) \quad (1.92)$$

whereas for the cosine function

$$\cos \pi x = \prod_{n=1}^{\infty}\left[1 - \frac{4x^2}{(2n-1)^2}\right] \quad (1.93)$$

An interesting result can be derived from (1.92) by setting $x = \frac{1}{2}$. That is,

$$1 = \frac{\pi}{2}\prod_{n=1}^{\infty}\left[1 - \frac{1}{(2n)^2}\right] = \frac{\pi}{2}\prod_{n=1}^{\infty}\frac{(2n-1)(2n+1)}{(2n)^2}$$

and by solving for $\pi/2$, we see that

$$\frac{\pi}{2} = \frac{2 \times 2}{1 \times 3} \cdot \frac{4 \times 4}{3 \times 5} \cdot \frac{6 \times 6}{5 \times 7} \cdots \quad (1.94)$$

which is Wallis's famous formula for $\pi/2$. In Section 2.2.4 we will again use (1.92) to derive another interesting relation between the sine function and the gamma function.

EXERCISES 1.7

1. If $\prod_{n=1}^{\infty}(1 + a_n)$ converges to the value $P \neq 0$, show that

$$\lim_{n \to \infty} a_n = 0$$

Hint: Consider the ratio $\displaystyle\lim_{n \to \infty} \frac{\prod_{k=1}^{n}(1 + a_k)}{\prod_{k=1}^{n-1}(1 + a_k)}$.

In problems 2–6, show that the infinite product converges by finding its value.

2. $\displaystyle\prod_{n=1}^{\infty}\left[1 - \frac{2}{(n+1)(n+2)}\right]$.

3. $\displaystyle\prod_{n=1}^{\infty}\left[1 + \frac{6}{(n+1)(2n+9)}\right]$.

4. $\displaystyle\prod_{n=1}^{\infty}\left(1 - \frac{1}{4n^2}\right)$.

5. $\prod_{n=1}^{\infty}\left[1+\dfrac{1}{(4n-1)(4n-3)}\right].$

6. $\prod_{n=1}^{\infty}\left[1+\dfrac{(-1)^{n+1}}{n}\right].$

7. Use (1.92) and (1.93) to verify the identity
$$2\sin x\cos x = \sin 2x.$$

8. Given the function $f(x) = \cos kx$, $-\pi \le x \le \pi$, where k is not an integer,

    (a) find its Fourier trigonometric series.
    (b) Letting $k = z$ in (a), and substituting $x = 0$ and $x = \pi$, obtain the series expansions
$$\csc \pi z = \dfrac{1}{\pi z} + \dfrac{2z}{\pi}\sum_{n=1}^{\infty}\dfrac{(-1)^n}{z^2-n^2}$$
$$\cot \pi z = \dfrac{1}{\pi z} + \dfrac{2z}{\pi}\sum_{n=1}^{\infty}\dfrac{1}{z^2-n^2}$$

    (c) Assume $0 < z < 1$ and integrate the series in (b) for $\cot \pi z$ from 0 to x, $0 < x < 1$, and show that
$$\log\dfrac{\sin \pi x}{\pi x} = \sum_{n=1}^{\infty}\log\left(1-\dfrac{x^2}{n^2}\right)$$
    so that
$$\sin \pi x = \pi x \prod_{n=1}^{\infty}\left(1-\dfrac{x^2}{n^2}\right)$$

    (d) From (c), deduce that
$$\sin x = x\prod_{n=1}^{\infty}\left(1-\dfrac{x^2}{n^2\pi^2}\right)$$

9. By using the results of problem 8, show that
$$\int_0^{\infty}\dfrac{x^{z-1}}{1+x}\,dx = \dfrac{\pi}{\sin \pi z}, \quad 0 < z < 1$$

Hint: Express the integral as a sum of two integrals, the first having $(0, 1)$ as the interval of integration and the second $(1, \infty)$. Then let $x = 1/t$ in the second integral, and use the geometric series for $(1 + x)^{-1}$.

2

The Gamma Function and Related Functions

2.1 Introduction

In the eighteenth century, L. Euler (1707–1783) concerned himself with the problem of interpolating between the numbers

$$n! = \int_0^\infty e^{-t} t^n \, dt, \qquad n = 0, 1, 2, \ldots$$

with nonintegral values of n. This problem led Euler in 1729 to the now famous *gamma function*, a generalization of the factorial function that gives meaning to $x!$ when x is any positive number. His result can be extended to certain negative numbers and even to complex numbers. The notation $\Gamma(x)$ that is now widely accepted for the gamma function is not due to Euler, however, but was introduced in 1809 by A. Legendre (1752–1833), who was also responsible for the *duplication formula* for the gamma function. Nearly 150 years after Euler's discovery of it, the theory concerning the gamma function was greatly expanded by means of the theory of entire functions developed by K. Weierstrass (1815–1897).

Because it is a generalization of $n!$, the gamma function has been examined over the years as a means of generalizing certain functions, operations, etc., that are commonly defined in terms of factorials. In addition to these applications, the gamma function is useful in the evaluation of many nonelementary integrals; the same is true of the related beta function, often called the Eulerian integral of the first kind. In 1771, forty-three years after discovering the gamma function, Euler discovered that the beta function is actually a particular combination of gamma functions.

The logarithmic derivative of the gamma function leads to the *digamma function*. Further differentiation of the digamma function produces the family of *polygamma functions*, all of which are also related to the *zeta function* of G. Riemann (1826–1866).

2.2 Gamma Function

One of the simplest but very important special functions is the *gamma function*. It appears occasionally by itself in physical applications (mostly in the form of some integral), but much of its importance stems from its usefulness in developing other functions such as *Bessel functions* (Chapter 6) and *hypergeometric functions* (Chapters 8–10), which have more direct physical application.

The gamma function has several equivalent definitions, most of which are due to Euler. To begin, we define it by*

$$\Gamma(x) = \lim_{n \to \infty} \frac{n! n^x}{x(x+1)(x+2) \cdots (x+n)} \qquad (2.1)$$

If x is not zero or a negative integer, it can be shown that the limit (2.1) exists.[†] It is apparent, however, that $\Gamma(x)$ cannot be defined at $x = 0, -1, -2, \ldots$, since the limit becomes infinite for any of these values. Let us formalize this last statement as a theorem.

Theorem 2.1. If $x = -n$ $(n = 0, 1, 2, \ldots)$, then $|\Gamma(x)| = \infty$, or equivalently,

$$\frac{1}{\Gamma(-n)} = 0, \qquad n = 0, 1, 2, \ldots$$

By setting $x = 1$ in Equation (2.1), we see that

$$\Gamma(1) = \lim_{n \to \infty} \frac{n! n}{1 \times 2 \times 3 \times \cdots \times n(n+1)} = \lim_{n \to \infty} \frac{n}{n+1}$$

---
*A variation of (2.1), called *Euler's infinite product* (see problem 43), was actually the starting point of Euler's work on the interpolation problem for $n!$.
[†]See E.D. Rainville, *Special Functions*. New York: Chelsea, 1960, p. 5.

from which we deduce the special value

$$\Gamma(1) = 1 \tag{2.2}$$

Other values of $\Gamma(x)$ are not so easily obtained, but the substitution of $x + 1$ for x in (2.1) leads to

$$\Gamma(x + 1) = \lim_{n \to \infty} \frac{n! n^{x+1}}{(x + 1)(x + 2) \cdots (x + n)(x + n + 1)}$$

$$= \lim_{n \to \infty} \frac{nx}{x + n + 1} \cdot \lim_{n \to \infty} \frac{n! n^x}{x(x + 1) \cdots (x + n)}$$

from which we deduce the *recurrence formula*

$$\Gamma(x + 1) = x\Gamma(x) \tag{2.3}$$

Equation (2.3) is the basic functional relation for the gamma function; it is in the form of a *difference equation*. While many of the special functions satisfy some linear *differential equation*, it has been shown that the gamma function does not satisfy any linear differential equation with rational coefficients.*

A direct connection between the gamma function and factorials can be obtained from (2.2) and (2.3). That is, if we combine these relations, we have

$$\Gamma(2) = 1 \times \Gamma(1) = 1$$
$$\Gamma(3) = 2 \times \Gamma(2) = 2 \times 1 = 2!$$
$$\Gamma(4) = 3 \times \Gamma(3) = 3 \times 2! = 3!$$
$$\vdots$$

and through mathematical induction it can be shown that

$$\Gamma(n + 1) = n!, \quad n = 0, 1, 2, \ldots \tag{2.4}$$

Thus the gamma function is a generalization of the factorial function from the domain of positive integers to the domain of all real numbers (except as noted in Theorem 2.1). Also, Equation (2.4) confirms a result which beginning algebra students often find puzzling to understand, viz., $0! = 1$.

It is sometimes considered a nuisance that $n!$ is not $\Gamma(n)$, but $\Gamma(n + 1)$. Because of this, some authors adopt the notation $x!$ for the gamma

---

*See R. Campbell, *Les intégrals Eulériennes et leurs applications*, Paris: Dunod, 1966, pp. 152–159.

function, whether or not x is an integer. C. Gauss (1777–1855) introduced the notation $\Pi(x)$, where $\Pi(x) = x!$, but this notation is seldom utilized. The symbol Γ, due to Legendre, is the most widely used today. We will not use the notation of Gauss, nor will we use the factorial notation except when dealing with nonnegative integer values.

2.2.1 Integral Representations

Our reason for using the limit definition (2.1) of the gamma function is mostly historical, but also that it defines the gamma function for negative values of x as well as positive values. The gamma function rarely appears in the form (2.1) in applications. Instead, it most often arises in the evaluation of certain integrals; for example, Euler was able to show that*

$$\Gamma(x) = \int_0^\infty e^{-t} t^{x-1} dt, \qquad x > 0 \tag{2.5}$$

This *integral representation* of $\Gamma(x)$ is the most common way in which the gamma function is now defined. Since integrals are fairly easy to manipulate, (2.5) is often preferred to (2.1) for developing properties of this function. Equation (2.5) is less general than (2.1), however, since the variable x is restricted in (2.5) to positive values. Lastly, we note that (2.5) is an improper integral, due to the infinite limit of integration and also because the factor t^{x-1} becomes infinite at $t = 0$ for values of x in the interval $0 < x < 1$. Nonetheless, the integral (2.5) is *uniformly convergent* for all $a \le x \le b$, where $0 < a \le b < \infty$.

Let us first establish the equivalence of (2.1) and (2.5) for positive values of x. To do so, we set

$$\begin{aligned} F(x) &= \int_0^\infty e^{-t} t^{x-1} dt \\ &= \lim_{n \to \infty} \int_0^n \left(1 - \frac{t}{n}\right)^n t^{x-1} dt, \qquad x > 0 \end{aligned} \tag{2.6}$$

where we are making the observation

$$e^{-t} = \lim_{n \to \infty} \left(1 - \frac{t}{n}\right)^n \tag{2.7}$$

Using successive integration by parts, after making the change of variable

---

*Legendre termed the right-hand side of (2.5) the *Eulerian integral of the second kind*.

$z = t/n$, we find

$$F(x) = \lim_{n \to \infty} n^x \int_0^1 (1-z)^n z^{x-1} dx$$

$$= \lim_{n \to \infty} n^x \left[(1-z)^n \frac{z^x}{x} \Big|_0^1 + \frac{n}{x} \int_0^1 (1-z)^{n-1} z^x dz \right]$$

$$= \cdots \qquad (2.8)$$

$$= \lim_{n \to \infty} n^x \left[\frac{n(n-1) \cdots 2 \times 1}{x(x+1) \cdots (x+n-1)} \int_0^1 z^{x+n-1} dz \right]$$

$$= \lim_{n \to \infty} \frac{n! n^x}{x(x+1)(x+2) \cdots (x+n)}$$

and thus we have shown that

$$F(x) = \int_0^\infty e^{-t} t^{x-1} dt = \Gamma(x), \qquad x > 0 \qquad (2.9)$$

It follows from the uniform convergence of the integral (2.5) that $\Gamma(x)$ is a continuous function for all $x > 0$ (see Theorem 1.19). To investigate the behavior of $\Gamma(x)$ as x approaches the value zero from the right, we use the recurrence formula (2.3) written in the form

$$\Gamma(x) = \frac{\Gamma(x+1)}{x}$$

Thus, we see that

$$\lim_{x \to 0^+} \Gamma(x) = \lim_{x \to 0^+} \frac{\Gamma(x+1)}{x} = +\infty \qquad (2.10)$$

Another consequence of the uniform convergence of the defining integral for $\Gamma(x)$ is that we may differentiate the function under the integral sign to obtain*

$$\Gamma'(x) = \int_0^\infty e^{-t} t^{x-1} \log t \, dt, \qquad x > 0 \qquad (2.11)$$

and

$$\Gamma''(x) = \int_0^\infty e^{-t} t^{x-1} (\log t)^2 dt, \qquad x > 0 \qquad (2.12)$$

---

*Actually, to completely justify the derivative relations (2.11) and (2.12) requires that we first establish the uniform convergence of the integrals in them. See Theorem 1.21 in Section 1.6.3.

The integrand in (2.12) is positive over the entire interval of integration, and thus it follows that $\Gamma''(x) > 0$. This implies that the graph of $y = \Gamma(x)$ is *concave upward* for all $x > 0$. While maxima and minima are ordinarily found by setting the derivative of the function to zero, here we make the observation that, since $\Gamma(1) = \Gamma(2) = 1$ and $\Gamma(x)$ is always concave upward, the gamma function has *only a minimum* on the interval $x > 0$. Moreover, the minimum occurs on the interval $1 < x < 2$. The exact position of the minimum was first computed by Gauss and found to be $x_0 = 1.4616...$, which leads to the minimum value $\Gamma(x_0) = 0.8856...$. Lastly, from the continuity of $\Gamma(x)$ and its concavity, we deduce that

$$\lim_{x \to +\infty} \Gamma(x) = +\infty \tag{2.13}$$

With this last result, we have determined the fundamental characteristics of the graph of the gamma function for $x > 0$ (see Fig. 2.1).

The gamma function is defined for negative values of x by Equation (2.1), but can be evaluated more conveniently by using the recurrence formula

$$\Gamma(x) = \frac{\Gamma(x+1)}{x}, \qquad x \neq 0, -1, -2, \ldots \tag{2.14}$$

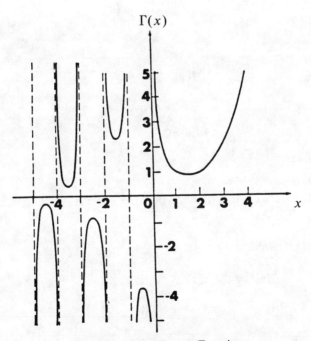

Figure 2.1 The Gamma Function

We are particularly interested in the behavior of the gamma function in the vicinity of the discontinuities at $x = 0, -1, -2, \ldots$. From the above expression, we immediately obtain

$$\lim_{x \to 0^-} \Gamma(x) = \lim_{x \to 0^-} \frac{\Gamma(x+1)}{x} = -\infty \qquad (2.15)$$

and

$$\lim_{x \to -1^+} \Gamma(x) = \lim_{x \to -1^+} \frac{\Gamma(x+1)}{x} = -\infty \qquad (2.16)$$

By replacing x with $x + 1$ in (2.14), we get

$$\Gamma(x+1) = \frac{\Gamma(x+2)}{x+1}$$

which leads to

$$\Gamma(x) = \frac{\Gamma(x+1)}{x} = \frac{\Gamma(x+2)}{x(x+1)}$$

Using this last expression, we find the limiting values

$$\lim_{x \to -1^-} \Gamma(x) = \lim_{x \to -1^-} \frac{\Gamma(x+2)}{x(x+1)} = +\infty \qquad (2.17)$$

and

$$\lim_{x \to -2^+} \Gamma(x) = \lim_{x \to -2^+} \frac{\Gamma(x+2)}{x(x+1)} = +\infty \qquad (2.18)$$

Continuing this process, we finally derive the formula

$$\Gamma(x) = \frac{\Gamma(x+k)}{x(x+1)(x+2)\cdots(x+k-1)}, \qquad k = 1, 2, 3, \ldots \qquad (2.19)$$

which defines the gamma function over the interval $-k < x < 0$, except for $x = -1, -2, -3, \ldots, -k+1$.

Example 1: Evaluate $\Gamma(-\tfrac{3}{2})$.

Solution: Making use of (2.19) with $k = 2$ yields*

$$\Gamma(-\tfrac{3}{2}) = (-\tfrac{2}{3})(-2)\Gamma(\tfrac{1}{2}) = \tfrac{4}{3}\sqrt{\pi}$$

---

*$\Gamma(\tfrac{1}{2}) = \sqrt{\pi}$. See Equation (2.23).

If we now assemble all the information we have on the gamma function for both positive and negative values of x, we obtain the graph of this function shown in Fig. 2.1. Values of $\Gamma(x)$ are commonly tabulated for the interval $1 \leq x \leq 2$, and other values of $\Gamma(x)$ can then be generated through use of the recurrence formulas.

In addition to

$$\Gamma(x) = \int_0^\infty e^{-t} t^{x-1}\, dt, \qquad x > 0$$

there are a variety of other integral representations of $\Gamma(x)$, most of which can be derived from that one by simple changes of variable. For example, if we set $t = u^2$ in the above integral, we get

$$\Gamma(x) = 2\int_0^\infty e^{-u^2} u^{2x-1}\, du, \qquad x > 0 \qquad (2.20)$$

whereas the substitution $t = \log(1/u)$ yields

$$\Gamma(x) = \int_0^1 \left(\log \frac{1}{u}\right)^{x-1} du, \qquad x > 0 \qquad (2.21)$$

A slightly more complicated relation can be derived by using the representation (2.20) and forming the product

$$\Gamma(x)\Gamma(y) = 2\int_0^\infty e^{-u^2} u^{2x-1}\, du \cdot 2\int_0^\infty e^{-v^2} v^{2y-1}\, dv$$

$$= 4\int_0^\infty \int_0^\infty e^{-(u^2+v^2)} u^{2x-1} v^{2y-1}\, du\, dv$$

The presence of the term $u^2 + v^2$ in the integrand suggests the change of coordinates

$$u = r\cos\theta, \qquad v = r\sin\theta$$

which leads to

$$\Gamma(x)\Gamma(y) = 4\int_0^{\pi/2} \int_0^\infty e^{-r^2} r^{2x-1} \cos^{2x-1}\theta\, r^{2y-1} \sin^{2y-1}\theta\, r\, dr\, d\theta$$

$$= 4\int_0^\infty e^{-r^2} r^{2(x+y)-1}\, dr \cdot \int_0^{\pi/2} \cos^{2x-1}\theta \sin^{2y-1}\theta\, d\theta$$

$$= 2\Gamma(x+y) \int_0^{\pi/2} \cos^{2x-1}\theta \sin^{2y-1}\theta\, d\theta$$

Finally, solving for the integral, we get the interesting relation

$$\int_0^{\pi/2} \cos^{2x-1}\theta \sin^{2y-1}\theta\, d\theta = \frac{\Gamma(x)\Gamma(y)}{2\Gamma(x+y)}, \qquad x > 0, \ y > 0 \qquad (2.22)$$

58 • Special Functions for Engineers and Applied Mathematicians

By setting $x = y = \tfrac{1}{2}$ in (2.22), we have

$$\int_0^{\pi/2} d\theta = \frac{\Gamma(\tfrac{1}{2})\Gamma(\tfrac{1}{2})}{2\Gamma(1)}$$

from which we deduce the special value

$$\Gamma(\tfrac{1}{2}) = \sqrt{\pi} \qquad (2.23)$$

Example 2: Evaluate $\int_0^\infty e^{-t^2} dt$.

Solution: By comparison with (2.20), we see that

$$\int_0^\infty e^{-t^2} dt = \tfrac{1}{2}\Gamma(\tfrac{1}{2}) = \tfrac{1}{2}\sqrt{\pi}$$

Example 3: Evaluate $\int_0^\infty x^4 e^{-x^3} dx$.

Solution: Let $t = x^3$, and then

$$\int_0^\infty x^4 e^{-x^3} dx = \frac{1}{3}\int_0^\infty e^{-t} t^{2/3} dt = \tfrac{1}{3}\Gamma(\tfrac{5}{3})$$

2.2.2 Legendre Duplication Formula

A formula involving gamma functions that is somewhat comparable to the double-angle formulas for trigonometric functions is the *Legendre duplication formula*

$$2^{2x-1}\Gamma(x)\Gamma(x + \tfrac{1}{2}) = \sqrt{\pi}\,\Gamma(2x) \qquad (2.24)$$

In order to derive this relation, we first set $y = x$ in (2.22) to get

$$\frac{\Gamma(x)\Gamma(x)}{2\Gamma(2x)} = \int_0^{\pi/2} \cos^{2x-1}\theta \sin^{2x-1}\theta\, d\theta$$

$$= 2^{1-2x}\int_0^{\pi/2} \sin^{2x-1} 2\theta\, d\theta$$

where we have used the double-angle formula for the sine function. Next we make the variable change $\phi = 2\theta$, which yields

$$\frac{\Gamma(x)\Gamma(x)}{2\Gamma(2x)} = 2^{-2x}\int_0^{\pi} \sin^{2x-1}\phi\, d\phi$$

$$= 2^{1-2x}\int_0^{\pi/2} \sin^{2x-1}\phi\, d\phi$$

$$= \frac{2^{1-2x}\Gamma(\tfrac{1}{2})\Gamma(x)}{2\Gamma(x + \tfrac{1}{2})}$$

where the last step results from (2.22). Simplification of this identity leads to (2.24).

An important special case of (2.24) occurs when $x = n$ ($n = 0, 1, 2, \ldots$), i.e.,

$$\Gamma(n + \tfrac{1}{2}) = \frac{(2n)!}{2^{2n} n!} \sqrt{\pi}, \qquad n = 0, 1, 2, \ldots \qquad (2.25)$$

the verification of which is left to the exercises (see problem 39).

Example 4: Compute $\Gamma(\tfrac{3}{2})$.

Solution: The substitution of $n = 1$ in (2.25) yields

$$\Gamma(\tfrac{3}{2}) = \Gamma(1 + \tfrac{1}{2}) = \frac{2!\sqrt{\pi}}{2^2 \times 1!} = \tfrac{1}{2}\sqrt{\pi}$$

2.2.3 The Weierstrass Infinite Product

Although it was originally found by Schlömilch in 1844, thirty-two years before Weierstrass's famous work on entire functions, Weierstrass is usually credited with the infinite-product definition of the gamma function

$$\frac{1}{\Gamma(x)} = x e^{\gamma x} \prod_{n=1}^{\infty} \left(1 + \frac{x}{n}\right) e^{-x/n} \qquad (2.26)$$

where γ is the *Euler-Mascheroni constant* defined by*

$$\gamma = \lim_{n \to \infty} \sum_{k=1}^{n} \frac{1}{k} - \log n = 0.577215\ldots \qquad (2.27)$$

We can derive this representation of $\Gamma(x)$ directly from (2.1) by first observing that

$$\begin{aligned}
\frac{1}{\Gamma(x)} &= \lim_{n \to \infty} \frac{x(x+1)(x+2) \cdots (x+n)}{n! n^x} \\
&= x \lim_{n \to \infty} n^{-x} \left[\frac{(x+1)}{1} \cdot \frac{(x+2)}{2} \cdots \frac{(x+n)}{n} \right] \\
&= x \lim_{n \to \infty} \exp[-(\log n) x] \prod_{k=1}^{n} \left(1 + \frac{x}{k}\right) \qquad (2.28)
\end{aligned}$$

where we have written $n^{-x} = \exp[-(\log n) x]$. Next, relying on properties

---
*The constant γ is commonly called (simply) *Euler's constant*.

of exponentials, we recognize the identity

$$\exp\left[\left(\sum_{k=1}^{n}\frac{1}{k}\right)x\right] = \prod_{k=1}^{n} e^{x/k}$$

Thus, if we multiply (2.28) by the left-hand side of this expression and divide by the right-hand side, we arrive at

$$\frac{1}{\Gamma(x)} = x \lim_{n\to\infty} \exp\left[\left(\sum_{k=1}^{n}\frac{1}{k} - \log n\right)x\right] \cdot \lim_{n\to\infty}\prod_{k=1}^{n}\left(1+\frac{x}{k}\right)e^{-x/k}$$

which reduces to (2.26).

An important identity involving the gamma function and sine function can now be derived by using (2.26). We begin with the product of gamma functions

$$\frac{1}{\Gamma(x)\Gamma(-x)} = xe^{\gamma x}\prod_{n=1}^{\infty}\left(1+\frac{x}{n}\right)e^{-x/n} \cdot (-x)e^{-\gamma x}\prod_{n=1}^{\infty}\left(1-\frac{x}{n}\right)e^{x/n}$$

or

$$\frac{1}{\Gamma(x)\Gamma(-x)} = -x^2 \prod_{n=1}^{\infty}\left(1-\frac{x^2}{n^2}\right) \qquad (2.29)$$

where we assume that x is nonintegral. Recalling Equation (1.92) in Section 1.7.2, which gives the infinite-product definition of the sine function, we have

$$\prod_{n=1}^{\infty}\left(1-\frac{x^2}{n^2}\right) = \frac{\sin\pi x}{\pi x} \qquad (2.30)$$

Comparison of (2.29) and (2.30) reveals that

$$\Gamma(x)\Gamma(-x) = -\frac{\pi}{x\sin\pi x} \qquad (x \text{ nonintegral}) \qquad (2.31)$$

Also, by writing the recurrence formula (2.3) in the form

$$-x\Gamma(-x) = \Gamma(1-x)$$

we deduce the identity

$$\Gamma(x)\Gamma(1-x) = \frac{\pi}{\sin\pi x} \qquad (x \text{ nonintegral}) \qquad (2.32)$$

Example 5: Evaluate the integral $\int_0^{\pi/2} \tan^{1/2}\theta\, d\theta$.

Solution: Making use of (2.22) and (2.32), we get

$$\int_0^{\pi/2} \tan^{1/2}\theta \, d\theta = \int_0^{\pi/2} \sin^{1/2}\theta \cos^{-1/2}\theta \, d\theta$$

$$= \frac{\Gamma(\tfrac{3}{4})\Gamma(\tfrac{1}{4})}{2\Gamma(1)}$$

$$= \frac{1}{2}\frac{\pi}{\sin(\pi/4)}$$

$$= \frac{\pi}{\sqrt{2}}$$

Remark: An *entire function* is one that is analytic for all finite values of its argument. Weierstrass was the first to show that any entire function (under appropriate restrictions) with an infinite number of zeros, such as $\sin x$ and $\cos x$, is essentially determined by its zeros. This result led to the infinite-product representations of such functions, and in particular, to the infinite-product representation of the gamma function.

2.2.4 Fractional-Order Derivatives

Besides generalizing the notion of factorials, the gamma function can be used in a variety of situations to generalize discrete processes into the continuum. Such generalizations are not new, however: mathematicians over the years have concerned themselves with this concept. In particular, the question concerning derivatives of nonintegral order was first raised by Leibniz in 1695, many years before Euler introduced the gamma function.

The general procedure for developing fractional derivatives is too involved for our purposes.* However, we can illustrate the concept by first recalling the familiar derivative formula from calculus,

$$D^n x^a = a(a-1)\cdots(a-n+1)x^{a-n}, \qquad a \geq 0 \qquad (2.33)$$

where $D^n = d^n/dx^n$. In terms of the gamma function, we can rewrite (2.33) as (see problem 10)

$$D^n x^a = \frac{\Gamma(a+1)}{\Gamma(a-n+1)} x^{a-n}$$

The right-hand side of this expression is meaningful for any real number n for which $\Gamma(a - n + 1)$ is defined. Hence, we will assume that the same is

---

*For a deeper discussion of fractional derivatives, see L. Debnath, Generalized Calculus and Its Applications, *Int. J. Math. Educ. Sci. Technol.*, **9**, No. 4, pp. 399–416 (1978).

true of the left-hand side and write

$$D^\nu x^a = \frac{\Gamma(a+1)}{\Gamma(a-\nu+1)} x^{a-\nu}, \quad a \geq 0 \qquad (2.34)$$

where ν is not restricted to integer values. Equation (2.34) provides a simple method of computing *fractional-order derivatives* of polynomials.

Example 6: Compute $D^{1/2} x^2$.

Solution: Directly from (2.34), we obtain

$$D^{1/2} x^2 = \frac{\Gamma(3)}{\Gamma(\tfrac{5}{2})} x^{3/2}$$

the simplification of which yields

$$D^{1/2} x^2 = \frac{8}{3\sqrt{\pi}} x^{3/2}$$

Generalization of the differentiation formula for $D^n x^{-a}$, which covers the case of negative exponents, is left to the exercises (see problem 52).

EXERCISES 2.2

1. Use Equation (2.1) directly to evaluate
   (a) $\Gamma(2)$. (b) $\Gamma(3)$.

In problems 2–7, give numerical values for the expressions.

2. $\Gamma(6)/\Gamma(3)$.

3. $\Gamma(7)/\Gamma(4)\Gamma(3)$.

4. $\Gamma(\tfrac{7}{2})$.

5. $\Gamma(-\tfrac{1}{2})$.

6. $\Gamma(-\tfrac{5}{2})/\Gamma(\tfrac{1}{2})$.

7. $\Gamma(\tfrac{8}{3})/\Gamma(\tfrac{2}{3})$.

In problems 8–14, verify the given identity.

8. $\Gamma(a+n) = a(a+1)(a+2) \cdots (a+n-1)\Gamma(a), \quad n = 1, 2, 3, \ldots$.

9. $\dfrac{\Gamma(n-a)}{\Gamma(-a)} = (-1)^n a(a-1)(a-2) \cdots (a-n+1), \quad n = 1, 2, 3, \ldots$.

10. $\dfrac{\Gamma(a)}{\Gamma(a-n)} = (a-1)(a-2) \cdots (a-n), \quad n = 1, 2, 3, \ldots$.

11. $\dfrac{\Gamma(k-n)}{\Gamma(-n)} = \begin{cases} \dfrac{(-1)^k n!}{(n-k)!}, & 0 \le k \le n \ (k, n \text{ non-negative integers}), \\ 0, & k > n. \end{cases}$

Hint: See problem 9.

12. $\dbinom{a}{n} = \dfrac{\Gamma(a+1)}{n!\,\Gamma(a-n+1)}$, $n = 0, 1, 2, \ldots$.

Hint: See problem 10.

13. $\dbinom{-\frac{1}{2}}{n} = \dfrac{(-1)^n (2n)!}{2^{2n}(n!)^2}$, $n = 0, 1, 2, \ldots$.

14. $\dbinom{-2k-1}{m} = (-1)^m \dfrac{(m+2k)!}{(2k)!\,m!}$, $k, m = 0, 1, 2, \ldots$.

15. In problems in electromagnetic theory it is quite common to come across products like

$$2 \times 4 \times 6 \times \cdots \times 2n \equiv (2n)!!$$

and

$$1 \times 3 \times 5 \times \cdots \times (2n+1) \equiv (2n+1)!!$$

Use these definitions of the !! notation to show that

(a) $(2n)!! = 2^n n!$,
(b) $(2n+1)!! = \dfrac{(2n+1)!}{2^n n!}$,
(c) $(-2n-1)!! = \dfrac{(-1)^n 2^n n!}{(2n)!}$,
(d) $(-1)!! = 1$.

Hint: See problem 10 for (c) and (d).

16. Prove that $\int_0^\infty e^{-t} t^{x-1}\, dt$ converges uniformly in $1 \le x \le 2$.

In problems 17–20, verify the given integral representation.

17. $\Gamma(x) = s^x \int_0^\infty e^{-st} t^{x-1}\, dt$, $x, s > 0$.

18. $\Gamma(x) = \int_{-\infty}^\infty \exp(xt - e^t)\, dt$, $x > 0$.

Hint: Let $u = e^t$.

19. $\Gamma(x) = \int_1^\infty e^{-t} t^{x-1}\, dt + \sum_{n=0}^\infty \dfrac{(-1)^n}{n!(x+n)}$, $x > 0$.

20. $\Gamma(x) = (\log b)^x \int_0^\infty t^{x-1} b^{-t}\, dt$, $x > 0$, $b > 1$.

Hint: Let $u = t \log b$.

In problems 21–29, use properties of the gamma function to obtain the result.

21. $\int_a^\infty e^{2ax-x^2}\, dx = \tfrac{1}{2}\sqrt{\pi}\, e^{a^2}$.

Hint: $2ax - x^2 = -(x-a)^2 + a^2$.

22. $\int_0^\infty e^{-2x} x^6\, dx = \dfrac{45}{8}$.

23. $\int_0^\infty \sqrt{x}\, e^{-x^3}\, dx = \dfrac{\sqrt{\pi}}{3}$.

24. $\int_0^1 \dfrac{du}{\sqrt{-\log u}} = \sqrt{\pi}$.

25. $\int_0^1 x^k (\log x)^n\, dx = \dfrac{(-1)^n n!}{(k+1)^{n+1}}$, $k > -1$, $n = 0, 1, 2, \ldots$.

26. $\int_0^{\pi/2} \cos^6\theta\, d\theta = \dfrac{5\pi}{32}$.

27. $\int_0^{\pi/2} \sin^3\theta \cos^2\theta\, d\theta = \dfrac{2}{15}$.

28. $\int_0^\pi \cos^4 x\, dx = \dfrac{3\pi}{8}$.

29. $\int_0^{\pi/2} \sin^{2n+1}\theta\, d\theta = \int_0^{\pi/2} \cos^{2n+1}\theta\, d\theta = \dfrac{2^{2n}(n!)^2}{(2n+1)!}$, $n = 0, 1, 2, \ldots$.

In problems 30–35, evaluate the integral in terms of the gamma function and simplify when possible.

30. $\int_0^\infty \dfrac{e^{-st}}{\sqrt{t}}\, dt$, $s > 0$.

31. $\int_0^\infty \dfrac{dx}{1 + x^4}$.

Hint: Let $x^2 = \tan\theta$.

32. $\int_0^{\pi/2} \sqrt{\sin 2x}\, dx$.

33. $\int_0^1 t^{x-1}(\log \tfrac{1}{t})^{y-1}\, dt$, $x, y > 0$.

34. $\int_0^{\pi/2} \cot^{1/2}\theta\, d\theta$.

35. $\int_0^\infty e^{-st^p} t^{x-1}\, dt$, $p, s, x > 0$.

36. Using the recurrence formula (2.3), deduce that

(a) $\Gamma(x) = \Gamma'(x+1) - x\Gamma'(x)$,

(b) $\Gamma(x) = \int_0^\infty e^{-t}(t-x) t^{x-1} \log t\, dt$, $x > 0$.

In problems 37 and 38, use the Euler formulas

$$\cos x = \frac{e^{ix} + e^{-ix}}{2}, \quad \sin x = \frac{e^{ix} - e^{-ix}}{2i}$$

and properties of the gamma function to derive the result. Assume that $b, x > 0$ and $-\frac{1}{2}\pi < a < \frac{1}{2}\pi$.

37. $\Gamma(x)\cos ax = b^x \int_0^\infty t^{x-1} e^{-bt\cos a} \cos(bt\sin a)\, dt.$

38. $\Gamma(x)\sin ax = b^x \int_0^\infty t^{x-1} e^{-bt\cos a} \sin(bt\sin a)\, dt.$

39. Based on the Legendre duplication formula, show that (for $n = 0, 1, 2, \ldots$)

(a) $\Gamma(n + \frac{1}{2}) = \dfrac{(2n)!\sqrt{\pi}}{2^{2n} n!},$

(b) $\Gamma(\frac{1}{2} - n) = \dfrac{(-1)^n 2^{2n-1}(n-1)!\sqrt{\pi}}{(2n-1)!},$

(c) $\Gamma(\frac{1}{2} + n)\Gamma(\frac{1}{2} - n) = (-1)^n \pi.$

40. Show that

$$\Gamma(3x) = \frac{1}{2\pi} 3^{3x-1/2} \Gamma(x)\Gamma(x + \tfrac{1}{3})\Gamma(x + \tfrac{2}{3})$$

41. Show that

$$|\Gamma'(x)|^2 \le \Gamma(x)\Gamma''(x), \quad x > 0$$

42. Show that

(a) $\Gamma(1 + x)\Gamma(1 - x) = \pi x \csc \pi x$ (x nonintegral),
(b) $\Gamma(\frac{1}{2} + x)\Gamma(\frac{1}{2} - x) = \pi \sec \pi x,\ x \ne n + \frac{1}{2},\ n = 0, 1, 2, \ldots\,.$

43. Derive Euler's infinite-product representation

$$\frac{1}{\Gamma(x)} = x \prod_{n=1}^\infty \frac{\left(1 + \dfrac{x}{n}\right)}{\left(1 + \dfrac{1}{n}\right)^x}$$

44. Derive the recurrence relation $\Gamma(x + 1) = x\Gamma(x)$, by use of the

(a) integral definition (2.5),
(b) Weierstrass infinite product (2.26).

45. A particle of mass m starts from rest at $r = 1$ and moves along a radial line toward the origin $r = 0$ under the reciprocal force law $f = -k/r$, where k is a positive constant. The energy equation of the particle is

given by

$$\tfrac{1}{2}m\left(\frac{dr}{dt}\right)^2 + k\log r = 0$$

(a) Show that the time required for the particle to reach the origin is $(m\pi/2k)^{1/2}$.

(b) If the particle starts from rest at $r = a$ ($a > 0$), the energy equation becomes

$$\tfrac{1}{2}m\left(\frac{dr}{dt}\right)^2 + k\log r = k\log a$$

Again find the time required for the particle to reach the origin.

46. Find the area enclosed by the curve $x^4 + y^4 = 1$.

47. Find the total arclength of the lemniscate $r^2 = a^2\cos 2\theta$.

48. Find the area inside the curve $x^{2/3} + y^{2/3} = 1$.

49. Find the volume in the first octant below the surface

$$x^{1/2} + y^{1/2} + z^{1/2} = 1$$

50. Compute the fractional-order derivatives

(a) $D^{1/2}c$, where c is constant,
(b) $D^{1/2}(3x^2 - 7x + 4)$,
(c) $D^{3/2}x^2$,
(d) $D^\nu x^\nu$, where ν is not a positive integer.

51. Show that
(a) $D^{1/2}(D^{1/2}x^2) = Dx^2$,
(b) $D^{-1/2}(D^{1/2}x^2) = x^2$,
(c) $D^\nu(D^\mu x^a) = D^{\nu+\mu}x^a$.

52. By generalizing the formula for $D^n x^{-a}$, show that

$$D^\nu x^{-a} = (-1)^\nu \frac{\Gamma(\nu + a)}{\Gamma(a)} x^{-(a+\nu)}, \qquad a > 0$$

2.3 Beta Function

A useful function of two variables is the *beta function**

$$B(x, y) = \int_0^1 t^{x-1}(1-t)^{y-1}\, dt, \qquad x > 0, \quad y > 0 \qquad (2.35)$$

---

*This is called the *Eulerian integral of the first kind*.

The utility of the beta function is often overshadowed by that of the gamma function, partly perhaps because it can be evaluated in terms of the gamma function. However, since it occurs so frequently in practice, a special designation for it is widely accepted.

If we make the change of variable $u = 1 - t$ in (2.35), we find

$$B(x, y) = \int_0^1 (1 - u)^{x-1} u^{y-1} \, du$$

from which we deduce the *symmetry property*

$$B(x, y) = B(y, x) \tag{2.36}$$

Another representation of the beta function results if we make the variable change $t = u/(1 + u)$, leading to

$$B(x, y) = \int_0^\infty \frac{u^{x-1}}{(1 + u)^{x+y}} \, du, \quad x > 0, \quad y > 0 \tag{2.37}$$

Finally, to show how the beta function is related to the gamma function, we set $t = \cos^2 \theta$ in (2.35) to find

$$B(x, y) = 2 \int_0^{\pi/2} \cos^{2x-1} \theta \sin^{2y-1} \theta \, d\theta$$

and hence from (2.22) we obtain the relation

$$B(x, y) = \frac{\Gamma(x) \Gamma(y)}{\Gamma(x + y)}, \quad x > 0, \quad y > 0 \tag{2.38}$$

Example 6: Evaluate the integral $I = \int_0^\infty x^{-1/2} (1 + x)^{-2} \, dx$.

Solution: By comparison with (2.37), we recognize

$$I = B\left(\tfrac{1}{2}, \tfrac{3}{2}\right)$$

$$= \frac{\Gamma\left(\tfrac{1}{2}\right) \Gamma\left(\tfrac{3}{2}\right)}{\Gamma(2)}$$

Hence, we deduce that

$$\int_0^\infty x^{-1/2} (1 + x)^{-2} \, dx = \frac{\pi}{2}$$

Example 7: Show that

$$\int_0^\infty \frac{\cos x}{x^p} \, dx = \frac{\pi}{2 \Gamma(p) \cos(p\pi/2)}, \quad 0 < p < 1$$

Solution: Making the observation (problem 17 in Exercises 2.2)

$$\frac{1}{x^p} = \frac{1}{\Gamma(p)} \int_0^\infty e^{-xt} t^{p-1} \, dt$$

it follows that

$$\int_0^\infty \frac{\cos x}{x^p} \, dx = \frac{1}{\Gamma(p)} \int_0^\infty \cos x \int_0^\infty e^{-xt} t^{p-1} \, dt \, dx$$

$$= \frac{1}{\Gamma(p)} \int_0^\infty t^{p-1} \int_0^\infty e^{-xt} \cos x \, dx \, dt$$

$$= \frac{1}{\Gamma(p)} \int_0^\infty \frac{t^p}{1 + t^2} \, dt$$

where we have reversed the order of integration. If we now let $u = t^2$, then

$$\int_0^\infty \frac{\cos x}{x^p} \, dx = \frac{1}{2\Gamma(p)} \int_0^\infty \frac{u^{\frac{1}{2}(p-1)}}{1 + u} \, du$$

$$= \frac{1}{2\Gamma(p)} B\left(\frac{1+p}{2}, \frac{1-p}{2}\right)$$

However (see problem 10),

$$B\left(\frac{1+p}{2}, \frac{1-p}{2}\right) = \pi \sec\left(\frac{p\pi}{2}\right)$$

and thus we have our result.

Example 7 illustrates one of the basic approaches we use in the evaluation of nonelementary integrals. That is, we replace part (or all) of the integrand by its series representation or integral representation and then interchange the order in which the operations are carried out.

EXERCISES 2.3

In problems 1–4, evaluate the beta function.

1. $B(\frac{2}{3}, \frac{1}{3})$.
2. $B(\frac{3}{4}, \frac{1}{4})$.
3. $B(\frac{1}{2}, 1)$.
4. $B(x, 1 - x), 0 < x < 1$.

In problems 5–10, verify the identity.

5. $B(x + 1, y) + B(x, y + 1) = B(x, y), \; x, y > 0$.

6. $B(x, y+1) = \dfrac{y}{x} B(x+1, y) = \dfrac{y}{x+y} B(x, y)$, $x, y > 0$.

7. $B(x, x) = 2^{1-2x} B(x, \tfrac{1}{2})$, $x > 0$.

8. $B(x, y) B(x+y, z) B(x+y+z, w) = \dfrac{\Gamma(x)\Gamma(y)\Gamma(z)\Gamma(w)}{\Gamma(x+y+z+w)}$,
$x, y, z, w > 0$.

9. $B(n, n) B(n + \tfrac{1}{2}, n + \tfrac{1}{2}) = \pi 2^{1-4n} n^{-1}$, $n = 1, 2, 3, \ldots$.

10. $B\left(\dfrac{1+p}{2}, \dfrac{1-p}{2}\right) = \pi \sec(p\pi/2)$, $0 < p < 1$.

In problems 11–18, use properties of the beta and gamma functions to evaluate the integral.

11. $\displaystyle\int_0^1 \sqrt{x(1-x)}\, dx$.

12. $\displaystyle\int_0^1 x^4 (1-x^2)^{-1/2}\, dx$.

13. $\displaystyle\int_0^\infty \dfrac{x}{(1+x^3)^2}\, dx$.

   Hint: Set $t = x^3/(1+x^3)$.

14. $\displaystyle\int_{-1}^1 \left(\dfrac{1+x}{1-x}\right)^{1/2} dx$.

   Hint: Set $x = 2t - 1$.

15. $\displaystyle\int_a^b (b-x)^{m-1}(x-a)^{n-1}\, dx$, where m, n are positive integers.

16. $\displaystyle\int_0^2 x^2 (2-x)^{-1/2}\, dx$.

17. $\displaystyle\int_0^a x^4 \sqrt{a^2 - x^2}\, dx$.

18. $\displaystyle\int_0^2 x\sqrt[3]{8 - x^3}\, dx$.

In problems 19–30, verify the integral formula.

19. $\displaystyle\int_0^\infty \dfrac{x^{p-1}}{1+x}\, dx = \pi \csc p\pi$, $0 < p < 1$.

20. $\displaystyle\int_0^\infty \dfrac{\sin x}{x^p}\, dx = \dfrac{\pi}{2\Gamma(p)\sin(p\pi/2)}$, $0 < p < 1$.

21. $\displaystyle\int_0^\infty \sin x^2\, dx = \dfrac{1}{2}\sqrt{\dfrac{\pi}{2}}$.

   Hint: Use problem 20.

22. $\displaystyle\int_0^\infty \cos x^2\, dx = \frac{1}{2}\sqrt{\frac{\pi}{2}}$.

23. $\displaystyle\int_0^{\pi/2} \tan^p x\, dx = \int_0^{\pi/2} \cot^p x\, dx = \frac{\pi}{2\cos(p\pi/2)},\ 0 < p < 1$.

24. $\displaystyle\int_0^\infty \frac{x^{p-1}\log x}{1+x}\, dx = -\pi^2 \csc p\pi \cot p\pi,\ 0 < p < 1$.

25. $\displaystyle\int_0^\infty \frac{x^{p-1}}{1+x^a}\, dx = \frac{\pi}{a\sin(p\pi/a)},\ 0 < p < a$.

26. $\displaystyle\int_0^\infty e^{-st}(1-e^{-t})^n\, dt = \frac{n!\,\Gamma(s)}{\Gamma(s+n+1)}$, where $s > 0,\ n = 0, 1, 2, \ldots$.

27. $\displaystyle\int_{-\infty}^\infty \frac{e^{2x}}{ae^{3x}+b}\, dx = \frac{2\pi}{3\sqrt{3}} a^{-2/3} b^{-1/3}$, where $a, b > 0$.

28. $\displaystyle\int_{-\infty}^\infty \frac{e^{2x}}{(e^{3x}+1)^2}\, dx = \frac{2\pi}{9\sqrt{3}}$.

Hint: Differentiate with respect to b in problem 27.

29. $\displaystyle\int_0^1 \frac{t^{x-1}+t^{y-1}}{(t+1)^{x+y}}\, dt = 2B(x, y)$, where $x, y > 0$.

30. $\displaystyle\int_0^1 \frac{t^{x-1}(1-t)^{y-1}}{(t+p)^{x+y}}\, dt = \frac{B(x, y)}{p^x(1+p)^{x+y}}$, where $x, y, p > 0$.

31. Using the notation of problem 15 in Exercises 2.2, show that

(a) $\displaystyle\int_{-1}^1 (1-x^2)^{1/2} x^{2n}\, dx = \begin{cases} \dfrac{\pi}{2}, & n = 0, \\ \pi\dfrac{(2n-1)!!}{(2n+2)!!}, & n = 1, 2, 3, \ldots. \end{cases}$

(b) $\displaystyle\int_{-1}^1 (1-x^2)^{-1/2} x^{2n}\, dx = \begin{cases} \pi, & n = 0, \\ \pi\dfrac{(2n-1)!!}{(2n)!!}, & n = 1, 2, 3, \ldots. \end{cases}$

32. Show that

$$\int_{-1}^1 (1-x^2)^n\, dx = 2^{2n+1}\frac{(n!)^2}{(2n+1)!},\quad n = 0, 1, 2, \ldots$$

33. The *incomplete beta function* is defined by

$$B_x(p, q) = \int_0^x t^{p-1}(1-t)^{q-1}\, dt,\quad 0 \le x \le 1,\ p, q > 0$$

(a) Show that

$$B_x(p,q) = x^p \Gamma(q) \sum_{n=0}^{\infty} \frac{(-1)^n x^n}{\Gamma(q-n)(p+n)n!}, \quad 0 \le x \le 1$$

(b) From (a), deduce that

$$\sum_{n=0}^{\infty} \frac{(-1)^n}{\Gamma(q-n)(p+n)n!} = \frac{\Gamma(p)}{\Gamma(p+q)}$$

2.4 Incomplete Gamma Function

Generalizing the Euler integral (2.5), we introduce the related function

$$\gamma(a,x) = \int_0^x e^{-t} t^{a-1} dt, \quad a > 0 \qquad (2.39)$$

called the *incomplete gamma function*. This function most commonly arises in probability theory, particularly those applications involving the chi-square distribution. It is customary to also introduce the companion function

$$\Gamma(a,x) = \int_x^{\infty} e^{-t} t^{a-1} dt, \quad a > 0 \qquad (2.40)$$

which is known as the *complementary incomplete gamma function*. Thus, it follows that

$$\gamma(a,x) + \Gamma(a,x) = \Gamma(a) \qquad (2.41)$$

Because of the close relationship between these two functions, the choice of using $\gamma(a,x)$ or $\Gamma(a,x)$ in practice is simply a matter of convenience.

By substituting the series representation for e^{-t} in (2.39), we get

$$\gamma(a,x) = \int_0^x \left(\sum_{n=0}^{\infty} \frac{(-1)^n}{n!} t^{n+a-1} \right) dt$$

and then, performing termwise integration, we are led to the series representation

$$\gamma(a,x) = x^a \sum_{n=0}^{\infty} \frac{(-1)^n x^n}{n!(n+a)}, \quad a > 0 \qquad (2.42)$$

It immediately follows from (2.41) that

$$\Gamma(a,x) = \Gamma(a) - x^a \sum_{n=0}^{\infty} \frac{(-1)^n x^n}{n!(n+a)}, \quad a > 0 \qquad (2.43)$$

2.4.1 Asymptotic Series

The integration of Equation (2.40) by parts gives us

$$\Gamma(a, x) = \int_x^\infty e^{-t} t^{a-1} \, dt$$

$$= -e^{-t} t^{a-1} \Big|_x^\infty + (a-1) \int_x^\infty e^{-t} t^{a-2} \, dt$$

$$= e^{-x} x^{a-1} + (a-1) \int_x^\infty e^{-t} t^{a-2} \, dt \qquad (2.44)$$

while continued integration by parts yields

$$\Gamma(a, x) = e^{-x} x^{a-1} + (a-1) e^{-x} x^{a-2} + (a-1)(a-2) \int_x^\infty e^{-t} t^{a-3} \, dt$$

and so on. Thus we generate the *asymptotic series**

$$\Gamma(a, x) \sim e^{-x} x^{a-1} \left[1 + \frac{a-1}{x} + \frac{(a-1)(a-2)}{x^2} + \cdots \right], \qquad x \to \infty \qquad (2.45)$$

which can be expressed as

$$\Gamma(a, x) \sim \Gamma(a) x^{a-1} e^{-x} \sum_{k=0}^\infty \frac{1}{\Gamma(a-k) x^k}, \qquad a > 0, \quad x \to \infty \qquad (2.46)$$

If we set $a = n + 1$ ($n = 0, 1, 2, \ldots$) in (2.46), we find that

$$\Gamma(n+1, x) = n! x^n e^{-x} \sum_{k=0}^n \frac{x^{-k}}{(n-k)!}, \qquad (2.47)$$

where the series truncates because $1/\Gamma(n+1-k) = 0$ for $k > n$ (Theorem 2.1). The change of variable $j = n - k$ further simplifies (2.47) to

$$\Gamma(n+1, x) = n! e^{-x} \sum_{j=0}^n \frac{x^j}{j!} \qquad (2.48)$$

or

$$\Gamma(n+1, x) = n! e^{-x} e_n(x), \qquad n = 0, 1, 2, \ldots \qquad (2.49)$$

where $e_n(x)$ denotes the first $n + 1$ terms of the Maclaurin series for e^x.†

---
*The asymptotic series (2.45) or (2.46) diverges for all finite x.
†For additional properties of the function $e_n(x)$, see problems 11 and 12 in Exercises 4.2.

By a similar analysis, it can be shown that

$$\gamma(n + 1, x) = n![1 - e^{-x}e_n(x)], \qquad n = 0, 1, 2, \ldots \qquad (2.50)$$

Remark: It is interesting to note that both (2.49) and (2.50) are valid representations for all $x > 0$, while the asymptotic series (2.46) [from which (2.49) and (2.50) were derived] diverges for all x.

EXERCISES 2.4

1. Show that
   (a) $\gamma(a + 1, x) = a\gamma(a, x) - x^a e^{-x}$,
   (b) $\Gamma(a + 1, x) = a\Gamma(a, x) + x^a e^{-x}$.

2. Show that
   (a) $\dfrac{d}{dx}[x^{-a}\Gamma(a, x)] = -x^{-a-1}\Gamma(a + 1, x)$,
   (b) $\dfrac{d^m}{dx^m}[x^{-a}\Gamma(a, x)] = (-1)^m x^{-a-m}\Gamma(a + m, x)$, $m = 1, 2, 3, \ldots$.

3. Show that
$$\Gamma(a)\Gamma(a + n, x) - \Gamma(a + n)\Gamma(a, x) = \Gamma(a + n)\gamma(a, x) - \Gamma(a)\gamma(a + n, x)$$

4. Verify the integral formula
$$\Gamma(a, xy) = y^a e^{-xy} \int_0^\infty e^{-yt}(t + x)^{a-1}\,dt, \qquad x, y > 0, \quad a > 1$$

5. Verify the integral representation
$$\gamma(a, x) = x^{a/2} \int_0^\infty e^{-t} t^{\frac{1}{2}a - 1} J_a(2\sqrt{xt})\,dt, \qquad a > 0$$

   where $J_a(z)$ is the *Bessel function* defined by (see Chapter 6)
$$J_a(z) = \sum_{n=0}^\infty \frac{(-1)^n (z/2)^{2n+a}}{n!\,\Gamma(n + a + 1)}$$

6. Formally derive the asymptotic series (2.46) by setting $y = 1$ in the result of problem 4 and using the binomial series
$$\left(1 + \frac{t}{x}\right)^{a-1} = \sum_{k=0}^\infty \binom{a-1}{k}\left(\frac{t}{x}\right)^k, \qquad x > t$$

2.5 Digamma and Polygamma Functions

Closely associated with the derivative of the gamma function is the *logarithmic-derivative function*, or *digamma function*, defined by*

$$\psi(x) \equiv \frac{d}{dx}\log \Gamma(x) = \frac{\Gamma'(x)}{\Gamma(x)}, \qquad x \neq 0, -1, -2, \ldots \qquad (2.51)$$

In order to find an infinite series representation of $\psi(x)$, we first take the natural logarithm of both sides of the Weierstrass infinite product

$$\frac{1}{\Gamma(x)} = xe^{\gamma x}\prod_{n=1}^{\infty}\left(1 + \frac{x}{n}\right)e^{-x/n}$$

which yields

$$-\log \Gamma(x) = \log x + \gamma x + \sum_{n=1}^{\infty}\left[\log\left(1 + \frac{x}{n}\right) - \frac{x}{n}\right], \qquad x > 0 \quad (2.52)$$

Then, negating both sides of (2.52) and differentiating the result with respect to x, we find

$$\psi(x) \equiv \frac{d}{dx}\log \Gamma(x) = -\frac{1}{x} - \gamma + \sum_{n=1}^{\infty}\left(\frac{1}{n} - \frac{1}{x+n}\right)$$

which we choose to write as

$$\psi(x) = -\gamma + \sum_{n=0}^{\infty}\left(\frac{1}{n+1} - \frac{1}{n+x}\right), \qquad x > 0 \qquad (2.53)$$

The restriction $x > 0$ follows from Equation (2.52).†

Noteworthy here is the special value

$$\psi(1) = \frac{\Gamma'(1)}{\Gamma(1)} = -\gamma \qquad (2.54)$$

and by recalling Equation (2.11), we see that

$$\Gamma'(1) = -\gamma = \int_0^{\infty} e^{-t}\log t\, dt \qquad (2.55)$$

Based upon Equation (2.51), it is clear that the digamma function has the same domain of definition as the gamma function. It has characteristics quite distinct from those of the gamma function, however, since it is related

---
*The function $\psi(x)$ is also commonly called the *psi function*.
†Actually, (2.53) is valid for all x except $x = 0, -1, -2, \ldots$, although we will not prove it.

to the derivative of $\Gamma(x)$. For example, unlike the gamma function, the function $\psi(x)$ crosses the x-axis. In fact, it has infinitely many zeros, corresponding to the extrema of $\Gamma(x)$, i.e., points where $\Gamma'(x) = 0$. For positive x the only extremum of the gamma function occurs at $x_0 = 1.4616\ldots$. Because x_0 corresponds to a minimum of $\Gamma(x)$, it follows that $\Gamma'(x)$ and $\psi(x)$ are both negative on the interval $0 < x < x_0$ and both positive for $x > x_0$. For large values of x, it can be shown that the digamma function is approximately equal to $\log x$ [see Equation (2.74) below]. The general characteristics of $\psi(x)$ for both positive and negative values of x are illustrated in Fig. 2.2.

The function $\psi(x)$ satisfies relations somewhat analogous to those for the gamma function, which can be derived by taking logarithmic derivatives of the latter. As an illustrative example, let us consider the recurrence formula

$$\Gamma(x + 1) = x\Gamma(x) \tag{2.56}$$

By taking the logarithm we have

$$\log \Gamma(x + 1) = \log x + \log \Gamma(x)$$

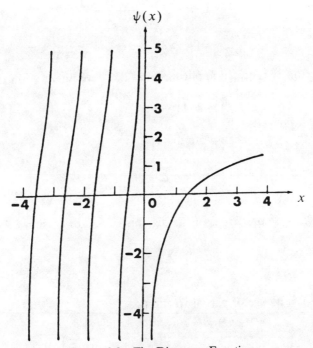

Figure 2.2 The Digamma Function

which upon differentiation yields
$$\frac{d}{dx}\log\Gamma(x+1) = \frac{1}{x} + \frac{d}{dx}\log\Gamma(x)$$
Thus,
$$\psi(x+1) = \psi(x) + \frac{1}{x} \tag{2.57}$$

Also, the logarithmic derivative of
$$\Gamma(x)\Gamma(1-x) = \pi\csc\pi x$$
results in the identity
$$\psi(1-x) - \psi(x) = \pi\cot\pi x \tag{2.58}$$
and finally, the logarithmic derivative of the Legendre duplication formula (2.24) leads to
$$\psi(x) + \psi(x + \tfrac{1}{2}) + 2\log 2 = 2\psi(2x) \tag{2.59}$$
The details of deriving (2.58) and (2.59) are left to the exercises.

If n denotes a positive integer, it follows from (2.57) that
$$\psi(n+1) = \psi(n) + \frac{1}{n}$$
$$= \psi(n-1) + \frac{1}{n-1} + \frac{1}{n}$$
$$= \psi(n-2) + \frac{1}{n-2} + \frac{1}{n-1} + \frac{1}{n}$$
and so forth. By repeated application of (2.57), we finally deduce that
$$\psi(n+1) = \psi(1) + 1 + \frac{1}{2} + \frac{1}{3} + \cdots + \frac{1}{n}$$
Since $\psi(1) = -\gamma$, we can write this as
$$\psi(n+1) = -\gamma + \sum_{k=1}^{n} \frac{1}{k}, \qquad n = 1, 2, 3, \ldots \tag{2.60}$$

Example 8: Use properties of the digamma function to sum the series
$$\sum_{n=2}^{\infty} \frac{1}{n^2 - 1}$$

Solution: By use of partial fractions,
$$\frac{1}{n^2 - 1} = \frac{1}{2}\left(\frac{1}{n-1} - \frac{1}{n+1}\right)$$

and therefore

$$\sum_{n=2}^{\infty} \frac{1}{n^2 - 1} = \frac{1}{2} \sum_{n=2}^{\infty} \left(\frac{1}{n-1} - \frac{1}{n+1} \right)$$

$$= \frac{1}{2} \sum_{k=0}^{\infty} \left(\frac{1}{k+1} - \frac{1}{k+3} \right)$$

where we have introduced the change of index $n - 2 = k$. Now, from Equations (2.53) and (2.60), it follows that

$$\sum_{n=2}^{\infty} \frac{1}{n^2 - 1} = \tfrac{1}{2}[\psi(3) + \gamma]$$

$$= \tfrac{1}{2}[-\gamma + 1 + \tfrac{1}{2} + \gamma]$$

or

$$\sum_{n=2}^{\infty} \frac{1}{n^2 - 1} = \frac{3}{4}$$

2.5.1 Integral Representations

Like the gamma function, the digamma function also has various integral representations. Let us start with the known relation

$$\Gamma'(x) = \int_0^\infty e^{-t} t^{x-1} \log t \, dt, \qquad x > 0 \qquad (2.61)$$

and replace $\log t$ with the *Frullani integral representation* (see Example 14 in Section 1.6.3)

$$\log t = \int_0^\infty \frac{e^{-u} - e^{-ut}}{u} \, du, \qquad t > 0 \qquad (2.62)$$

Hence,

$$\Gamma'(x) = \int_0^\infty e^{-t} t^{x-1} \left(\int_0^\infty \frac{e^{-u} - e^{-ut}}{u} \, du \right) dt$$

$$= \int_0^\infty \int_0^\infty e^{-t} t^{x-1} \left(\frac{e^{-u} - e^{-ut}}{u} \right) dt \, du$$

where we have reversed the order of integration. Next, splitting the inside integral into a sum of integrals, and recalling the integral relation (see problem 17 in Exercises 2.2)

$$\int_0^\infty e^{-t(u+1)} t^{x-1} \, dt = \frac{\Gamma(x)}{(u+1)^x}, \qquad x > 0 \qquad (2.63)$$

we see that

$$\Gamma'(x) = \int_0^\infty \frac{1}{u}\left[e^{-u}\int_0^\infty e^{-t}t^{x-1}\,dt - \int_0^\infty e^{-t(u+1)}t^{x-1}\,dt\right]du$$

$$= \int_0^\infty \frac{1}{u}\left[e^{-u}\Gamma(x) - \frac{\Gamma(x)}{(u+1)^x}\right]du$$

Finally, division of this last result by $\Gamma(x)$ leads to the desired integral relation

$$\psi(x) = \int_0^\infty \frac{1}{u}\left[e^{-u} - (u+1)^{-x}\right]du, \qquad x > 0 \qquad (2.64)$$

Another integral representation can be derived by first writing (2.64) as

$$\psi(x) = \int_0^\infty \frac{e^{-u}}{u}\,du - \int_0^\infty \frac{(u+1)^{-x}}{u}\,du$$

and then making the substitution $u + 1 = e^t$ in the second integral to get

$$\psi(x) = \int_0^\infty \frac{e^{-u}}{u}\,du - \int_0^\infty \frac{e^{-t(x-1)}}{e^t - 1}\,dt$$

Combining the last two integrals once again as a single integral yields

$$\psi(x) = \int_0^\infty \left(\frac{e^{-t}}{t} - \frac{e^{-t(x-1)}}{e^t - 1}\right)dt, \qquad x > 0 \qquad (2.65)$$

Remark: Although (2.64) is a convergent integral, it is not technically correct to write it as the difference of two integrals, since each integral by itself is divergent. We are simply using a mathematical gimmick here in order to formally derive (2.65), which happens also to be a convergent integral.

2.5.2 Asymptotic Series for ψ and Γ

Our next task is to derive asymptotic series for both the digamma and gamma functions. We begin with the integral representation

$$\psi(x+1) = \int_0^\infty \left(\frac{e^{-t}}{t} - \frac{e^{-xt}}{e^t - 1}\right)dt \qquad (2.66)$$

which comes from (2.65) with x replaced by $x + 1$. We then rewrite (2.66) in the form

$$\psi(x+1) = \int_0^\infty \left(\frac{e^{-t}}{t} - \frac{e^{-xt}}{t} + \frac{e^{-xt}}{t} - \frac{e^{-xt}}{e^t - 1}\right)dt$$

$$= \int_0^\infty \frac{e^{-t} - e^{-xt}}{t}\,dt + \int_0^\infty \left(\frac{1}{t} - \frac{1}{e^t - 1}\right)e^{-xt}\,dt$$

$$= \log x + I \qquad (2.67)$$

where we recognize the Frullani integral (2.62) and define

$$I = \int_0^\infty \left(\frac{1}{t} - \frac{1}{e^t - 1}\right) e^{-xt} dt, \qquad x > 0 \qquad (2.68)$$

In order to perform the integration in (2.68), we need to represent the function $(e^t - 1)^{-1}$ in a series and integrate termwise. Since this function is not defined at $t = 0$, it does not have a Maclaurin series about this point. However, the related function $t(e^t - 1)^{-1}$ and all its derivatives are well defined at $t = 0$, so we write

$$\frac{t}{e^t - 1} = \sum_{n=0}^{\infty} B_n \frac{t^n}{n!}, \qquad |t| < \infty \qquad (2.69)$$

where

$$B_n = \frac{d^n}{dt^n}\left[t(e^t - 1)^{-1}\right]\bigg|_{t=0}, \qquad n = 0, 1, 2, \ldots \qquad (2.70)$$

The constants B_n are called the *Bernoulli numbers*;* the first few are found to be

$$\begin{aligned} B_0 &= 1 \\ B_1 &= -\tfrac{1}{2} \\ B_2 &= \tfrac{1}{6} \\ B_3 &= 0 \\ B_4 &= -\tfrac{1}{30} \\ &\vdots \end{aligned} \qquad (2.71)$$

All Bernoulli numbers with odd index, except B_1, are zero. To show this, we simply replace t by $-t$ in (2.69) and then subtract the result from (2.69) itself, finding

$$\frac{t}{e^t - 1} - \frac{-t}{e^{-t} - 1} = -t = \sum_{n=0}^{\infty} \left[1 - (-1)^n\right] B_n \frac{t^n}{n!}$$

and by equating coefficients of like powers of t, we see that $B_1 = -\tfrac{1}{2}$ and $B_3 = B_5 = B_7 = \cdots = 0$.

If we divide both sides of (2.69) by t, we get

$$\frac{1}{e^t - 1} = \sum_{n=0}^{\infty} B_n \frac{t^{n-1}}{n!}$$

$$= \frac{1}{t} - \frac{1}{2} + \sum_{n=2}^{\infty} B_n \frac{t^{n-1}}{n!}$$

---

*The Bernoulli numbers are named after Jacob Bernoulli (1654–1705), who first introduced them.

and since all odd B_n are zero for n greater than one, we replace n by $2n$ in the sum to obtain the result

$$\frac{1}{e^t - 1} = \frac{1}{t} - \frac{1}{2} + \sum_{n=1}^{\infty} B_{2n} \frac{t^{2n-1}}{(2n)!} \tag{2.72}$$

Hence, the substitution of (2.72) into (2.68) gives us

$$I = \int_0^{\infty} \left(\frac{1}{t} - \frac{1}{t} + \frac{1}{2} - \sum_{n=1}^{\infty} B_{2n} \frac{t^{2n-1}}{(2n)!} \right) e^{-xt} dt$$

$$= \frac{1}{2} \int_0^{\infty} e^{-xt} dt - \sum_{n=1}^{\infty} \frac{B_{2n}}{(2n)!} \int_0^{\infty} e^{-xt} t^{2n-1} dt$$

Evaluating the above integrals in terms of gamma functions, the expression for I becomes

$$I = \frac{1}{2x} - \frac{1}{2} \sum_{n=1}^{\infty} \frac{B_{2n}}{n} \frac{1}{x^{2n}} \tag{2.73}$$

and this in turn, substituted into (2.67), leads to the asymptotic series

$$\psi(x+1) \sim \log x + \frac{1}{2x} - \frac{1}{2} \sum_{n=1}^{\infty} \frac{B_{2n}}{n} \frac{1}{x^{2n}}, \qquad x \to \infty \tag{2.74}$$

Unlike many of the asymptotic series that we derive, (2.74) converges for all $x > 0$.

In statistical mechanics, probability theory, and so forth, it often happens that we are dealing with large factorials, or gamma functions with large arguments. To facilitate the computations involving such expressions it is helpful to have an accurate asymptotic formula from which to approximate $\Gamma(x)$. Our approach to finding such a formula will be to first find a suitable asymptotic relation for $\log \Gamma(x+1)$ and then exponentiate this result.

Since, by definition,

$$\psi(x+1) = \frac{d}{dx} \log \Gamma(x+1)$$

it follows that the indefinite integral of (2.74) leads to the asymptotic series

$$\log \Gamma(x+1) \sim C + (x + \tfrac{1}{2}) \log x - x + \frac{1}{2} \sum_{n=1}^{\infty} \frac{B_{2n}}{n(2n-1)} \frac{1}{x^{2n-1}} \tag{2.75}$$

where C is a constant of integration. In order to evaluate C, we would normally need to know the exact behavior of the series (2.75) for some value of x. However, by allowing $x \to \infty$, we can eliminate the series in (2.75), and thus we see that

$$C = \lim_{x \to \infty} \left[\log \Gamma(x+1) - (x + \tfrac{1}{2})\log x + x \right]$$

$$= \lim_{x \to \infty} \left[\log\left(\frac{\Gamma(x+1)}{x^{x+\frac{1}{2}}} \right) + x \right] \qquad (2.76)$$

Now, by defining

$$K = e^C = \lim_{x \to \infty} \frac{\Gamma(x+1)}{x^{x+\frac{1}{2}}} e^x \qquad (2.77)$$

we have the limit relation

$$\lim_{x \to \infty} \Gamma(x+1) = K \lim_{x \to \infty} \left(e^{-x} x^{x+\frac{1}{2}} \right) \qquad (2.78)$$

The constant K can be determined by substituting (2.78) into the Legendre duplication formula written as

$$\sqrt{\pi} = \lim_{x \to \infty} \frac{2^{2x-1} \Gamma(x) \Gamma(x + \tfrac{1}{2})}{\Gamma(2x)} \qquad (2.79)$$

The result is $K = \sqrt{2\pi}$ (see problem 21), and therefore (2.78) leads to the asymptotic formula

$$\Gamma(x+1) \sim \sqrt{2\pi x}\, x^x e^{-x}, \qquad x \to \infty \qquad (2.80)$$

In particular, if we set $x = n$, where n is a large positive integer, we get the well-known expression

$$n! \sim \sqrt{2\pi n}\, n^n e^{-n}, \qquad n \gg 1 \qquad (2.81)$$

called *Stirling's formula*.*

It is interesting to note that Stirling's formula is remarkably accurate even for small values of n. For example, when $n = 6$ we find $6! \simeq 710.08$, an error of only 1.4% from the exact value of 720. Of course, for larger values of n the formula is even more accurate.

Our original intent was to find an asymptotic *series* for the gamma function, and to do this we substitute $K = \sqrt{2\pi}$ into (2.77), which identifies

$$C = \log K = \tfrac{1}{2} \log 2\pi \qquad (2.82)$$

---

*Equation (2.81), which is a special case of the asymptotic series for the gamma function, was published in 1730 by James Stirling (1692–1770).

Then, returning to the series (2.75), we have

$$\log \Gamma(x+1) \sim \tfrac{1}{2}\log 2\pi + (x+\tfrac{1}{2})\log x - x + \frac{1}{2}\sum_{n=1}^{\infty} \frac{B_{2n}}{n(2n-1)}\frac{1}{x^{2n-1}},$$
$$x \to \infty \quad (2.83)$$

This last expression is called *Stirling's series*. It represents a convergent series for $\log \Gamma(x+1)$ for all positive values of x. Moreover, the absolute value of the error incurred in using this series to evaluate $\log \Gamma(x+1)$ is less than the absolute value of the first term neglected in the series.

Although Stirling's series is valid for all positive x, it is used primarily for evaluating the gamma function for large arguments. We can eliminate the logarithm terms by exponentiating both sides to get (retaining only the first few terms of the series)

$$\Gamma(x+1) \sim \sqrt{2\pi}\, x^{x+\tfrac{1}{2}} e^{-x} \exp\left(\frac{1}{12x} - \frac{1}{360x^3} + \cdots\right)$$

or

$$\Gamma(x+1) \sim \sqrt{2\pi x}\, x^{x} e^{-x}\left(1 + \frac{1}{12x} + \frac{1}{288x^2} + \cdots\right), \quad x \to \infty$$
$$(2.84)$$

In this final step, we have replaced the last exponential function by the first few terms of its Maclaurin series.

Finally, if we set $x = n$, where n is a large positive integer, and retain only the first two terms of (2.84), we get a more accurate version of Stirling's formula [Equation (2.81)]:

$$n! \sim \sqrt{2\pi n}\, n^{n} e^{-n}\left(1 + \frac{1}{12n}\right), \quad n \gg 1 \quad (2.85)$$

Here we find for $n = 6$ that $6! \simeq 719.94$, which has an error of only $8.3 \times 10^{-3}\%$. Perhaps even more remarkable is that if we let $n = 1, 2, 3, \ldots$, we calculate from (2.85) the values

$$1! \simeq 0.99898$$
$$2! \simeq 1.99896$$
$$3! \simeq 5.99833$$
$$\vdots$$

and thus conclude that (2.85) is accurate enough for many applications for *all* positive integers.

2.5.3 Polygamma Functions

By repeated differentiation of the digamma function

$$\psi(x) = \frac{d}{dx}\log\Gamma(x) \qquad (2.86)$$

we form the family of *polygamma functions*

$$\psi^{(m)}(x) = \frac{d^{m+1}}{dx^{m+1}}\log\Gamma(x), \qquad m = 1,2,3,\ldots \qquad (2.87)$$

Recalling Equation (2.53),

$$\psi(x) = -\gamma + \sum_{n=0}^{\infty}\left(\frac{1}{n+1} - \frac{1}{n+x}\right) \qquad (2.88)$$

we readily determine the representation

$$\psi^{(m)}(x) = (-1)^{m+1}m!\sum_{n=0}^{\infty}\frac{1}{(n+x)^{m+1}}, \qquad m = 1,2,3,\ldots \qquad (2.89)$$

Of special interest is the evaluation of (2.89) when $x = 1$, i.e.,

$$\psi^{(m)}(1) = (-1)^{m+1}m!\sum_{n=0}^{\infty}\frac{1}{(n+1)^{m+1}}$$

$$= (-1)^{m+1}m!\sum_{n=1}^{\infty}\frac{1}{n^{m+1}}$$

or

$$\psi^{(m)}(1) = (-1)^{m+1}m!\zeta(m+1), \qquad m = 1,2,3,\ldots \qquad (2.90)$$

where

$$\zeta(p) = \sum_{n=1}^{\infty}\frac{1}{n^p}, \qquad p > 1 \qquad (2.91)$$

is the *Riemann zeta function* (see Section 2.5.4). The evaluation of $\psi^{(m)}(x)$ for other values of x also leads to the zeta function (see problems 29 and 30).

Although (2.88) and (2.89) are valid representations of $\psi(x)$ and $\psi^{(m)}(x)$, respectively, for all values of x except $x = 0, -1, -2, \ldots$, they are not the most convenient series to use for computational purposes, particularly in the neighborhood of $x = 1$. Instead, it may be preferable to have power-series expansions for such calculations.

To begin, we seek a power series of the form

$$\log\Gamma(x+1) = \sum_{n=0}^{\infty}c_n x^n \qquad (2.92)$$

where we choose $\log\Gamma(x+1)$ instead of $\log\Gamma(x)$ so that we can expand

about $x = 0$. The constants in this Maclaurin expansion are defined by
$$c_0 = \log \Gamma(x + 1)|_{x=0} = \log \Gamma(1) = 0$$
$$c_1 = \frac{d}{dx} \log \Gamma(x + 1)\bigg|_{x=0} = \psi(1) = -\gamma \qquad (2.93)$$
and for $n \geq 2$,
$$c_n = \frac{1}{n!} \frac{d^n}{dx^n} \log \Gamma(x + 1)\bigg|_{x=0} = \frac{1}{n!} \psi^{(n-1)}(1)$$
which in view of (2.90), becomes
$$c_n = \frac{(-1)^n}{n} \zeta(n), \qquad n = 2, 3, 4, \ldots \qquad (2.94)$$

Hence, the substitution of (2.93) and (2.94) into (2.92) yields the result
$$\log \Gamma(x + 1) = -\gamma x + \sum_{n=2}^{\infty} \frac{(-1)^n \zeta(n)}{n} x^n, \qquad -1 < x \leq 1 \quad (2.95)$$
where the interval of convergence is shown.

Termwise differentiation of (2.95) is permitted, and leads to
$$\psi(x + 1) = \frac{d}{dx} \log \Gamma(x + 1)$$
$$= -\gamma + \sum_{n=2}^{\infty} (-1)^n \zeta(n) x^{n-1}$$
or, by making a change of index,
$$\psi(x + 1) = -\gamma + \sum_{n=1}^{\infty} (-1)^{n+1} \zeta(n + 1) x^n, \qquad -1 < x < 1 \quad (2.96)$$

This last series no longer converges at the endpoint $x = 1$ as was the case in (2.95). Continued differentiation of (2.96) finally leads to the following relation for $m = 1, 2, 3, \ldots$ (see problem 23):
$$\psi^{(m)}(x + 1) = (-1)^{m+1} \sum_{n=0}^{\infty} (-1)^n \frac{(m + n)!}{n!} \zeta(m + n + 1) x^n,$$
$$-1 < x < 1 \quad (2.97)$$
which also converges for $-1 < x < 1$.

Both the digamma and polygamma functions are used at times for summing series, particularly those series involving rational functions with the power of the denominator at least two greater than that in the numerator. In such cases, the infinite series can be expressed as a finite sum of digamma or polygamma functions by the use of partial-fraction expansions (see Example 8). Of course, the values of the digamma and polygamma functions must usually be obtained from tables.*

2.5.4 Riemann Zeta Function

The *Riemann zeta function*

$$\zeta(x) = \sum_{n=1}^{\infty} \frac{1}{n^x}, \qquad x > 1 \tag{2.98}$$

first arose in Section 1.2.2 as a series that is useful in proving convergence or divergence of other series by means of a comparison test. We also found that the zeta function is closely related to the logarithm of the gamma function and to the polygamma functions. Although the zeta function was known to Euler, it was Riemann in 1859 who established most of its properties, which now are very important in the field of number theory, among others. Thus it bears his name.

An interesting relation for the zeta function can be derived by first making the observation

$$\zeta(x)(1 - 2^{-x}) = 1 + \frac{1}{2^x} + \frac{1}{3^x} + \frac{1}{4^x} + \cdots$$
$$- \left(\frac{1}{2^x} + \frac{1}{4^x} + \frac{1}{6^x} + \cdots \right)$$

where all terms are eliminated from (2.98) in which n is a multiple of 2. Therefore we deduce that

$$\zeta(x)(1 - 2^{-x}) = \sum_{n=1}^{\infty} \frac{1}{(2n-1)^x} \tag{2.99}$$

One of the advantages of (2.99) is that, using it, $\zeta(x)$ can be computed to the same accuracy as given by (2.98), but with only half as many terms. Similarly, the product

$$\zeta(x)(1 - 2^{-x})(1 - 3^{-x}) = 1 + \frac{1}{3^x} + \frac{1}{5^x} + \frac{1}{7^x} + \cdots$$
$$- \left(\frac{1}{3^x} + \frac{1}{9^x} + \frac{1}{15^x} + \cdots \right) \tag{2.100}$$

eliminates all terms from (2.99) in which n is a multiple of 3. Continuing in this fashion, it can eventually be shown that the infinite product over all prime numbers greater than one leads to

$$\zeta(x)(1 - 2^{-x})(1 - 3^{-x}) \cdots (1 - P^{-x}) \cdots = 1 \tag{2.101}$$

where P denotes a prime number. Hence, we have *Euler's infinite-product*

---

*See M. Abramowitz and I.A. Stegun (Eds.), *Handbook of Mathematical Tables*, New York: Dover, 1965, Chapter 6.

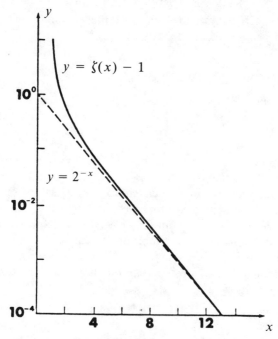

Figure 2.3 The Graphs of $\zeta(x) - 1$ and 2^{-x} (Note the logarithmic scale on the vertical axis.)

representation

$$\zeta(x) = \prod_{P=2}^{\infty} (1 - P^{-x})^{-1}, \qquad P \text{ prime} \qquad (2.102)$$

It can readily be shown that the zeta function has the integral representation (see problem 17)

$$\zeta(x) = \frac{1}{\Gamma(x)} \int_0^\infty \frac{t^{x-1}}{e^t - 1} \, dt, \qquad x > 1 \qquad (2.103)$$

Also, by using complex-variable methods, it can be shown that*

$$\zeta(1 - x) = 2^{1-x} \pi^{-x} \cos(\tfrac{1}{2}\pi x) \Gamma(x) \zeta(x) \qquad (2.104)$$

which is the famous formula of Riemann. Other relations involving this function, as well as some special values, are taken up in the exercises.

The graph of $\zeta(x) - 1$ is shown in Fig. 2.3 for $x > 1$. For comparison, the dotted line is the graph of 2^{-x}.

---

*See E.T. Whittaker and G.N. Watson, *A Course of Modern Analysis*, Cambridge: Cambridge U.P., 1965, p. 269.

EXERCISES 2.5

1. Show that
$$\psi(x) - \psi(y) = \sum_{n=0}^{\infty}\left(\frac{1}{y+n} - \frac{1}{x+n}\right)$$

2. Take the logarithmic derivative of $\Gamma(x)\Gamma(1-x) = \pi \csc \pi x$ to deduce the identity
$$\psi(1-x) - \psi(x) = \pi \cot \pi x$$

3. By taking the logarithmic derivative of the Legendre duplication formula
$$2^{2x-1}\Gamma(x)\Gamma(x+\tfrac{1}{2}) = \sqrt{\pi}\,\Gamma(2x)$$

   (a) deduce that
$$\psi(x) + \psi(x+\tfrac{1}{2}) + 2\log 2 = 2\psi(2x)$$

   (b) From (a), deduce that $\psi(\tfrac{1}{2}) = -\gamma - 2\log 2$.
   (c) For $n = 1, 2, 3, \ldots$, show that
$$\psi(n+\tfrac{1}{2}) = -\gamma - 2\log 2 + 2\sum_{k=1}^{n}(2k-1)^{-1}$$

4. Derive the formula
$$3\psi(3) = \psi(x) + \psi(x+\tfrac{1}{3}) + \psi(x+\tfrac{2}{3}) + 3\log 3$$

   Hint: Recall problem 40 in Exercises 2.2.

In problems 5–8, verify the given relation.

5. $\psi(n+1) = -\gamma + \sum_{k=1}^{\infty}\dfrac{n}{k(k+n)}$, $n = 0, 1, 2, \ldots$.

6. $\lim\limits_{n \to \infty}[\psi(x+n) - \log n] = 0$.

7. $\psi(\tfrac{1}{2}+p) = \psi(\tfrac{1}{2}-p) + \pi \tan \pi p$.

8. $\exp[\psi(x)] = x\prod\limits_{n=0}^{\infty}\left(1 + \dfrac{1}{x+n}\right)e^{-1/(x+n)}$.

9. Show that
   (a) $\gamma = -\displaystyle\int_0^1 \log\left(\log\frac{1}{t}\right) dt$.
   (b) $\gamma = \dfrac{1}{2} + 2\displaystyle\int_0^{\infty}\dfrac{t\,dt}{(1+t^2)(e^{2\pi t}-1)}$.

88 • Special Functions for Engineers and Applied Mathematicians

10. Starting with Equation (2.55), use integration by parts followed by a change of variable to show that

$$\gamma = \int_0^1 \frac{1 - e^{-t} - e^{-1/t}}{t}\, dt$$

11. Derive the Maclaurin series expansion

$$t \coth t = \sum_{m=0}^{\infty} B_{2m} \frac{(2t)^{2m}}{(2m)!}$$

Hint: First show that

$$\coth t = \frac{e^t + e^{-t}}{e^t - e^{-t}} = \frac{1}{e^{2t} - 1} + \frac{1}{1 - e^{-2t}}$$

12. Starting with the infinite product representation

$$\sin x = x \prod_{n=1}^{\infty} \left(1 - \frac{x^2}{n^2 \pi^2}\right)$$

(a) show that the logarithmic derivative leads to

$$x \cot x = 1 - 2 \sum_{n=1}^{\infty} \frac{(x/n\pi)^2}{1 - x^2/n^2\pi^2}, \qquad -\pi < x < \pi$$

(b) From (a), deduce that

$$x \cot x = 1 - 2 \sum_{m=1}^{\infty} \zeta(2m) \left(\frac{x}{\pi}\right)^{2m}$$

13. By using the identity $\coth ix = -i \cot x$ ($i^2 = -1$) and the result of problem 11,

(a) deduce that

$$x \cot x = 1 + \sum_{m=1}^{\infty} B_{2m} \frac{(-1)^m}{(2m)!} (2x)^{2m}$$

(b) Comparing the result of (a) with that of problem 12(b), deduce the relation

$$\zeta(2m) = \frac{(2\pi)^{2m}(-1)^{m-1}}{2(2m)!} B_{2m}, \qquad m = 1, 2, 3, \ldots$$

14. Show that
    (a) $\zeta(2) = \pi^2/6$,   (b) $\zeta(4) = \pi^4/90$,   (c) $\zeta(6) = \pi^6/945$.

    Hint: Use problem 13(b).

15. Show that
$$\gamma = \sum_{n=2}^{\infty} \frac{(-1)^n}{n} \zeta(n)$$

16. The total energy radiated by a blackbody (*Stefan-Boltzmann law*) is proportional to the integral
$$I = \int_0^{\infty} \frac{x^3}{e^x - 1} dx$$
    Show that $I = \pi^4/15$.

    Hint: Observe that $(1 - e^{-x})^{-1} = \sum_{n=0}^{\infty} e^{-nx}$ and use problem 14.

17. Starting with the observation
$$\frac{1}{n^x} = \frac{1}{\Gamma(x)} \int_0^{\infty} e^{-nt} t^{x-1} dt, \quad x > 1, \quad n = 1, 2, 3, \ldots$$
    sum over all values of n to deduce that
$$\zeta(x) = \frac{1}{\Gamma(x)} \int_0^{\infty} \frac{t^{x-1}}{e^t - 1} dt, \quad x > 1$$

18. Show that ($p > 1$)
$$\int_0^{\infty} t^{p-1} \left(\frac{1}{e^t - 1} - \frac{1}{e^t + 1} \right) dt = 2^{1-p} \Gamma(p) \zeta(p)$$

19. Using the results of problems 17 and 18, deduce that ($p > 1$)
$$\frac{1}{\Gamma(p)} \int_0^{\infty} \frac{t^{p-1}}{e^t + 1} dt = \sum_{n=1}^{\infty} \frac{(-1)^{n-1}}{n^p} = (1 - 2^{1-p}) \zeta(p)$$

20. By expressing $\log(1 + x)$ in its Maclaurin series, show that
$$\int_0^1 \frac{\log(1 + x)}{x} dx = \frac{\pi^2}{12}$$

    Hint: See problems 14 and 19.

21. By substituting the limit expression

$$\lim_{x \to \infty} \Gamma(x+1) = K \lim_{x \to \infty} \left(e^{-x} x^{x+\frac{1}{2}} \right)$$

into the Legendre duplication formula written in the form

$$\sqrt{\pi} = \lim_{x \to \infty} \frac{2^{2x-1} \Gamma(x) \Gamma(x + \frac{1}{2})}{\Gamma(2x)}$$

deduce that $K = \sqrt{2\pi}$.

22. Use problem 21 to establish that

$$\lim_{x \to \infty} x^{b-a} \frac{\Gamma(x+a+1)}{\Gamma(x+b+1)} = 1$$

23. Show that the mth derivative of Equation (2.96) leads to

$$\psi^{(m)}(x+1) = (-1)^{m+1} \sum_{n=0}^{\infty} (-1)^n \frac{(m+n)!}{n!} \zeta(m+n+1) x^n$$

24. Show that

(a) $\psi'(1) = \pi^2/6$,   (b) $\psi'(2) = \dfrac{\pi^2}{6} - 1$,

(c) $\psi'(\tfrac{1}{2}) = \pi^2/2$,   (d) $\psi'''(2) = \dfrac{\pi^4}{15} - 6$.

25. Write the sum of the series in terms of the digamma and polygamma functions and evaluate:

(a) $\displaystyle\sum_{n=1}^{\infty} \frac{1}{n(n+1)}$,   (b) $\displaystyle\sum_{n=0}^{\infty} \frac{1}{(n+2)(n+4)}$,

(c) $\displaystyle\sum_{n=1}^{\infty} \frac{1}{n(n+1)^2}$,   (d) $\displaystyle\sum_{n=1}^{\infty} \frac{1}{n(4n^2-1)}$.

26. Show that ($m = 1, 2, 3, \ldots$)

$$\psi^{(m)}(x) = (-1)^{m+1} \int_0^{\infty} \frac{t^m e^{-xt}}{1 - e^{-t}} \, dt$$

27. Derive the asymptotic series

$$\psi'(x+1) \sim \frac{1}{x} - \frac{1}{2x^2} + \sum_{n=1}^{\infty} B_{2n} x^{-(2n+1)}, \qquad x \to \infty$$

Note: This series diverges for all x.

28. Use the first *four* terms of the series in problem 27 (including the terms outside the summation) to approximate $\psi'(4)$, and compare with the exact value $\psi'(4) = \pi^2/6 - \frac{49}{36}$.

29. For $k = 2, 3, 4, \ldots,$ show that

$$\psi^{(m)}(k) = (-1)^{m+1} m! \left[\zeta(m+1) - \sum_{n=1}^{k-1} \frac{1}{n^{m+1}} \right]$$

30. For $k = 2, 3, 4, \ldots,$ show that

$$\psi^{(m)}(k - \tfrac{1}{2})$$
$$= (-1)^{m+1} m! \left[(2^{m+1} - 1)\zeta(m+1) - 2^{m+1} \sum_{n=1}^{k-1} \frac{1}{(2n-1)^{m+1}} \right]$$

3

Other Functions Defined by Integrals

3.1 Introduction

In addition to the gamma function, there are numerous other special functions whose primary definition involves an integral. Some of these other functions were introduced in Chapter 2 along with the gamma function, and in the present chapter we wish to consider several other such functions defined by integrals.

The *error function* derives its name from its importance in the theory of errors, but it also occurs in probability theory and in certain heat-conduction problems on infinite domains. The closely related *Fresnel integrals*, which are fundamental in the theory of optics, can be derived directly from the error function. A special case of the incomplete gamma function (Section 2.4) leads to the *exponential integral* and related functions—the *logarithmic integral*, which is important in analysis and number theory, and the *sine* and *cosine integrals*, which arise in Fourier-transform theory.

Elliptic integrals first arose in the problems associated with computing the arc length of an ellipse and a lemniscate (a curve in the shape of a figure eight). Some early results concerning elliptic integrals were discovered by L. Euler and J. Landen, but virtually the whole theory of these integrals was developed by Legendre over a period spanning 40 years. The inverses of the elliptic integrals, called *elliptic functions*, were independently introduced in 1827 by C.G.J. Jacobi (1802–1859) and N.H. Abel (1802–1829). Many of the properties of elliptic functions, however, had already been developed as early as 1809 by Gauss. Elliptic functions have the distinction of being doubly periodic, with one real period and one imaginary period. Among other areas of application, the elliptic functions are important in solving the pendulum problem (Section 3.4.2).

3.2 The Error Function and Related Functions

The *error function* is defined by the integral

$$\text{erf}(x) = \frac{2}{\sqrt{\pi}} \int_0^x e^{-t^2}\, dt, \qquad -\infty < x < \infty \tag{3.1}$$

This function is encountered in probability theory, the theory of errors, the theory of heat conduction, and various branches of mathematical physics. By representing the exponential function in (3.1) in terms of its power-series expansion, we have

$$\text{erf}(x) = \frac{2}{\sqrt{\pi}} \int_0^x \sum_{n=0}^{\infty} \frac{(-1)^n}{n!} t^{2n}\, dt$$

from which we deduce (termwise integration of power series is permitted)

$$\text{erf}(x) = \frac{2}{\sqrt{\pi}} \sum_{n=0}^{\infty} \frac{(-1)^n x^{2n+1}}{n!(2n+1)}, \qquad |x| < \infty \tag{3.2}$$

Examination of the series (3.2) reveals that the error function is an *odd function*, i.e.,

$$\text{erf}(-x) = -\text{erf}(x) \tag{3.3}$$

Also, we see that

$$\text{erf}(0) = \frac{2}{\sqrt{\pi}} \int_0^0 e^{-t^2}\, dt = 0 \tag{3.4}$$

and by using properties of the gamma function, we find that (in the limit)

$$\text{erf}(\infty) = \frac{2}{\sqrt{\pi}} \int_0^{\infty} e^{-t^2}\, dt = \frac{\Gamma(\frac{1}{2})}{\sqrt{\pi}} = 1 \tag{3.5}$$

The graph of $\text{erf}(x)$ is shown in Fig. 3.1.

In some applications it is useful to introduce the *complementary error function*

$$\text{erfc}(x) = \frac{2}{\sqrt{\pi}} \int_x^{\infty} e^{-t^2}\, dt \tag{3.6}$$

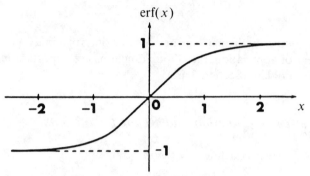

Figure 3.1 The Error Function

Clearly it follows that

$$\text{erfc}(x) = \frac{2}{\sqrt{\pi}} \int_0^\infty e^{-t^2}\, dt - \frac{2}{\sqrt{\pi}} \int_0^x e^{-t^2}\, dt$$

from which we deduce

$$\text{erfc}(x) = 1 - \text{erf}(x) \qquad (3.7)$$

Hence, all properties of erfc(x) can be derived from those of erf(x).

Example 1: Find the Laplace transform of $f(t) = \text{erfc}(t^{-1/2})$.

Solution: The Laplace transform is defined by

$$\mathscr{L}\{\text{erfc}(t^{-1/2}); s\} = \int_0^\infty e^{-st}\, \text{erfc}(t^{-1/2})\, dt$$

$$= \int_0^\infty e^{-st} \frac{2}{\sqrt{\pi}} \int_{t^{-1/2}}^\infty e^{-u^2}\, du\, dt$$

If we interpret this last expression as an iterated integral, we can then interchange the order of integration (see Fig. 3.2). Hence,

$$\mathscr{L}\{\text{erfc}(t^{-1/2}); s\} = \frac{2}{\sqrt{\pi}} \int_0^\infty e^{-u^2} \int_{u^{-2}}^\infty e^{-st}\, dt\, du$$

$$= \frac{2}{s\sqrt{\pi}} \int_0^\infty e^{-u^2 - su^{-2}}\, du$$

and by calling upon the integral formula (see problem 6)

$$\int_0^\infty e^{-a^2 x^2 - b^2 x^{-2}}\, dx = \frac{\sqrt{\pi}}{2a} e^{-2ab}, \qquad a > 0, \quad b \geq 0$$

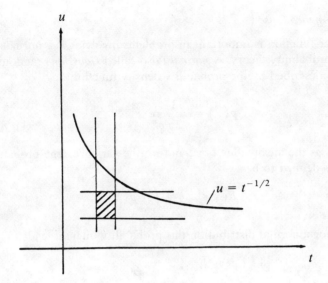

Figure 3.2

we deduce that

$$\mathscr{L}\{\mathrm{erfc}(t^{-1/2}); s\} = \frac{1}{s} e^{-2\sqrt{s}}, \qquad s > 0$$

3.2.1 Asymptotic Series

An *asymptotic series* for the complementary error function can be obtained through repeated integration by parts. To obtain this series, we first observe that integration by parts leads to

$$\int_x^\infty e^{-t^2}\, dt = \frac{e^{-x^2}}{2x} - \frac{1}{2}\int_x^\infty \frac{e^{-t^2}}{t^2}\, dt$$

and by integrating by parts again, we get

$$\int_x^\infty e^{-t^2}\, dt = \frac{e^{-x^2}}{2x} - \frac{e^{-x^2}}{2^2 x^2} + \frac{1 \times 3}{x^2} \int_x^\infty \frac{e^{-t^2}}{t^4}\, dt$$

Continuing this process indefinitely, we finally derive the asymptotic series

$$\mathrm{erfc}(x) \sim \frac{e^{-x^2}}{\sqrt{\pi}\, x}\left[1 + \sum_{n=1}^{\infty} (-1)^n \frac{1 \times 3 \times \cdots \times (2n-1)}{(2x^2)^n}\right], \qquad x \to \infty$$

(3.8)

3.2.2 Applications

The error function is important in problems involving the normal distribution in probability theory. A *normal* (also called *Gaussian*) random variable x is one described by the probability density function

$$p(x) = \frac{1}{\sqrt{2\pi}\,\sigma} e^{-(x-m)^2/2\sigma^2} \tag{3.9}$$

where m is the mean value of x, and σ^2 the variance. The probability that $x \leq X$ is defined to be

$$P(x \leq X) = \int_{-\infty}^{X} p(x)\,dx \tag{3.10}$$

Hence, for a normal distribution this probability integral leads to

$$P(x \leq X) = \frac{1}{\sqrt{2\pi}\,\sigma} \int_{-\infty}^{X} e^{-(x-m)^2/2\sigma^2}\,dx$$

$$= \frac{1}{\sqrt{2\pi}\,\sigma} \int_{-\infty}^{\infty} e^{-(x-m)^2/2\sigma^2}\,dx - \frac{1}{\sqrt{2\pi}\,\sigma} \int_{X}^{\infty} e^{-(x-m)^2/2\sigma^2}\,dx$$

$$= 1 - \tfrac{1}{2}\mathrm{erfc}\left(\frac{X-m}{\sqrt{2}\,\sigma}\right) \tag{3.11}$$

where the last step is obtained after making the change of variable $t = (x-m)/\sqrt{2}\,\sigma$. By using (3.7), we can rewrite (3.11) in the form

$$P(x \leq X) = \frac{1}{2}\left[1 + \mathrm{erf}\left(\frac{X-m}{\sqrt{2}\,\sigma}\right)\right] \tag{3.12}$$

As we expect, the probability (3.12) approaches unity in the limit as $X \to \infty$.

Another application involving the error function concerns the problem of heat flow in the infinite medium $-\infty < x < \infty$ when the initial distribution of temperature $f(x)$ is known and the region is free of any heat sources. Physically, this problem might represent the linear flow of heat in a very long slender rod whose lateral surface is insulated. The problem is mathematically characterized by

$$\frac{\partial^2 u}{\partial x^2} = a^{-2}\frac{\partial u}{\partial t}, \quad -\infty < x < \infty, \quad t > 0$$

$$u(x,0) = f(x), \quad -\infty < x < \infty \tag{3.13}$$

where a^2 is a physical constant. The formal solution of (3.13) for any piecewise smooth function f is known to be*

$$u(x,t) = \frac{1}{2a\sqrt{\pi t}} \int_{-\infty}^{\infty} f(\xi) e^{-(x-\xi)^2/4a^2 t} d\xi \qquad (3.14)$$

which can be verified by direct substitution into (3.13).

As a specific example, let us suppose the initial temperature distribution in (3.13) is prescribed by

$$f(x) = \begin{cases} T_0, & |x| < 1 \\ 0, & |x| > 1 \end{cases} \qquad (3.15)$$

where T_0 is constant. The substitution of (3.15) into (3.14) yields

$$u(x,t) = \frac{T_0}{2a\sqrt{\pi t}} \int_{-1}^{1} e^{-(x-\xi)^2/4a^2 t} d\xi \qquad (3.16)$$

which, following the change of variable $z = (x - \xi)/2a\sqrt{t}$, becomes

$$u(x,t) = \frac{T_0}{2\sqrt{\pi}} \int_{(x-1)/2a\sqrt{t}}^{(x+1)/2a\sqrt{t}} e^{-z^2} dz \qquad (3.17)$$

Finally, evaluating (3.17), we obtain (see problem 1)

$$u(x,t) = \tfrac{1}{2} T_0 \left[\mathrm{erf}\left(\frac{x+1}{2a\sqrt{t}}\right) - \mathrm{erf}\left(\frac{x-1}{2a\sqrt{t}}\right) \right] \qquad (3.18)$$

Physical intuition suggests that $u(x,t) \to 0$ as $t \to \infty$, which we leave to the reader to verify. Also, it is interesting to note that the solution (3.18) is a continuous function for all x and for all $t > 0$, even though the input temperature distribution (3.15) is discontinuous.

3.2.3 Fresnel Integrals

Closely associated with the error function are the *Fresnel integrals*

$$C(x) = \int_0^x \cos(\tfrac{1}{2}\pi t^2) dt \qquad (3.19)$$

and

$$S(x) = \int_0^x \sin(\tfrac{1}{2}\pi t^2) dt \qquad (3.20)$$

These integrals come up in various branches of physics and engineering, such as in diffraction theory and the theory of vibrations, among others.

---
*See D. Powers, *Boundary Value Problems*, 2nd ed., New York: Academic, 1979, p. 134.

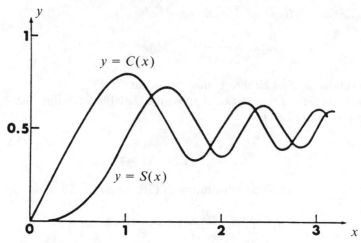

Figure 3.3 The Fresnel Integrals

From definition, we have the immediate results
$$C(0) = S(0) = 0 \tag{3.21}$$
The derivatives of these functions are
$$C'(x) = \cos(\tfrac{1}{2}\pi x^2), \qquad S'(x) = \sin(\tfrac{1}{2}\pi x^2) \tag{3.22}$$
and thus we deduce that both $C(x)$ and $S(x)$ are oscillatory. Namely, $C(x)$ has extrema at the points where $x^2 = 2n + 1$ ($n = 0, 1, 2, \ldots$), and $S(x)$ has extrema where $x^2 = 2n$ ($n = 1, 2, 3, \ldots$). The largest maxima occur first and are found to be $C(1) = 0.77989\ldots$ and $S(\sqrt{2}) = 0.71397\ldots$. For $x \to \infty$, we can use the integral formulas (see problem 19)
$$\int_0^\infty \cos t^2 \, dt = \int_0^\infty \sin t^2 \, dt = \frac{1}{2}\sqrt{\frac{\pi}{2}} \tag{3.23}$$
to obtain the results
$$C(\infty) = S(\infty) = \tfrac{1}{2} \tag{3.24}$$
The graphs of $C(x)$ and $S(x)$ for positive x are shown in Fig. 3.3.

To derive the relation between the Fresnel integrals and the error function, we start with
$$\operatorname{erf}(z) = \frac{2}{\sqrt{\pi}} \int_0^z e^{-u^2} \, du \tag{3.25}$$
where z may be real or complex.* Substituting $z = (i\pi/2)^{1/2} x$ and $u =$

---

*For a discussion of the error function with complex argument, see N.N. Lebedev, *Special Functions and Their Applications*, New York: Dover, 1972, Chapter 2.

$(i\pi/2)^{1/2}t$ into (3.25) leads to

$$\text{erf}\left[\left(\frac{i\pi}{2}\right)^{1/2} x\right] = (2i)^{1/2} \int_0^x e^{-i\pi t^2/2} \, dt$$

$$= (2i)^{1/2}\left[\int_0^x \cos(\tfrac{1}{2}\pi t^2) \, dt - i\int_0^x \sin(\tfrac{1}{2}\pi t^2) \, dt\right]$$

from which it follows that

$$(2i)^{-1/2}\text{erf}\left[(i\pi/2)^{1/2} x\right] = C(x) - iS(x). \qquad (3.26)$$

Other properties of these functions are taken up in the exercises.

EXERCISES 3.2

1. Show that
   (a) $\int_{-a}^{a} e^{-t^2} \, dt = \sqrt{\pi}\, \text{erf}(a)$,
   (b) $\int_a^b e^{-t^2} \, dt = \tfrac{1}{2}\sqrt{\pi}\,[\text{erf}(b) - \text{erf}(a)]$,
   (c) $\dfrac{d}{dx}\text{erf}(x) = \dfrac{2}{\sqrt{\pi}} e^{-x^2}$.

2. Evaluate
   (a) erfc(0),
   (b) erfc(∞).

3. Establish the relations (see Section 2.4)
   (a) $\text{erf}(x) = \dfrac{1}{\sqrt{\pi}}\gamma(\tfrac{1}{2}, x^2)$,
   (b) $\text{erfc}(x) = \dfrac{1}{\sqrt{\pi}}\Gamma(\tfrac{1}{2}, x^2)$.

4. Show that $(s \geq 0)$

$$\int_0^\infty e^{-st - t^2/4} \, dt = \sqrt{\pi}\, e^{s^2} \text{erfc}(s)$$

   Hint: Write $\tfrac{1}{4}t^2 + st = (\tfrac{1}{2}t + s)^2 - s^2$ and make the change of variable $u = \tfrac{1}{2}t + s$.

5. Show that $(s > 0)$

$$\int_0^\infty e^{-st}\text{erf}(t) \, dt = \frac{1}{s} e^{\frac{1}{4}s^2}\text{erfc}(\tfrac{1}{2}s)$$

   Hint: Reverse the order of integration.

6. Considering the integral

$$I(b) = \int_0^\infty e^{-a^2x^2 - b^2x^{-2}}\,dx, \qquad a > 0, \quad b \geq 0$$

as a function of the parameter b,

(a) show that I satisfies the first-order linear differential equation (DE)

$$\frac{dI}{db} + 2aI = 0$$

(b) Evaluate $I(0)$ directly from the integral.
(c) Solve the DE in (a) subject to the initial condition in (b) to deduce the result

$$I(b) = \frac{\sqrt{\pi}}{2a} e^{-2ab}$$

7. Considering the integral

$$I(a) = \int_{-\infty}^\infty \frac{e^{-a^2x^2}}{x^2 + b^2}\,dx, \qquad a \geq 0, \quad b > 0$$

as a function of the parameter a,

(a) show that I satisfies the first-order linear DE

$$\frac{dI}{da} - 2ab^2 I = -2\sqrt{\pi}$$

(b) Evaluate $I(0)$ directly from the integral.
(c) Solve the DE in (a) subject to the initial condition in (b) to deduce that

$$I(a) = \frac{\pi}{b} e^{a^2 b^2} \operatorname{erfc}(ab)$$

8. Use the result of problem 7 to show that

$$\int_0^{\pi/2} e^{-a^2 \tan^2 x}\,dx = \int_0^{\pi/2} e^{-a^2 \cot^2 x}\,dx = \frac{\pi}{2} e^{a^2} \operatorname{erfc}(a), \qquad a \geq 0$$

9. Solve the problem described by (3.13) when

$$f(x) = \begin{cases} 0, & x < 0 \\ T_0, & x > 0 \end{cases}$$

10. The temperature distribution in a very long rod, initially at zero temperature and subject to a time-varying heat reservoir at one end, is governed by

$$\frac{\partial^2 u}{\partial x^2} = a^{-2}\frac{\partial u}{\partial t}, \qquad 0 < x < \infty, \quad t > 0$$

$$u(x,0) = 0, \qquad 0 < x < \infty$$
$$u(0,t) = f(t), \qquad t > 0$$

with formal solution

$$u(x,t) = \frac{x}{2a\sqrt{\pi}} \int_0^t \frac{f(\tau)}{(t-\tau)^{3/2}} \exp\left[-\frac{x^2}{4a^2(t-\tau)}\right] d\tau$$

Solve this problem when $f(t) = T_1$ (constant).

11. If the boundary condition in problem 10 is

$$u(0,t) = f(t) = \begin{cases} T_1, & 0 < t < b \\ 0, & t \geq b \end{cases}$$

show that the subsequent temperature distribution is given by

$$u(x,t) = \begin{cases} T_1 \operatorname{erfc}(x/2a\sqrt{t}), & 0 < t < b \\ T_1[\operatorname{erf}(x/2a\sqrt{t-b}) - \operatorname{erf}(x/2a\sqrt{t})], & t \geq b \end{cases}$$

Verify that $u(x,t)$ is continuous at $t = b$.

12. If the boundary condition in problem 10 is modified to

$$\frac{\partial u}{\partial x}(0,t) = -f(t)$$

the formal solution becomes

$$u(x,t) = \frac{a}{\sqrt{\pi}} \int_0^t \frac{f(\tau)}{\sqrt{t-\tau}} \exp\left[-\frac{x^2}{4a^2(t-\tau)}\right] d\tau$$

(a) For the special case $f(t) = K$ (constant), show that

$$u(x,t) = K\left[2a(t/\pi)^{1/2} e^{-x^2/4a^2 t} - x\operatorname{erfc}(x/2a\sqrt{t})\right]$$

(b) What is the temperature in the rod at the end $x = 0$ as a function of time?

13. Use integration by parts to obtain

$$\int \operatorname{erf}(x)\, dx = x\operatorname{erf}(x) + \frac{1}{\sqrt{\pi}} e^{-x^2} + C$$

102 • Special Functions for Engineers and Applied Mathematicians

In problems 14–16, derive the integral representation.

14. $\mathrm{erf}(x) = \dfrac{2}{\pi} \displaystyle\int_0^\infty e^{-t^2} \dfrac{\sin(2xt)}{t}\, dt.$

   Hint: Write $\sin(2xt)$ in a power series.

15. $[\mathrm{erf}(x)]^2 = 1 - \dfrac{4}{\pi} \displaystyle\int_0^1 \dfrac{e^{-x^2(1+t^2)}}{1+t^2}\, dt.$

   Hint: Write the expansion on the left as a double integral and transform to polar coordinates.

16. $[\mathrm{erfc}(x)]^2 = \dfrac{4}{\sqrt{\pi}} e^{-2x^2} \displaystyle\int_0^\infty e^{-t^2 - 2\sqrt{2}\,xt} \mathrm{erf}(t)\, dt, \; x > 0.$

   Hint: Use the result of problem 4.

17. Show that the Fresnel integrals satisfy

   (a) $C(-x) = -C(x)$.
   (b) $S(-x) = -S(x)$.

18. Obtain the series representations

   (a) $C(x) = \displaystyle\sum_{n=0}^\infty \dfrac{(-1)^n (\pi/2)^{2n}}{(2n)!(4n+1)} x^{4n+1}.$

   (b) $S(x) = \displaystyle\sum_{n=0}^\infty \dfrac{(-1)^n (\pi/2)^{2n+1}}{(2n+1)!(4n+3)} x^{4n+3}.$

19. Establish the integral formula (see problem 6)

$$\int_0^\infty e^{-a^2 t^2}\, dt = \dfrac{\sqrt{\pi}}{2a}.$$

Then, writing $a = (1-i)/\sqrt{2}$ and separating into real and imaginary parts, deduce that

$$\int_0^\infty \cos t^2\, dt = \int_0^\infty \sin t^2\, dt = \dfrac{1}{2}\sqrt{\dfrac{\pi}{2}}.$$

20. Establish the integral formula

$$\dfrac{2}{\sqrt{\pi}} \int_0^x e^{-a^2 t^2}\, dt = \dfrac{1}{a} \mathrm{erf}(ax).$$

Then, following the suggestion in problem 19 and using the asymptotic series (3.8), derive the asymptotic series

$$C(x) \sim \frac{1}{2} - \frac{1}{\pi x}\left[B(x)\cos(\tfrac{1}{2}\pi x^2) - A(x)\sin(\tfrac{1}{2}\pi x^2)\right], \qquad x \to \infty$$

$$S(x) \sim \frac{1}{2} - \frac{1}{\pi x}\left[A(x)\cos(\tfrac{1}{2}\pi x^2) + B(x)\sin(\tfrac{1}{2}\pi x^2)\right], \qquad x \to \infty$$

where $A(x)$ and $B(x)$ are each asymptotic series related to (3.8).

3.3 The Exponential Integral and Related Functions

The *exponential integral* is defined by*

$$\text{Ei}(x) = \int_{-\infty}^{x} \frac{e^t}{t}\, dt, \qquad x \neq 0 \tag{3.27}$$

Another definition that is often given results from the replacement of x by $-x$ and t by $-t$, which leads to

$$-\text{Ei}(-x) \equiv E_1(x) = \int_{x}^{\infty} \frac{e^{-t}}{t}\, dt, \qquad x > 0 \tag{3.28}$$

The exponential integral (3.27) or (3.28) is encountered in several areas, including antenna theory and some astrophysical problems. Also, many integrals of a more complicated nature can be expressed in terms of the exponential integrals.

Comparison of (3.28) with Equation (2.40) in Section 2.4 reveals that $E_1(x)$ is related to the incomplete gamma functions according to

$$E_1(x) = \Gamma(0, x) = \lim_{a \to 0}\left[\Gamma(a) - \gamma(a, x)\right] \tag{3.29}$$

Thus, properties of $E_1(x)$ can be deduced from those of the incomplete gamma functions. For example, from the series for $\gamma(a, x)$ [see Equation (2.42) in Section 2.4], we have

$$\begin{aligned}
E_1(x) &= \lim_{a \to 0}\left[\Gamma(a) - x^a \sum_{n=0}^{\infty} \frac{(-1)^n x^n}{n!(n + a)}\right] \\
&= \lim_{a \to 0}\left[\frac{a\Gamma(a) - x^a}{a}\right] - \sum_{n=1}^{\infty} \frac{(-1)^n x^n}{n!\, n}
\end{aligned} \tag{3.30}$$

Using the recurrence formula for the gamma function and L'Hôpital's rule,

---

*Technically, Equation (3.27) does not define $\text{Ei}(x)$ for $x > 0$ unless we interpret the integral as its Cauchy principal value, i.e.,

$$\int_{-\infty}^{x} \frac{e^t}{t}\, dt = \lim_{\varepsilon \to 0^+}\left[\int_{-\infty}^{-\varepsilon} \frac{e^t}{t}\, dt + \int_{\varepsilon}^{x} \frac{e^t}{t}\, dt\right]$$

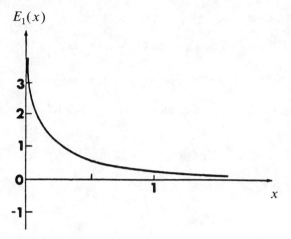

Figure 3.4 The Exponential Integral $E_1(x)$

it follows that*

$$\lim_{a \to 0}\left[\frac{a\Gamma(a) - x^a}{a}\right] = \lim_{a \to 0}\left[\Gamma'(a+1) - x^a \log x\right]$$
$$= -\gamma - \log x \quad (3.31)$$

Hence we have derived the series representation

$$E_1(x) = -\gamma - \log x - \sum_{n=1}^{\infty} \frac{(-1)^n x^n}{n!n}, \quad x > 0 \quad (3.32)$$

Equation (3.32) illustrates the logarithmic behavior of $E_1(x)$ for small arguments, i.e.,

$$E_1(x) \sim -\log x, \quad x \to 0^+ \quad (3.33)$$

For large arguments, we can use Equation (2.46) in Section 2.4 to deduce that†

$$E_1(x) \sim \frac{e^{-x}}{x}, \quad x \to \infty \quad (3.34)$$

The graph of $E_1(x)$ for positive x is shown in Fig. 3.4.

---

*Recall from Equation (2.55) in Section 2.5 that $\Gamma'(1) = -\gamma$.
†The complete asymptotic series for $E_1(x)$ was developed in Example 8 in Section 1.4.1

3.3.1 Logarithmic Integral

Closely related to the exponential integral is the *logarithmic integral*

$$\text{li}(x) = \int_0^x \frac{dt}{\log t}, \qquad x \neq 1 \qquad (3.35)$$

Setting $u = \log t$, we see that (3.35) becomes

$$\text{li}(x) = \int_{-\infty}^{\log x} \frac{e^u}{u} \, du$$

and thus deduce that

$$\text{li}(x) = \text{Ei}(\log x) = -E_1(-\log x), \qquad 0 < x < 1 \qquad (3.36)$$

By using (3.36), we can immediately deduce properties of $\text{li}(x)$ from those developed for the exponential integrals. In particular, Equation (3.32) leads to

$$\text{li}(x) = \gamma + \log(-\log x) + \sum_{n=1}^{\infty} \frac{(\log x)^n}{n! n}, \qquad 0 < x < 1 \qquad (3.37)$$

The graph of $\text{li}(x)$ is shown in Fig. 3.5.

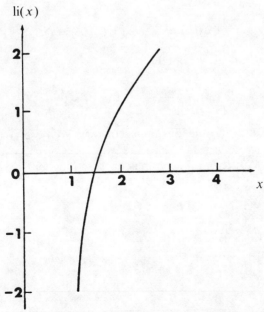

Figure 3.5 The Logarithmic Integral

3.3.2 Sine and Cosine Integrals

Another set of special functions that are related to the exponential integral are the *sine integral* and *cosine integral* defined, respectively, by

$$\text{Si}(x) = \int_0^x \frac{\sin t}{t}\, dt, \qquad x > 0 \tag{3.38}$$

and

$$\text{Ci}(x) = \int_\infty^x \frac{\cos t}{t}\, dt, \qquad x > 0 \tag{3.39}$$

To relate these integrals to the exponential integral requires complex-variable theory, and thus we omit the derivation.*

It is convenient in some applications to introduce another sine integral defined by

$$\text{si}(x) = -\int_x^\infty \frac{\sin t}{t}\, dt \tag{3.40}$$

which is related to (3.38) by (see problem 6)

$$\text{Si}(x) = \frac{\pi}{2} + \text{si}(x) \tag{3.41}$$

Special values of these functions include (see problem 7)

$$\text{Si}(0) = 0, \qquad \text{Si}(\infty) = \frac{\pi}{2} \tag{3.42a}$$

$$\text{Ci}(0^+) = -\infty, \qquad \text{Ci}(\infty) = 0 \tag{3.42b}$$

Also, taking derivatives, we obtain

$$\text{Si}'(x) = \frac{\sin x}{x}, \qquad \text{Ci}'(x) = \frac{\cos x}{x} \tag{3.43}$$

which shows that both functions are oscillatory. We observe that $\text{Si}(x)$ has extrema at $x = n\pi$ ($n = 0, 1, 2, \ldots$), while $\text{Ci}(x)$ has extrema at $x = (n + \tfrac{1}{2})\pi$ ($n = 0, 1, 2, \ldots$). The graphs of these functions are shown in Fig. 3.6.

EXERCISES 3.3

1. Show that ($x > 0$)

$$E_1(x) = e^{-x} \int_0^\infty \frac{e^{-xt}}{1+t}\, dt$$

---

*See N.N. Lebedev, *Special Functions and Their Applications*, New York: Dover, 1972, pp. 33–37.

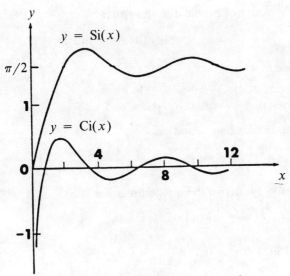

Figure 3.6 The Sine and Cosine Integrals

2. Derive the asymptotic formula
$$\text{Ei}(x) \sim \frac{e^x}{x} \sum_{n=0}^{\infty} \frac{n!}{x^n}, \qquad x \to \infty$$

3. From the result of problem 2, show that
$$E_1(x) \sim \frac{e^{-x}}{x} \sum_{n=0}^{\infty} \frac{(-1)^n n!}{x^n}, \qquad x \to \infty$$

4. Derive the asymptotic formula
$$\text{li}(x) \sim \frac{x}{\log x} \sum_{n=0}^{\infty} \frac{n!}{(\log x)^n}, \qquad x \to \infty$$

5. Let $f(t) = \int_0^\infty \frac{\sin tx}{x} dx$, $t > 0$
   (a) By taking the Laplace transform of both sides, show that
   $$\mathcal{L}\{f(t); s\} = \int_0^\infty (x^2 + s^2)^{-1} dx$$
   (b) Evaluate the integral in (a), and by taking the inverse Laplace transform, deduce the value $f(t) = \pi/2$.

6. Using the result of problem 5, show that
$$\text{Si}(x) = \frac{\pi}{2} + \text{si}(x)$$

7. Show that
(a) $\text{Si}(\infty) = \dfrac{\pi}{2}$,   (b) $\text{Ci}(\infty) = 0$,
(c) $\text{Si}(0) = 0$,   (d) $\text{Ci}(0^+) = -\infty$.

8. Derive the series representation
$$\text{Si}(x) = \sum_{n=0}^{\infty} \frac{(-1)^n x^{2n+1}}{(2n+1)(2n+1)!}$$

In problems 9–14, derive the integral relation.

9. $\displaystyle\int_0^\infty e^{-st} E_1(t)\, dt = \frac{1}{s} \log(1+s),\ s > 0.$

10. $\displaystyle\int_0^\infty e^{-st} \text{Si}(t)\, dt = \frac{1}{s} \tan^{-1}\left(\frac{1}{s}\right),\ s > 0.$

11. $\displaystyle\int_0^\infty e^{-st} \text{si}(t)\, dt = -\frac{1}{s} \tan^{-1}(s),\ s > 0.$

12. $\displaystyle\int_0^\infty e^{-st} \text{Ci}(t)\, dt = -\frac{1}{2s} \log(1+s^2),\ s > 0.$

13. $\displaystyle\int_0^\infty \cos x\, \text{Ci}(x)\, dx = \int_0^\infty \sin x\, \text{si}(x)\, dx = -\frac{\pi}{4}.$

14. $\displaystyle\int_0^\infty [\text{Ci}(x)]^2\, dx = \int_0^\infty [\text{si}(x)]^2\, dx = \frac{\pi}{2}.$

In problems 15–20, express the given integral in terms of $\text{Si}(x)$ and/or $\text{Ci}(x)$.

15. $\displaystyle\int_a^b \frac{\sin t}{t}\, dt.$

16. $\displaystyle\int_a^b \frac{\sin t^2}{t}\, dt.$
    Hint: Let $t^2 = u.$

17. $\displaystyle\int_a^b \frac{\cos t^2}{t}\, dt.$

18. $\displaystyle\int_a^b \frac{\sin t}{t^3}\, dt.$
    Hint: Use integration by parts.

19. $\displaystyle\int_0^b \left(\frac{1-\cos t}{t}\right) \sin at\, dt.$

20. $\displaystyle\int_2^3 \frac{\cos 2t}{1-t^2}\, dt.$
    Hint: Start with partial fractions.

3.4 Elliptic Integrals

The parametric equations for an elliptic arc are given by ($b > a$)
$$\left. \begin{array}{l} x = a\cos\theta \\ y = b\sin\theta \end{array} \right\},\quad 0 \le \theta \le \phi \tag{3.44}$$

Using the formula for arclength from calculus, we find the length of the

elliptic arc (3.44) leads to the integral
$$L = \int_0^\phi \sqrt{a^2\sin^2\theta + b^2\cos^2\theta}\, d\theta \tag{3.45}$$
which can also be expressed in the form
$$L = b\int_0^\phi \sqrt{1 - e^2\sin^2\theta}\, d\theta \tag{3.46}$$
where e is the eccentricity of the ellipse defined by
$$e = \frac{1}{b}\sqrt{b^2 - a^2} \tag{3.47}$$

The integral in (3.46) cannot be evaluated in terms of elementary functions. Because of its origin, it is called an *elliptic integral*.

There are three classifications of elliptic integrals, called *elliptic integrals of the first, second, and third kinds*, and defined respectively by
$$F(m,\phi) = \int_0^\phi \frac{d\theta}{\sqrt{1 - m^2\sin^2\theta}}, \qquad 0 < m < 1 \tag{3.48}$$
$$E(m,\phi) = \int_0^\phi \sqrt{1 - m^2\sin^2\theta}\, d\theta, \qquad 0 < m < 1 \tag{3.49}$$
and
$$\Pi(m,\phi,a) = \int_0^\phi \frac{d\theta}{\sqrt{1 - m^2\sin^2\theta}\,(1 + a^2\sin^2\theta)}, \qquad 0 < m < 1, \quad a \neq m, 0 \tag{3.50}$$

The parameter ϕ is called the *amplitude*, and m the *modulus*. When $\phi = \pi/2$ we refer to (3.48)–(3.50) as *complete elliptic integrals*, and they are often given the special designations
$$K(m) = \int_0^{\pi/2} \frac{d\theta}{\sqrt{1 - m^2\sin^2\theta}}, \qquad 0 < m < 1 \tag{3.51}$$
$$E(m) = \int_0^{\pi/2} \sqrt{1 - m^2\sin^2\theta}\, d\theta, \qquad 0 < m < 1 \tag{3.52}$$
and
$$\Pi(m,a) = \int_0^{\pi/2} \frac{d\theta}{\sqrt{1 - m^2\sin^2\theta}\,(1 + a^2\sin^2\theta)}, \qquad 0 < m < 1, \quad a \neq m, 0 \tag{3.53}$$

Sometimes these integrals are designated by simply the letters K, E, and Π.

Some of the importance connected with these integrals lies in the following theorem, which we state without proof.*

---

*For a proof of Theorem 3.1, see F. Bowman, *Introduction to Elliptic Functions with Applications*, New York: Dover, 1961.

Theorem 3.1. If $R(x, y)$ is a rational function in x and y, and $P(x)$ is a polynomial of degree four or less, then the integral

$$\int R\left(x, \sqrt{P(x)}\right) dx$$

can always be expressed in terms of elliptic integrals.

3.4.1 Limiting Values and Series Representations

For the limiting case $m \to 0$, we find that (3.51) leads to

$$K(0) = \int_0^{\pi/2} d\theta = \frac{\pi}{2} \tag{3.54}$$

and similarly,

$$E(0) = \frac{\pi}{2} \tag{3.55}$$

In the other limiting case where $m \to 1$ we obtain the results (see problem 1)

$$K(1) = \infty \tag{3.56}$$
$$E(1) = 1 \tag{3.57}$$

We can generate an infinite-series representation for K and E by first expanding the integrands in (3.51) and (3.52) in binomial series and then using termwise integration. For example,

$$(1 - m^2 \sin^2\theta)^{-1/2} = \sum_{n=0}^{\infty} \binom{-\frac{1}{2}}{n} (-1)^n m^{2n} \sin^{2n}\theta \tag{3.58}$$

and hence, by using the integral formula (see problem 16)

$$\int_0^{\pi/2} \sin^{2n}\theta \, d\theta = \frac{\pi}{2} (-1)^n \binom{-\frac{1}{2}}{n} \tag{3.59}$$

we deduce the series representation

$$K(m) = \frac{\pi}{2} \sum_{n=0}^{\infty} \binom{-\frac{1}{2}}{n}^2 m^{2n} \tag{3.60}$$

In the same fashion, it follows that

$$E(m) = \frac{\pi}{2} \sum_{n=0}^{\infty} \binom{\frac{1}{2}}{n} \binom{-\frac{1}{2}}{n} m^{2n} \tag{3.61}$$

3.4.2 The Pendulum Problem

A mass μ is suspended from the end of a rod of constant length b (whose weight is negligible). Summing forces (see Fig. 3.7) makes it clear that the weight component $\mu g \cos\phi$ acting in the normal direction to the path is offset by the force of restraint in the rod.* Therefore, the only weight

---
*Here g is the gravitational constant.

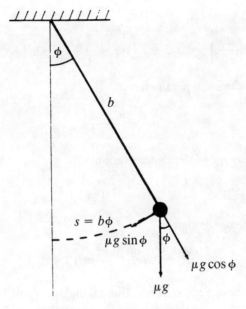

Figure 3.7 Swinging Pendulum

component contributing to the motion is $\mu g \sin\phi$, which acts in the direction of the tangent to the path. If we denote the arclength of the path by s, then Newton's second law of motion ($F = ma$) leads to

$$\mu \frac{d^2 s}{dt^2} = -\mu g \sin\phi$$

where the minus sign signifies that the tangential force component opposes the motion for increasing s. The arclength s of a circle of radius b is related to the central angle ϕ through the formula $s = b\phi$, and so the equation of motion (after simplification) becomes

$$\frac{d^2\phi}{dt^2} + k^2 \sin\phi = 0 \tag{3.62}$$

where $k^2 = g/b$.

Equation (3.62) is nonlinear and cannot be solved in terms of elementary functions. To solve it, we first note that it is equivalent to

$$\frac{1}{2}\left(\frac{d\phi}{dt}\right)^2 - k^2\cos\phi = C \tag{3.63}$$

i.e., (3.62) is the derivative of (3.63). (The constant C is proportional to the energy of the system.) Solving (3.63) for $(d\phi/dt)^2$, we have

$$\left(\frac{d\phi}{dt}\right)^2 = 2C + 2k^2\cos\phi = 2C + 2k^2 - 4k^2\sin^2(\tfrac{1}{2}\phi)$$

or

$$\left(\frac{d\phi}{dt}\right)^2 = 2(C + k^2)\left[1 - \frac{2k^2}{C + k^2}\sin^2(\tfrac{1}{2}\phi)\right] \quad (3.64)$$

If we now introduce the parameter

$$m^2 = \frac{C + k^2}{2k^2} \quad (3.65)$$

and make the change of dependent variable

$$y = \frac{1}{m}\sin(\tfrac{1}{2}\phi) \quad (3.66)$$

the chain rule demands that

$$\frac{d\phi}{dt} = 2m\frac{dy}{dt}(1 - m^2 y)^{-1/2} \quad (3.67)$$

Upon making these replacements and taking the (positive) square root, (3.64) becomes

$$\frac{dy}{dt} = k\sqrt{(1 - y^2)(1 - m^2 y^2)} \quad (3.68)$$

If we assume the position of the pendulum is $\phi = 0$ at time $t = 0$ and position $\phi = \Phi$ at time $t = T$, then the separation of variables applied to (3.68) leads to $[Y = (1/m)\sin(\tfrac{1}{2}\Phi)]$

$$kT = \int_0^Y \frac{dy}{\sqrt{(1 - y^2)(1 - m^2 y^2)}} \quad (3.69)$$

This integral is another form of the elliptic integral of the first kind $F(m, Y)$, as can be verified by making the substitution $y = \sin x$ (see problem 4).

Equation (3.69) gives the total time of motion of the pendulum in terms of an elliptic integral. If we wish to solve explicitly for the angle of motion $\Phi = 2\sin^{-1}(mY)$, we need to define an inverse function for $F(m, Y)$. Such a function exists and is called a *Jacobian elliptic function*. If in general we set

$$u = F(m, \phi) \quad (3.70)$$

then we can define three elliptic functions by the relations

$$\operatorname{sn} u = \sin\phi \quad (3.71)$$

$$\operatorname{cn} u = \cos\phi \quad (3.72)$$

and
$$\text{dn } u = \sqrt{1 - m^2 \sin^2 \phi} \tag{3.73}$$

These elliptic functions belong to the class of *doubly periodic functions* with one real period and one imaginary period. In this respect, they have characteristics of both the circular and hyperbolic functions. Much of the theory of elliptic functions is couched in the language of complex variables, and thus we will not pursue their general theory. Some elementary properties, however, are taken up in the exercises.

EXERCISES 3.4

1. Show that
   (a) $K(1) = \infty$,   (b) $E(1) = 1$.

2. Show that
$$\lim_{m \to 0} \frac{(K - E)}{m^2} = \frac{\pi}{4}$$

3. Verify that
$$F(m, \phi + \pi) - F(m, \phi) = 2K$$

In problems 4–9, derive the integral relation.

4. $\displaystyle\int_0^x \frac{dy}{\sqrt{(1 - y^2)(1 - m^2 y^2)}} = F(m, x).$

5. $\displaystyle\int_0^x \sqrt{\frac{1 - m^2 t^2}{1 - t^2}}\, dt = E(m, x).$

6. $\displaystyle\int_0^\phi \frac{\sin^2 x}{\sqrt{1 - m^2 \sin^2 x}}\, dx = \frac{1}{m^2}[F(m, \phi) - E(m, \phi)].$

7. $\displaystyle\int_0^{\pi/2} \frac{dx}{\sqrt{2 - \cos x}} = \frac{2}{\sqrt{3}}\left[K\left(\sqrt{\tfrac{2}{3}}\right) - F\left(\sqrt{\tfrac{2}{3}}, \pi/4\right)\right].$

8. $\displaystyle\int_0^2 (4 - x^2)^{-1/2}(9 - x^2)^{-1/2}\, dx = \tfrac{1}{3}K(\tfrac{2}{3}).$

9. $\displaystyle\int_0^1 (1 + x^2)^{-1/2}(1 + 2x^2)^{-1/2}\, dx = \frac{1}{\sqrt{2}}\left[K\left(\frac{1}{\sqrt{2}}\right) - F\left(\frac{1}{\sqrt{2}}, \frac{\pi}{4}\right)\right].$

10. Show that
$$\frac{1}{2\pi}\int_0^\pi \frac{\cos \theta\, d\theta}{(a^2 + b^2 + z^2 - 2ab\cos\theta)^{1/2}} = \frac{(ab)^{-1/2}}{k\pi}\left[\left(1 - \frac{k^2}{2}\right)K - E\right]$$

where
$$k^2 = \frac{4ab}{(a+b)^2 + z^2}$$

11. Find the perimeter of the ellipse $8x^2 + 9y^2 = 72$.
12. Find the area enclosed by one loop of the curve $y^2 = 1 - 4\sin^2 x$.
13. Find the arclength of the lemniscate $r^2 = \cos 2\theta$, $0 \leq \theta \leq \pi/2$.
14. Find the length of the curve $y = \sin x$, $0 \leq x \leq \pi/3$.
15. Find the surface area of a right circular cylinder of radius r intercepted by a sphere of radius a ($a > r$) whose center lies on the cylinder.
16. Show that ($n = 0, 1, 2, \ldots$)
$$\int_0^{\pi/2} \sin^{2n}\theta \, d\theta = \frac{\pi}{2}(-1)^n \binom{-\frac{1}{2}}{n}$$

Hint: See problem 17(a) in Exercises 1.2.

17. Show that
    (a) $\text{sn}(0) = 0$.   (b) $\text{cn}(0) = \text{dn}(0) = 1$.

18. Verify the identities
    (a) $\text{sn}^2 u + \text{cn}^2 u = 1$,
    (b) $m^2 \text{sn}^2 u + \text{dn}^2 u = 1$,
    (c) $\text{dn}^2 u - m^2 \text{cn}^2 u = 1 - m^2$.

19. Derive the derivative relations
    (a) $\frac{d}{du} \text{sn}\, u = \text{cn}\, u \,\text{dn}\, u$,
    (b) $\frac{d}{du} \text{cn}\, u = -\text{sn}\, u \,\text{dn}\, u$,
    (c) $\frac{d}{du} \text{dn}\, u = -m^2 \,\text{sn}\, u \,\text{cn}\, u$.

20. Show that
    (a) $\text{sn}(u + 4K) = \text{sn}\, u$,
    (b) $\text{cn}(u + 4K) = \text{cn}\, u$,
    (c) $\text{dn}(u + 2K) = \text{dn}\, u$.

21. Verify the addition formulae
    (a) $\text{sn}(u+v) = \dfrac{\text{sn}\, u \,\text{cn}\, v \,\text{dn}\, v + \text{sn}\, v \,\text{cn}\, u \,\text{dn}\, u}{1 - m^2 \text{sn}^2 u \,\text{sn}^2 v}$,
    (b) $\text{cn}(u+v) = \dfrac{\text{cn}\, u \,\text{cn}\, v - \text{sn}\, u \,\text{dn}\, u \,\text{sn}\, v \,\text{dn}\, v}{1 - m^2 \text{sn}^2 u \,\text{sn}^2 v}$,
    (c) $\text{dn}(u+v) = \dfrac{\text{dn}\, u \,\text{dn}\, v - m^2 \text{sn}\, u \,\text{cn}\, u \,\text{sn}\, v \,\text{cn}\, v}{1 - m^2 \text{sn}^2 u \,\text{sn}^2 v}$.

22. Show that
   (a) $\lim_{m \to 1} \operatorname{sn} u = \tanh u$,
   (b) $\lim_{m \to 1} \operatorname{cn} u = \operatorname{sech} u$,
   (c) $\lim_{m \to 1} \operatorname{dn} u = \operatorname{sech} u$.

4

Legendre Polynomials and Related Functions

4.1 Introduction

The *Legendre polynomials* are closely associated with physical phenomena for which spherical geometry is important. In particular, these polynomials first arose in the problem of expressing the Newtonian potential of a conservative force field in an infinite series involving the distance variables of two points and their included central angle (see Section 4.2). Other similar problems dealing with either gravitational potentials or electrostatic potentials also lead to Legendre polynomials, as do certain steady-state heat-conduction problems in spherical-shaped solids, and so forth.

There exists a whole class of polynomial sets which have many properties in common, and for which the Legendre polynomials represent the simplest example. Each polynomial set satisfies several recurrence formulas, is involved in numerous integral relationships, and forms the basis for series expansions resembling Fourier trigonometric series where the sines and cosines are replaced by members of the polynomial set. Because of all the similarities in these polynomial sets, and because the Legendre polynomials are the simplest such set, our development of the properties associated with the Legendre polynomials will be more extensive than similar developments in Chapter 5, where we introduce other polynomial sets.

In addition to the Legendre polynomials, we will present a brief discussion of the *Legendre functions of the second kind* and *associated Legendre functions*. The Legendre functions of the second kind arise as a second solution set of Legendre's differential equation, and the associated functions are related to derivatives of the Legendre polynomials.

4.2 The Generating Function

Among other areas of application, the subject of potential theory is concerned with the forces of attraction due to the presence of a gravitational field. Central to the discussion of problems of gravitational attraction is *Newton's law of universal gravitation*:

> "Every particle of matter in the universe attracts every other particle with a force whose direction is that of the line joining the two, and whose magnitude is directly as the product of their masses and inversely as the square of their distance from each other."

The force field generated by a single particle is usually considered to be *conservative*. That is, there exists a potential function V such that the gravitational force F at a point of free space (i.e., free of point masses) is related to the potential function according to

$$F = -\nabla V \tag{4.1}$$

where the minus sign is conventional. If r denotes the distance between a point mass and a point of free space, the potential function can be shown to have the form*

$$V(r) = \frac{k}{r} \tag{4.2}$$

where k is a constant whose numerical value does not concern us. Because of spherical symmetry of the gravitational field, the potential function V depends only upon the radial distance r.

Valuable information on the properties of potentials like (4.2) may be inferred from developments of the potential function into power series of certain types. In 1785, A.M. Legendre published his "Sur l'attraction des sphéroïdes," in which he developed the gravitational potential (4.2) in a power series involving the ratio of two distance variables. He found that the coefficients appearing in this expansion were polynomials that exhibited interesting properties.

In order to obtain Legendre's results, let us suppose that a particle of mass m is located at point P, which is a units from the origin of our

---

*See O.D. Kellogg, *Foundations of Potential Theory*, New York: Dover, 1953, Chapter III.

118 • **Special Functions for Engineers and Applied Mathematicians**

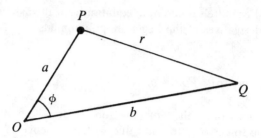

Figure 4.1

coordinate system (see Fig. 4.1). Let the point Q represent a point of free space r units from P and b units from the origin O. For the sake of definiteness, let us assume $b > a$. Then, from the law of cosines, we find the relation

$$r^2 = a^2 + b^2 - 2ab\cos\phi \tag{4.3}$$

where ϕ is the central angle between the rays $\overline{OP}$ and $\overline{OQ}$. By rearranging the terms and factoring out b^2, it follows that

$$r^2 = b^2\left[1 - 2\frac{a}{b}\cos\phi + \left(\frac{a}{b}\right)^2\right], \quad a < b \tag{4.4}$$

For notational simplicity, we introduce the parameters

$$t = \frac{a}{b}, \quad x = \cos\phi \tag{4.5}$$

and thus, upon taking the square root,

$$r = b(1 - 2xt + t^2)^{1/2} \tag{4.6}$$

Finally, the substitution of (4.6) into (4.2) leads to the expression

$$V = \frac{k}{b}(1 - 2xt + t^2)^{-1/2}, \quad 0 < t < 1 \tag{4.7}$$

for the potential function. For reasons that will soon be clear, we refer to the function $w(x,t) = (1 - 2xt + t^2)^{-1/2}$ as the *generating function* of the Legendre polynomials. Our task at this point is to develop $w(x,t)$ in a power series in the variable t.

4.2.1 Legendre Polynomials

From Example 6 in Section 1.3.2, we recall the binomial series

$$(1 - u)^{-1/2} = \sum_{n=0}^{\infty} \binom{-\frac{1}{2}}{n}(-1)^n u^n, \quad |u| < 1 \tag{4.8}$$

Hence, by setting $u = t(2x - t)$, we find that

$$w(x,t) = (1 - 2xt + t^2)^{-1/2}$$
$$= \sum_{n=0}^{\infty} \binom{-\frac{1}{2}}{n}(-1)^n t^n (2x - t)^n \quad (4.9)$$

which is valid for $|2xt - t^2| < 1$. For $|t| < 1$, it follows that $|x| \leq 1$. The factor $(2x - t)^n$ is simply a finite binomial series, and thus (4.9) can further be expressed as

$$w(x,t) = \sum_{n=0}^{\infty} \binom{-\frac{1}{2}}{n}(-1)^n t^n \sum_{k=0}^{n} \binom{n}{k}(-1)^k (2x)^{n-k} t^k$$

or

$$w(x,t) = \sum_{n=0}^{\infty} \sum_{k=0}^{n} \binom{-\frac{1}{2}}{n}\binom{n}{k}(-1)^{n+k}(2x)^{n-k} t^{n+k} \quad (4.10)$$

Since our goal is to obtain a power series involving powers of t to a single index, the change of indice $n \to n - k$ is suggested. Thus, recalling Equation (1.18) in Section 1.2.3, i.e.,

$$\sum_{n=0}^{\infty} \sum_{k=0}^{n} A_{n-k,k} = \sum_{n=0}^{\infty} \sum_{k=0}^{[n/2]} A_{n-2k,k}$$

we see that (4.10) can be written in the equivalent form

$$w(x,t) = \sum_{n=0}^{\infty} \left\{ \sum_{k=0}^{[n/2]} \binom{-\frac{1}{2}}{n-k}\binom{n-k}{k}(-1)^n (2x)^{n-2k} \right\} t^n \quad (4.11)$$

The innermost summation in (4.11) is of finite length and therefore represents a polynomial in x, which happens to be of degree n. If we denote this polynomial by the symbol

$$P_n(x) = \sum_{k=0}^{[n/2]} \binom{-\frac{1}{2}}{n-k}\binom{n-k}{k}(-1)^n (2x)^{n-2k} \quad (4.12)$$

then (4.11) leads to the intended result

$$w(x,t) = \sum_{n=0}^{\infty} P_n(x) t^n, \quad |x| \leq 1, \quad |t| < 1 \quad (4.13)$$

where $w(x,t) = (1 - 2xt + t^2)^{-1/2}$.

The polynomials $P_n(x)$ are called the *Legendre polynomials* in honor of their discoverer. By recognizing that [see Equation (1.27) in Section 1.2 and

Equation (2.25) in Section 2.2.2]

$$\binom{-\frac{1}{2}}{n} = (-1)^n \binom{n - \frac{1}{2}}{n}$$

$$= (-1)^n \frac{\Gamma(n + \frac{1}{2})}{n!\Gamma(\frac{1}{2})}$$

$$= \frac{(-1)^n (2n)!}{2^{2n}(n!)^2} \quad (4.14)$$

it follows that the product of binomial coefficients in (4.12) is

$$\binom{-\frac{1}{2}}{n-k}\binom{n-k}{k} = \frac{(-1)^{n-k}(2n-2k)!}{2^{2n-2k}(n-k)!k!(n-2k)!} \quad (4.15)$$

and hence, (4.12) becomes

$$P_n(x) = \sum_{k=0}^{[n/2]} \frac{(-1)^k (2n - 2k)! x^{n-2k}}{2^n k!(n-k)!(n-2k)!} \quad (4.16)$$

The first few Legendre polynomials are listed in Table 4.1.

Making an observation, we note that when n is an even number the polynomial $P_n(x)$ is an even function, and when n is odd the polynomial is an odd function. Therefore,

$$P_n(-x) = (-1)^n P_n(x), \quad n = 0, 1, 2, \ldots \quad (4.17)$$

The graphs of $P_n(x)$, $n = 0, 1, 2, 3, 4$, are sketched in Fig. 4.2 over the interval $-1 \le x \le 1$.

Returning now to Equation (4.7) with $x = \cos\phi$ and $t = a/b$, we find that the potential function has the series expansion

$$V = \frac{k}{b} \sum_{n=0}^{\infty} P_n(\cos\phi)\left(\frac{a}{b}\right)^n, \quad a < b \quad (4.18)$$

In terms of the argument $\cos\phi$, the Legendre polynomials can be expressed

Table 4.1 Legendre polynomials

$P_0(x) = 1$
$P_1(x) = x$
$P_2(x) = \frac{1}{2}(3x^2 - 1)$
$P_3(x) = \frac{1}{2}(5x^3 - 3x)$
$P_4(x) = \frac{1}{8}(35x^4 - 30x^2 + 3)$
$P_5(x) = \frac{1}{8}(63x^5 - 70x^3 + 15x)$

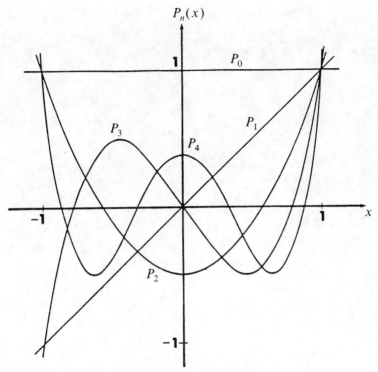

Figure 4.2 Graph of $P_n(x)$, $n = 0, 1, 2, 3, 4$

as trigonometric polynomials of the form shown in Table 4.2 (see problem 3).

In Fig. 4.3 the first few polynomials $P_n(\cos \phi)$ are plotted as a function of the angle ϕ.

4.2.2 Special Values and Recurrence Formulas

The Legendre polynomials are rich in recurrence relations and identities. Central to the development of many of these is the *generating-function*

Table 4.2 Legendre trigonometric polynomials.

$$P_0(\cos \phi) = 1$$
$$P_1(\cos \phi) = \cos \phi$$
$$P_2(\cos \phi) = \tfrac{1}{2}(3 \cos^2 \phi - 1)$$
$$= \tfrac{1}{4}(3 \cos 2\phi + 1)$$
$$P_3(\cos \phi) = \tfrac{1}{2}(5 \cos^3 \phi - 3 \cos \phi)$$
$$= \tfrac{1}{8}(5 \cos 3\phi + 3 \cos \phi)$$

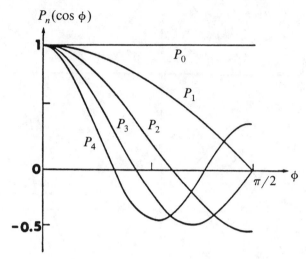

Figure 4.3 Graph of $P_n(\cos\phi)$, $n = 0, 1, 2, 3, 4$

relation

$$(1 - 2xt + t^2)^{-1/2} = \sum_{n=0}^{\infty} P_n(x)t^n, \quad |x| \leq 1, \quad |t| < 1 \quad (4.19)$$

Special values of the Legendre polynomials can be derived directly from (4.19) by substituting particular values for x. For example, the substitution of $x = 1$ yields

$$(1 - 2t + t^2)^{-1/2} = (1 - t)^{-1} = \sum_{n=0}^{\infty} P_n(1)t^n \quad (4.20)$$

However, we recognize that $(1 - t)^{-1}$ is the sum of a geometric series, so that (4.20) is equivalent to

$$\sum_{n=0}^{\infty} t^n = \sum_{n=0}^{\infty} P_n(1)t^n \quad (4.21)$$

Hence, from the uniqueness theorem of power series (Theorem 1.13), we can compare like coefficients of t^n in (4.21) to deduce the result

$$P_n(1) = 1, \quad n = 0, 1, 2, \ldots \quad (4.22)$$

Also, from (4.17) we see that

$$P_n(-1) = (-1)^n, \quad n = 0, 1, 2, \ldots \quad (4.23)$$

Legendre Polynomials and Related Functions • 123

The substitution of $x = 0$ into (4.19) leads to

$$(1 + t^2)^{-1/2} = \sum_{n=0}^{\infty} P_n(0)t^n \qquad (4.24)$$

but the term on the left-hand side has the binomial series expansion

$$(1 + t^2)^{-1/2} = \sum_{n=0}^{\infty} \binom{-\frac{1}{2}}{n} t^{2n} \qquad (4.25)$$

Comparing terms of the series on the right in (4.24) and (4.25), we note that (4.25) has only *even* powers of t. Thus we conclude that $P_n(0) = 0$ for $n = 1, 3, 5, \ldots$, or equivalently,

$$P_{2n+1}(0) = 0, \qquad n = 0, 1, 2, \ldots \qquad (4.26)$$

Since all odd terms in (4.24) are zero, we can replace n by $2n$ in the series and compare with (4.25), from which we deduce

$$P_{2n}(0) = \binom{-\frac{1}{2}}{n} = \frac{(-1)^n (2n)!}{2^{2n}(n!)^2}, \qquad n = 0, 1, 2, \ldots \qquad (4.27)$$

where we are recalling (4.14).

Remark: Actually, (4.26) could have been deduced from the fact that $P_{2n+1}(x)$ is an odd (continuous) function, and therefore must necessarily pass through the origin. (Why?)

In order to obtain the desired recurrence relations, we first make the observation that the function $w(x, t) = (1 - 2xt + t^2)^{-1/2}$ satisfies the derivative relation

$$(1 - 2xt + t^2)\frac{\partial w}{\partial t} + (t - x)w = 0 \qquad (4.28)$$

Direct substitution of the series (4.13) for $w(x, t)$ into (4.28) yields

$$(1 - 2xt + t^2)\sum_{n=0}^{\infty} nP_n(x)t^{n-1} + (t - x)\sum_{n=0}^{\infty} P_n(x)t^n = 0$$

Carrying out the indicated multiplications and simplifying gives us

$$\sum_{n=0}^{\infty} nP_n(x)t^{n-1} - 2x \underbrace{\sum_{n=0}^{\infty} nP_n(x)t^n}_{n \to n-1}$$
$$+ \underbrace{\sum_{n=0}^{\infty} nP_n(x)t^{n+1}}_{n \to n-2} + \underbrace{\sum_{n=0}^{\infty} P_n(x)t^{n+1}}_{n \to n-2} - x \underbrace{\sum_{n=0}^{\infty} P_n(x)t^n}_{n \to n-1} = 0$$

$$(4.29)$$

We now wish to change indices so that powers of t are the same in each summation. We accomplish this by leaving the first sum in (4.29) as it is, replacing n with $n-1$ in the second and last sums, and replacing n with $n-2$ in the remaining sums; thus, (4.29) becomes

$$\sum_{n=0}^{\infty} nP_n(x)t^{n-1} - 2x\sum_{n=1}^{\infty}(n-1)P_{n-1}(x)t^{n-1} + \sum_{n=2}^{\infty}(n-2)P_{n-2}(x)t^{n-1}$$
$$+ \sum_{n=2}^{\infty} P_{n-2}(x)t^{n-1} - x\sum_{n=1}^{\infty} P_{n-1}(x)t^{n-1} = 0$$

Finally, combining all summations, we have

$$\sum_{n=2}^{\infty}\left[nP_n(x) - 2x(n-1)P_{n-1}(x) + (n-2)P_{n-2}(x)\right.$$
$$\left. + P_{n-2}(x) - xP_{n-1}(x)\right]t^{n-1} + P_1(x) - xP_0(x) = 0 \quad (4.30)$$

But $P_1(x) - xP_0(x) = x - x = 0$, and the validity of (4.30) demands that the coefficient of t^{n-1} be zero for all x. Hence, after simplification we arrive at

$$nP_n(x) - (2n-1)xP_{n-1}(x) + (n-1)P_{n-2}(x) = 0, \qquad n = 2,3,4,\ldots$$

or, replacing n by $n+1$, we obtain the more conventional form

$$(n+1)P_{n+1}(x) - (2n+1)xP_n(x) + nP_{n-1}(x) = 0 \quad (4.31)$$

where $n = 1,2,3,\ldots$.

We refer to (4.31) as a *three-term recurrence formula*, since it forms a connecting relation between three successive Legendre polynomials. One of the primary uses of (4.31) in computations is to produce higher-order Legendre polynomials from lower-order ones by expressing them in the form

$$P_{n+1}(x) = \left(\frac{2n+1}{n+1}\right)xP_n(x) - \left(\frac{n}{n+1}\right)P_{n-1}(x) \quad (4.32)$$

where $n = 1,2,3,\ldots$. In practice, (4.32) is generally preferred to (4.16) in making computer calculations when several polynomials are involved.*

---

*Actually, to avoid excessive roundoff error in making computer calculations, Equation (4.32) should be rewritten in the form

$$P_{n+1}(x) = 2xP_n(x) - P_{n-1}(x) - \frac{xP_n(x) - P_{n-1}(x)}{n+1}.$$

A relation similar to (4.31) involving derivatives of the Legendre polynomials can be derived in the same fashion by first making the observation that $w(x,t)$ satisfies

$$(1 - 2xt + t^2)\frac{\partial w}{\partial x} - tw = 0 \qquad (4.33)$$

where this time the differentiation is with respect to x. Substituting the series for $w(x,t)$ directly into (4.33) leads to

$$(1 - 2xt + t^2)\sum_{n=0}^{\infty} P_n'(x)t^n - \sum_{n=0}^{\infty} P_n(x)t^{n+1} = 0$$

or, after carrying out the multiplications,

$$\sum_{n=0}^{\infty} P_n'(x)t^n - 2x\underbrace{\sum_{n=0}^{\infty} P_n'(x)t^{n+1}}_{n \to n-1} + \underbrace{\sum_{n=0}^{\infty} P_n'(x)t^{n+2}}_{n \to n-2} - \underbrace{\sum_{n=0}^{\infty} P_n(x)t^{n+1}}_{n \to n-1} = 0 \qquad (4.34)$$

Next, making an appropriate change of index in each summation, we get

$$\sum_{n=2}^{\infty} \left[P_n'(x) - 2xP_{n-1}'(x) + P_{n-2}'(x) - P_{n-1}(x) \right] t^n = 0 \qquad (4.35)$$

where all terms outside this summation add to zero. Thus, by equating the coefficient of t^n to zero in (4.35), we find

$$P_n'(x) - 2xP_{n-1}'(x) + P_{n-2}'(x) - P_{n-1}(x) = 0, \qquad n = 2, 3, 4, \ldots$$

or, by a change of index,

$$P_{n+1}'(x) - 2xP_n'(x) + P_{n-1}'(x) - P_n(x) = 0 \qquad (4.36)$$

for $n = 1, 2, 3, \ldots$.

Certain combinations of (4.21) and (4.36) can lead to further recurrence relations. For example, suppose we first differentiate (4.31), i.e.,

$$(n+1)P_{n+1}'(x) - (2n+1)P_n(x) - (2n+1)xP_n'(x) + nP_{n-1}'(x) = 0 \qquad (4.37)$$

From (4.36) we find

$$P_{n-1}'(x) = P_n(x) + 2xP_n'(x) - P_{n+1}'(x) \qquad (4.38a)$$

$$P_{n+1}'(x) = P_n(x) + 2xP_n'(x) - P_{n-1}'(x) \qquad (4.38b)$$

and the successive replacement of $P_{n-1}'(x)$ and $P_{n+1}'(x)$ in (4.37) by (4.38a) and (4.38b) leads to the two relations

$$P'_{n+1}(x) - xP'_n(x) = (n+1)P_n(x) \tag{4.39a}$$

$$xP'_n(x) - P'_{n-1}(x) = nP_n(x) \tag{4.39b}$$

The addition of (4.39a) and (4.39b) yields the more symmetric formula

$$P'_{n+1}(x) - P'_{n-1}(x) = (2n+1)P_n(x) \tag{4.40}$$

Finally, replacing n by $n-1$ in (4.39a) and then eliminating the term $P'_{n-1}(x)$ by use of (4.39b), we obtain

$$(1-x^2)P'_n(x) = nP_{n-1}(x) - nxP_n(x) \tag{4.41}$$

This last relation allows us to express the *derivative* of a Legendre polynomial in terms of Legendre polynomials.

4.2.3 Legendre's Differential Equation

All the recurrence relations that we have derived thus far involve successive Legendre polynomials. We may well wonder if any relation exists between derivatives of the Legendre polynomials and Legendre polynomials of the same index. The answer is in the affirmative, but to derive this relation we must consider second derivatives of the polynomials.

By taking the derivative of both sides of (4.41), we get

$$\frac{d}{dx}\left[(1-x^2)P'_n(x)\right] = nP'_{n-1}(x) - nP_n(x) - nxP'_n(x)$$

and then, using (4.39b) to eliminate $P'_{n-1}(x)$, we arrive at the derivative relation

$$\frac{d}{dx}\left[(1-x^2)P'_n(x)\right] + n(n+1)P_n(x) = 0 \tag{4.42}$$

which holds for $n = 0, 1, 2, \ldots$. Expanding the product term in (4.42) yields

$$(1-x^2)P''_n(x) - 2xP'_n(x) + n(n+1)P_n(x) = 0 \tag{4.43}$$

and thus we deduce that the Legendre polynomial $y = P_n(x)$ ($n = 0, 1, 2, \ldots$) is a solution of the linear second-order DE

$$(1-x^2)y'' - 2xy' + n(n+1)y = 0 \tag{4.44}$$

called *Legendre's differential equation*.*

Perhaps the most natural way in which Legendre polynomials arise in practice is as solutions of Legendre's equation. In such problems the basic

---

*In Section 4.6 we will discuss other solutions of Legendre's equation.

model is generally a partial differential equation. Solving the partial DE by the separation-of-variables technique leads to a system of ordinary DEs, and sometimes one of these is Legendre's DE. This is precisely the case, for example, in solving for the steady-state temperature distribution (independent of the azimuthal angle) in a solid sphere. We will delay any further discussion of such problems, however, until Chapter 7.

Remark: Any function $f_n(x)$ that satisfies Legendre's equation, i.e.,

$$(1 - x^2)f_n''(x) - 2xf_n'(x) + n(n + 1)f_n(x) = 0$$

will also satisfy *all* previous recurrence formulas given above, provided that $f_n(x)$ is properly normalized. Consequently, any further solutions of Legendre's equation can be selected in such a way that they automatically satisfy the whole set of recurrence relations already derived. The set of solutions $Q_n(x)$ introduced in Section 4.6 is a case in point.

EXERCISES 4.2

1. Use the series (4.16) to determine $P_n(x)$ directly for the specific cases $n = 0, 1, 2, 3, 4$, and 5.

2. Given that $P_0(x) = 1$ and $P_1(x) = x$, use the recurrence formula (4.33) to determine $P_2(x)$, $P_3(x)$, and $P_4(x)$.

3. Verify that
    (a) $P_0(\cos\phi) = 1$.
    (b) $P_1(\cos\phi) = \cos\phi$.
    (c) $P_2(\cos\phi) = \frac{1}{4}(3\cos 2\phi + 1)$.
    (d) $P_3(\cos\phi) = \frac{1}{8}(5\cos 3\phi + 3\cos\phi)$.

4. Given the function $w(x, t) = (1 - 2xt + t^2)^{-1/2}$,
    (a) show that $w(-x, -t) = w(x, t)$.
    (b) Use the result in (a) and the generating function relation (4.19) to deduce that (for $n = 0, 1, 2, \ldots$)

    $$P_n(-x) = (-1)^n P_n(x).$$

5. Verify the special values ($n = 0, 1, 2, \ldots$)
    (a) $P_n'(1) = \frac{1}{2}n(n + 1)$,   (b) $P_n'(-1) = (-1)^{n-1}\frac{1}{2}n(n + 1)$.

6. Verify the special values ($n = 0, 1, 2, \ldots$)
    (a) $P_{2n}'(0) = 0$.   (b) $P_{2n+1}'(0) = \dfrac{(-1)^n(2n + 1)}{2^{2n}}\dbinom{2n}{n}$.

7. Establish the generating-function relation

$$(1 - 2xt + t^2)^{-1} = \sum_{n=0}^{\infty} U_n(x)t^n, \qquad |t| < 1, \quad |x| \leq 1$$

where $U_n(x)$ is the nth *Chebyshev polynomial of the second kind*[*] defined by

$$U_n(x) = \sum_{k=0}^{[n/2]} \frac{(-1)^k (n-k)!}{k!(n-2k)!} (2x)^{n-2k}$$

8. Given the generating function $w(x, t) = (1 - 2xt + t^2)^{-1}$,

   (a) show that it satisfies the identity

$$(1 - 2xt + t^2)\frac{\partial w}{\partial t} + 2(t - x)w = 0$$

   (b) Substitute the series in problem 7 into the identity in (a) and derive the recurrence formula (for $n = 1, 2, 3, \ldots$)

$$U_{n+1}(x) - 2xU_n(x) + U_{n-1}(x) = 0$$

9. Show that the generating function in problem 8 also satisfies the identity

$$(1 - 2xt + t^2)\frac{\partial w}{\partial x} - 2tw = 0$$

   (a) and deduce the relation (for $n = 1, 2, 3, \ldots$)

$$U'_{n+1}(x) - 2xU'_n(x) + U'_{n-1}(x) - 2U_n(x) = 0$$

   (b) Show that (a) can be obtained directly from problem 8(b) by differentiation.

10. Using the results of problems 7–9, show that

    (a) $(1 - x^2)U'_n(x) = -nxU_n(x) + (n + 1)U_{n-1}(x)$,

    (b) $(1 - x^2)U''_n(x) - 3xU'_n(x) + n(n + 2)U_n(x) = 0$.

11. Using the Cauchy product of two power series (Section 1.3.3), show that

$$\frac{e^{xt}}{1 - t} = \sum_{n=0}^{\infty} e_n(x)t^n, \qquad |t| < 1$$

where $e_n(x)$ is the polynomial equal to the first $n + 1$ terms of the

---
[*]We will discuss these polynomials further in Section 5.4.2.

Maclaurin series for e^x, i.e.,

$$e_n(x) = \sum_{k=0}^{n} \frac{x^k}{k!}$$

12. Given the generating function $w(x,t) = e^{xt}/(1-t)$,

    (a) show that it satisfies the identity

    $$(1-t)\frac{\partial w}{\partial t} - [x(1-t) + 1]w = 0$$

    (b) Substitute the series in problem 11 into the identity in (a) and derive the recurrence formula ($n = 1, 2, 3, \ldots$)

    $$(n+1)e_{n+1}(x) - (n+1+x)e_n(x) + xe_{n-1}(x) = 0$$

    (c) Show directly from the series definition of $e_n(x)$ that

    $$e_n'(x) = e_{n-1}(x), \qquad n = 1, 2, 3, \ldots$$

13. Using the results of problems 11 and 12, show that $y = e_n(x)$ is a solution of the second-order linear DE

    $$xy'' - (x+n)y' + ny = 0$$

14. Make the change of variable $x = \cos\phi$ in the DE

    $$\frac{1}{\sin\phi}\frac{d}{d\phi}\left(\sin\phi\frac{dy}{d\phi}\right) + n(n+1)y = 0$$

    and show that it reduces to Legendre's DE (4.44).

15. Determine the values of n for which $y = P_n(x)$ is a solution of

    (a) $(1-x^2)y'' - 2xy' + n(n+1)y = 0$, $y(0) = 0$, $y(1) = 1$,
    (b) $(1-x^2)y'' - 2xy' + n(n+1)y = 0$, $y'(0) = 0$, $y(1) = 1$.

16. When a tightly stretched string is rotating with uniform angular speed ω about its rest position along the x-axis, the DE governing the displacements of the string in the vertical plane is approximately

    $$\frac{d}{dx}[T(x)y'] + \rho\omega^2 y = 0$$

    where $T(x)$ is the tension in the string and ρ the linear density (constant) of the string. If $T(x) = 1 - x^2$ and the boundary condition $y(-1) = y(1)$ is prescribed, determine the two lowest possible critical speeds ω. What shape does the string assume in the vertical plane in each case?

    Hint: Assume that $\rho\omega^2 = n(n+1)$.

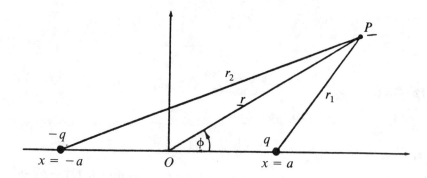

17. An electric dipole consists of electric charges q and $-q$ located along the x-axis as shown in the figure above. The potential induced at point P due to the charges is known to be $(r > a)$

$$V = kq\left(\frac{1}{r_1} - \frac{1}{r_2}\right)$$

where k is a constant. Express the potential in terms of the coordinates r and ϕ and show that it leads to an infinite series involving Legendre polynomials. Also show that if only the first nonzero term of the series is retained, the dipole potential is

$$V \simeq \frac{2akq}{r^2}\cos\phi, \qquad r \gg a$$

18. The electrostatic potential induced at point P for the array of charges shown in the figure below is given by $(r > a)$

$$V = kq\left(\frac{1}{r_1} + \frac{1}{r_2} - \frac{2}{r}\right)$$

where k is constant. Expressing V entirely in terms of r and ϕ, show

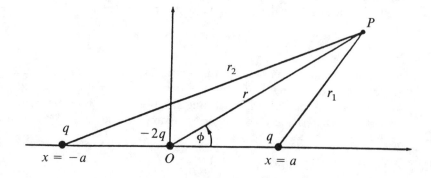

that the first nonzero term of the resulting series is

$$V \simeq \frac{kqa^2}{2r^3}(3\cos 2\phi + 1), \qquad r \gg a$$

19. Show that the even and odd Legendre polynomials have the series representations (for $n = 0, 1, 2, \ldots$)

    (a) $P_{2n}(x) = \dfrac{(-1)^n}{2^{2n-1}} \displaystyle\sum_{k=0}^{n} \dfrac{(-1)^k (2n + 2k - 1)!}{(2k)!(n + k - 1)!(n - k)!} x^{2k}$,

    (b) $P_{2n+1}(x) = \dfrac{(-1)^n}{2^{2n}} \displaystyle\sum_{k=0}^{n} \dfrac{(-1)^k (2n + 2k + 1)!}{(2k + 1)!(n + k)!(n - k)!} x^{2k+1}$.

20. Derive the identity ($n = 0, 1, 2, \ldots$)

    $$(1 - x^2) P_n'(x) = (n + 1)[x P_n(x) - P_{n+1}(x)]$$

21. Show that

    (a) $\displaystyle\sum_{k=0}^{n} (2k + 1) P_k(x) = P_{n+1}'(x) + P_n'(x)$,

    (b) $(1 - x) \displaystyle\sum_{k=0}^{n} (2k + 1) P_k(x) = (n + 1)[P_n(x) - P_{n+1}(x)]$.

22. Show that

    (a) $\displaystyle\sum_{n=0}^{\infty} [x P_n'(x) - n P_n(x)] t^n = t^2 (1 - 2xt + t^2)^{-3/2}$,

    (b) $\displaystyle\sum_{n=0}^{\infty} \sum_{k=0}^{[n/2]} (2n - 4k + 1) P_{n-2k}(x) t^n = (1 - 2xt + t^2)^{-3/2}$.

23. Using the result of problem 22, deduce that

    $$x P_n'(x) - n P_n(x) = \sum_{k=0}^{[\frac{1}{2}(n-2)]} (2n - 4k - 3) P_{n-2-2k}(x)$$

24. Show that

    $$P_n'(x) = \sum_{k=0}^{[\frac{1}{2}(n-1)]} (2n - 4k - 1) P_{n-1-2k}(x)$$

25. Show that

    $$\sum_{n=0}^{\infty} (2n + 1) P_n(x) t^n = (1 - t^2)(1 - 2xt + t^2)^{-3/2}$$

4.3 Other Representations of the Legendre Polynomials

For each n, the Legendre polynomials can be defined either by the series

$$P_n(x) = \sum_{k=0}^{[n/2]} \frac{(-1)^k (2n-2k)!}{2^n k!(n-k)!(n-2k)!} x^{n-2k} \qquad (4.45)$$

or by the recurrence formula

$$P_{n+1}(x) = \left(\frac{2n+1}{n+1}\right) x P_n(x) - \left(\frac{n}{n+1}\right) P_{n-1}(x) \qquad (4.46)$$

where $P_0(x) = 1$ and $P_1(x) = x$. In some situations, however, it is advantageous to have other representations from which further properties of the polynomials are more readily found.

4.3.1 Rodrigues's Formula

A representation of the Legendre polynomials involving differentiation is given by the *Rodrigues formula*

$$P_n(x) = \frac{1}{2^n n!} \frac{d^n}{dx^n}\left[(x^2-1)^n\right], \qquad n = 0, 1, 2, \ldots \qquad (4.47)$$

In order to verify (4.47), we start with the binomial series

$$(x^2 - 1)^n = \sum_{k=0}^{n} \frac{(-1)^k n!}{k!(n-k)!} x^{2n-2k}$$

and differentiate n times. Noting that

$$\frac{d^n}{dx^n} x^m = \begin{cases} \dfrac{m!}{(m-n)!} x^{m-n}, & n \le m \\ 0, & n > m \end{cases}$$

we infer

$$\frac{d^n}{dx^n}\left[(x^2-1)^n\right] = \sum_{k=0}^{[n/2]} \frac{(-1)^k n!(2n-2k)!}{k!(n-k)!(n-2k)!} x^{n-2k}$$
$$= 2^n n! P_n(x)$$

from which (4.47) now follows.

4.3.2 Laplace Integral Formula

An integral representation of $P_n(x)$ is given by

$$P_n(x) = \frac{1}{\pi} \int_0^{\pi} \left[x + (x^2-1)^{1/2} \cos\phi\right]^n d\phi, \qquad n = 0, 1, 2, \ldots \qquad (4.48)$$

which is called the *Laplace integral formula*. This relation is easily verified for $n = 0$ and $n = 1$, but more difficult to prove in the general case.

Let us call the integral I and expand the integrand in a finite binomial series to get

$$\begin{aligned}
I &= \frac{1}{\pi} \int_0^\pi \left[x + (x^2 - 1)^{1/2} \cos\phi \right]^n d\phi \\
&= \frac{1}{\pi} \int_0^\pi \sum_{k=0}^n \binom{n}{k} x^{n-k} (x^2 - 1)^{k/2} \cos^k\phi \, d\phi \\
&= \sum_{k=0}^n \binom{n}{k} x^{n-k} (x^2 - 1)^{k/2} \frac{1}{\pi} \int_0^\pi \cos^k\phi \, d\phi
\end{aligned} \quad (4.49)$$

The residual integral in (4.49) can be shown to satisfy

$$\frac{1}{\pi} \int_0^\pi \cos^k\phi \, d\phi = 0, \qquad k = 1, 3, 5, \ldots \quad (4.50)$$

and for even values of k we set $k = 2j$ to find

$$\begin{aligned}
\frac{1}{\pi} \int_0^\pi \cos^k\phi \, d\phi &= \frac{2}{\pi} \int_0^{\pi/2} \cos^{2j}\phi \, d\phi \\
&= \frac{(2j)!}{2^{2j}(j!)^2}, \qquad j = 0, 1, 2, \ldots
\end{aligned} \quad (4.51)$$

The verification of (4.50) and (4.51) is left to the exercises (see problems 5 and 6). Thus, all odd terms in (4.49) are zero, and by setting $k = 2j$ and using (4.51), we see that

$$I = \sum_{j=0}^{[n/2]} \frac{n! x^{n-2j} (x^2 - 1)^j}{2^{2j} (n - 2j)! (j!)^2} \quad (4.52)$$

What remains now is to show that (4.52) is a series representation of $P_n(x)$, and this we leave also to the exercises (problem 7).

4.3.3 Some Bounds on $P_n(x)$

One of the uses of the Laplace integral formula (4.48) is to establish some inequalities for the Legendre polynomials which furnish certain bounds on them. Of particular interest is the interval $|x| \leq 1$, but since the integrand in (4.48) is not real for this restriction on x, we first rewrite (4.48) in the form ($i^2 = -1$)

$$P_n(x) = \frac{1}{\pi} \int_0^\pi \left[x + i(1 - x^2)^{1/2} \cos\phi \right]^n d\phi, \qquad |x| \leq 1 \quad (4.53)$$

Now, using the fact that the absolute value of an integral is less than or equal to the integral of the absolute value of the integrand, we get

$$|P_n(x)| \leq \frac{1}{\pi} \int_0^\pi \left| x + i(1 - x^2)^{1/2} \cos\phi \right|^n d\phi \qquad (4.54)$$

From the algebra of complex numbers, it is known that $|a + ib| = (a^2 + b^2)^{1/2}$, and thus for $|x| \leq 1$ it follows that

$$\left| x + i(1 - x^2)^{1/2} \cos\phi \right|^n = [x^2 + (1 - x^2)\cos^2\phi]^{n/2}$$
$$= (\cos^2\phi + x^2 \sin^2\phi)^{n/2}$$
$$\leq (\cos^2\phi + \sin^2\phi)^{n/2}$$
$$\leq 1$$

Returning now to (4.54), we have shown that

$$|P_n(x)| \leq \frac{1}{\pi} \int_0^\pi d\phi$$

or

$$|P_n(x)| \leq 1, \qquad |x| \leq 1, \quad n = 0, 1, 2, \ldots \qquad (4.55)$$

which is our intended result. The equality in (4.55) holds only when $x = \pm 1$.

Another inequality, less obvious and more difficult to prove, is given by

$$P_n(x) < \left[\frac{\pi}{2n(1 - x^2)} \right]^{1/2}, \qquad |x| < 1, \quad n = 1, 2, 3, \ldots \qquad (4.56)$$

Again the Legendre integral representation is used to derive this inequality, although we will not do so here (see problem 10).

EXERCISES 4.3

1. Using Rodrigues's formula (4.47), derive the identities ($n = 1, 2, 3, \ldots$)
   (a) $(n + 1)P_{n+1}(x) = (2n + 1)xP_n(x) - nP_{n-1}(x)$,
   (b) $P_n'(x) = xP_{n-1}'(x) + nP_{n-1}(x)$,
   (c) $xP_n'(x) = nP_n(x) + P_{n-1}'(x)$,
   (d) $P_{n+1}'(x) - P_{n-1}'(x) = (2n + 1)P_n(x)$.

2. Representing $P_n(x)$ by Rodrigues's formula (4.47), show that
$$\int_{-1}^1 P_n(x)\,dx = 0, \qquad n = 1, 2, 3, \ldots$$

3. Using Rodrigues's formula (4.47) and integration by parts, show that
$$\int_{-1}^{1}[P_n(x)]^2\,dx = \frac{2}{2n+1}, \quad n=0,1,2,\ldots$$

4. By defining $v = (x^2-1)^n$, show that
   (a) $(1-x^2)\dfrac{dv}{dx} + 2nxv = 0.$
   (b) Differentiating the result in (a) $n+1$ times and defining $u = v^{(n)}$, show that u satisfies Legendre's equation
   $$(1-x^2)u'' - 2xu' + n(n+1)u = 0$$

5. Verify that
$$\frac{1}{\pi}\int_0^\pi \cos^{2n+1}\theta\,d\theta = 0, \quad n=0,1,2,\ldots$$

6. Verify that
   (a) $\dfrac{1}{\pi}\int_0^\pi \cos^{2n}\theta\,d\theta = \dfrac{2}{\pi}\int_0^{\pi/2}\cos^{2n}\theta\,d\theta.$
   (b) Using properties of the gamma function, show that
   $$\frac{2}{\pi}\int_0^{\pi/2}\cos^{2n}\theta\,d\theta = \frac{(2n)!}{2^{2n}(n!)^2}, \quad n=0,1,2,\ldots$$

7. Show that the generating function for the Legendre polynomials can be written in the form
   (a) $(1-2xt+t^2)^{-1/2} = (1-xt)^{-1}\left[1 - \dfrac{t^2(x^2-1)}{(1-xt)^2}\right]^{-1/2}$
   (b) Using the result in (a), expand the expression on the right in powers of t. Then, by comparing your result with Equation (4.19) in Section 4.2.1, deduce that
   $$P_n(x) = \sum_{k=0}^{[n/2]} \frac{n!x^{n-2k}(x^2-1)^k}{2^{2k}(n-2k)!(k!)^2}$$

8. (*Jordan inequality*) If $0 \le \phi \le \pi/2$, show that
$$\sin\phi \ge \frac{2\phi}{\pi}$$

Hint: Prove that $(\sin\phi)/\phi$ is a decreasing function on the given interval by showing its derivative is always negative. Hence, the minimum value occurs at $\phi = \pi/2$.

9. Derive the inequality
$$1 - y < e^{-y}, \quad y > 0$$

10. By using the Laplace integral formula (4.48), show that for $|x| < 1$,

    (a) $|P_n(x)| \le \dfrac{2}{\pi} \int_0^{\pi/2} [1 - (1 - x^2)\sin^2\phi]^{n/2} d\phi.$

    (b) Show that application of the Jordan inequality (problem 8) reduces (a) to
    $$|P_n(x)| \le \dfrac{2}{\pi} \int_0^{\pi/2} \left[1 - \dfrac{4\phi^2(1 - x^2)}{\pi^2}\right]^{n/2} d\phi$$

    (c) Making use of the inequality in problem 9 together with an appropriate change of variables, show that
    $$|P_n(x)| < \dfrac{2}{[2n(1 - x^2)]^{1/2}} \int_0^\infty e^{-t^2} dt$$
    and from this result, deduce that ($n = 1, 2, 3, \ldots$)
    $$|P_n(x)| < \left[\dfrac{\pi}{2n(1 - x^2)}\right]^{1/2}, \quad |x| < 1$$

11. Starting with the identity
$$(1 - x^2)P_n'(x) = nP_{n-1}(x) - nxP_n(x)$$
show that
$$|P_n'(x)| \le \dfrac{n}{1 - |x|}, \quad |x| < 1, \quad n = 1, 2, 3, \ldots$$

12. Starting with the identities
$$P_n(x) = xP_{n-1}(x) + \dfrac{x^2 - 1}{n} P_{n-1}'(x)$$
$$P_n'(x) = xP_{n-1}'(x) + nP_{n-1}(x)$$

    (a) show that (for $n = 1, 2, 3, \ldots$)
    $$\dfrac{1 - x^2}{n^2}[P_n'(x)]^2 + [P_n(x)]^2 = \dfrac{1 - x^2}{n^2}[P_{n-1}'(x)]^2 + [P_{n-1}(x)]^2$$

    (b) From (a), establish the inequality
    $$\dfrac{1 - x^2}{n^2}[P_n'(x)]^2 + [P_n(x)]^2 \le 1, \quad |x| \le 1$$

    (c) From (b), deduce that
    $$|P_n(x)| \le 1, \quad |x| \le 1$$

4.4 Legendre Series

In this section we wish to show how to represent certain functions by series of Legendre polynomials, called *Legendre series*. Because the general term in such series is a polynomial, we can interpret a Legendre series as some generalization of a power series for which the general term is also a polynomial, viz., $(x - a)^n$. However, to develop a given function f in a power series requires that the function f be at least continuous and differentiable in the interval of convergence. In the case of Legendre series we make no such requirement. In fact, many functions of practical interest exhibiting (finite) discontinuities may be represented by convergent Legendre series. Legendre series are only one member of a fairly large and special class of series collectively referred to as *generalized Fourier series*, all of which have many properties in common. In Section 1.5 we encountered *Fourier trigonometric series*, which are perhaps the best known members of this class, and in the following chapters we will come across several other members of this general class. Besides their obvious mathematical interest, it turns out that the applications of generalized Fourier series are very extensive—so much so, in fact, that they involve almost every facet of applied mathematics.

4.4.1 Orthogonality

Although we have already derived many identitities associated with the Legendre polynomials, none of these is so fundamental and far-reaching in practice as is the *orthogonality property*

$$\int_{-1}^{1} P_n(x) P_k(x)\, dx = 0, \qquad k \neq n \qquad (4.57)$$

Remark: It is sometimes helpful to think of (4.57) as a generalization of the scalar (dot) product of vector analysis. In fact, much of the following discussion has a vector analog in three-dimensional vector space.

To prove (4.57), we first take note of the fact that both $P_k(x)$ and $P_n(x)$ satisfy Legendre's DE (4.42), and thus we write

$$\frac{d}{dx}\left[(1 - x^2) P_k'(x)\right] + k(k + 1) P_k(x) = 0 \qquad (4.58a)$$

$$\frac{d}{dx}\left[(1 - x^2) P_n'(x)\right] + n(n + 1) P_n(x) = 0 \qquad (4.58b)$$

If we multiply the first of these equations by $P_n(x)$ and the second by

$P_k(x)$, subtract the results, and integrate from -1 to 1, we find

$$\int_{-1}^{1} P_n(x) \frac{d}{dx}[(1-x^2)P_k'(x)] \, dx - \int_{-1}^{1} P_k(x) \frac{d}{dx}[(1-x^2)P_n'(x)] \, dx$$
$$+ [k(k+1) - n(n+1)] \int_{-1}^{1} P_n(x) P_k(x) \, dx = 0 \quad (4.59)$$

On integrating the first integral above by parts, we have

$$\int_{-1}^{1} P_n(x) \frac{d}{dx}[(1-x^2)P_k'(x)] \, dx = P_n(x)(1-x^2)P_k'(x)\Big|_{-1}^{1}$$
$$- \int_{-1}^{1} (1-x^2) P_n'(x) P_k'(x) \, dx \quad (4.60a)$$

and similarly for the second integral,

$$\int_{-1}^{1} P_k(x) \frac{d}{dx}[(1-x^2)P_n'(x)] \, dx = - \int_{-1}^{1} (1-x^2) P_n'(x) P_k'(x) \, dx \quad (4.60b)$$

and therefore the difference of these two integrals is clearly zero. Hence, (4.59) reduces to

$$[k(k+1) - n(n+1)] \int_{-1}^{1} P_n(x) P_k(x) \, dx = 0$$

and since $k \neq n$ by hypothesis, the result (4.57) follows immediately.

When $k = n$, the situation is different. Let us define

$$A_n = \int_{-1}^{1} [P_n(x)]^2 \, dx \quad (4.61)$$

and replace one of the $P_n(x)$ in (4.61) by use of the identity [replace n with $n-1$ in (4.32)]

$$P_n(x) = \frac{2n-1}{n} x P_{n-1}(x) - \frac{n-1}{n} P_{n-2}(x) \quad (4.62)$$

to get

$$A_n = \int_{-1}^{1} P_n(x) \left[\frac{2n-1}{n} x P_{n-1}(x) - \frac{n-1}{n} P_{n-2}(x) \right] dx$$
$$= \frac{2n-1}{n} \int_{-1}^{1} x P_n(x) P_{n-1}(x) \, dx - \frac{n-1}{n} \int_{-1}^{1} P_n(x) P_{n-2}(x) \, dx \quad (4.63)$$

The second integral above vanishes because of the orthogonality property (4.57). To further simplify (4.63), we rewrite (4.62) in the form

$$xP_n(x) = \frac{1}{2n+1}[(n+1)P_{n+1}(x) + nP_{n-1}(x)]$$

and substitute it into (4.63), from which we deduce

$$A_n = \left(\frac{2n-1}{n}\right)\left(\frac{n+1}{2n+1}\right)\int_{-1}^{1} P_{n+1}(x)P_{n-1}(x)\,dx$$
$$+ \frac{2n-1}{2n+1}\int_{-1}^{1}[P_{n-1}(x)]^2\,dx$$

or

$$A_n = \frac{2n-1}{2n+1}A_{n-1}, \qquad n = 2,3,4,\ldots \qquad (4.64)$$

Equation (4.64) is simply a recurrence formula for A_n. Using the fact that

$$A_0 = \int_{-1}^{1}[P_0(x)]^2\,dx = \int_{-1}^{1}dx = 2$$

and

$$A_1 = \int_{-1}^{1}[P_1(x)]^2\,dx = \int_{-1}^{1}x^2\,dx = \tfrac{2}{3}$$

Equation (64) yields

$$A_2 = \tfrac{3}{5} \times \tfrac{1}{3} \times 2 = \tfrac{2}{5}$$
$$A_3 = \tfrac{5}{7} \times \tfrac{3}{5} \times \tfrac{1}{3} \times 2 = \tfrac{2}{7}$$

while in general it can be verified by mathematical induction that

$$A_n = \frac{2n-1}{2n+1} \times \frac{2n-3}{2n-1} \times \frac{2n-5}{2n-3} \times \cdots \times \frac{1}{3} \times 2 = \frac{2}{2n+1},$$
$$n = 0,1,2,\ldots \qquad (4.65)$$

Thus, we have derived the important result

$$\int_{-1}^{1}[P_n(x)]^2\,dx = \frac{2}{2n+1}, \qquad n = 0,1,2,\ldots \qquad (4.66)$$

4.4.2 Finite Legendre Series

Because of the special properties associated with Legendre polynomials, it may be useful in certain situations to represent arbitrary polynomials as linear combinations of Legendre polynomials. For example, if $q_m(x)$ denotes an arbitrary polynomial of degree m, then, since $P_0(x)$, $P_1(x), \ldots, P_m(x)$ are all polynomials of degree m or less, we might expect to find a representation of the form*

$$q_m(x) = c_0 P_0(x) + c_1 P_1(x) + \cdots + c_m P_m(x) \qquad (4.67)$$

Let us illustrate with a simple example.

Example 1: Express x^2 in a series of Legendre polynomials.

Solution: We write

$$\begin{aligned} x^2 &= c_0 P_0(x) + c_1 P_1(x) + c_2 P_2(x) \\ &= c_0 + c_1 x + c_2 \tfrac{1}{2}(3x^2 - 1) \\ &= \left(c_0 - \tfrac{1}{2}c_2\right) + c_1 x + \tfrac{3}{2}c_2 x^2 \end{aligned}$$

Now equating like coefficients, we see that

$$c_0 - \tfrac{1}{2}c_2 = 0, \qquad c_1 = 0, \qquad \tfrac{3}{2}c_2 = 1$$

from which we deduce $c_0 = \tfrac{1}{3}$, $c_1 = 0$, and $c_2 = \tfrac{2}{3}$. Hence,

$$x^2 = \tfrac{1}{3}P_0(x) + \tfrac{2}{3}P_2(x)$$

When the polynomial $q_m(x)$ is of a high degree, solving a system of simultaneous equations for the c's as we did in Example 1 is very tedious. A more systematic procedure can be developed by using the orthogonality property (4.57). We begin by writing (4.67) in the form

$$q_m(x) = \sum_{n=0}^{m} c_n P_n(x) \qquad (4.68)$$

Next, we multiply both sides of (4.68) by $P_k(x)$, $0 \le k \le m$, and integrate the result termwise (which is justified because the series is finite) from -1 to 1 to get

$$\int_{-1}^{1} q_m(x) P_k(x)\, dx = \sum_{n=0}^{m} c_n \int_{-1}^{1} P_n(x) P_k(x)\, dx \qquad \overset{0\ (n \ne k)}{} \qquad (4.69)$$

Because of the orthogonality property (4.57), each term of the series in

---

*Two polynomials can be equated if and only if they are of the same degree.

(4.69) vanishes except the term corresponding to $n = k$, and here we find

$$\int_{-1}^{1} q_m(x) P_k(x)\, dx = c_k \int_{-1}^{1} [P_k(x)]^2\, dx$$

$$= c_k \left(\frac{2}{2k+1}\right)$$

where the last step is a consequence of (4.66). Hence, we deduce that (changing the dummy index back to n)

$$c_n = (n + \tfrac{1}{2}) \int_{-1}^{1} q_m(x) P_n(x)\, dx, \qquad n = 0, 1, 2, \ldots, m \qquad (4.70)$$

Remark: If the polynomial $q_m(x)$ in (4.70) is even (odd), then only those c_n with even (odd) suffixes are nonzero, due to the even-odd property of the Legendre polynomials (see problems 25 and 26).

As a consequence of the fact that a polynomial of degree m can be expressed as a Legendre series involving only $P_m(x)$ and lower-order Legendre polynomials, we have the following theorem.*

Theorem 4.1. If $q_m(x)$ is a polynomial of degree m and $m < r$, then

$$\int_{-1}^{1} q_m(x) P_r(x)\, dx = 0, \qquad m < r$$

Proof: Since $q_m(x)$ is a polynomial of degree m, we can write

$$q_m(x) = \sum_{n=0}^{m} c_n P_n(x)$$

Then, multiplying both sides of this expression by $P_r(x)$ and integrating from -1 to 1, we get

$$\int_{-1}^{1} q_m(x) P_r(x)\, dx = \sum_{n=0}^{m} c_n \underbrace{\int_{-1}^{1} P_n(x) P_r(x)\, dx}_{0}$$

The largest value of n is m, and since $m < r$, the right-hand side is zero for each n [due to the orthogonality property (4.57)], and the theorem is proved. ∎

---

*Theorem 4.1 says that $P_r(x)$ is orthogonal to *every* polynomial of degree less than r.

4.4.3 Infinite Legendre Series

In some applications we will find it necessary to represent a function f, other than a polynomial, as a linear combination of Legendre polynomials. Such a representation will lead to an *infinite series* of the general form

$$f(x) = \sum_{n=0}^{\infty} c_n P_n(x) \qquad (4.71)$$

where the coefficients can be formally derived by a process similar to the derivation of (4.70), leading to

$$c_n = \left(n + \tfrac{1}{2}\right) \int_{-1}^{1} f(x) P_n(x)\, dx, \qquad n = 0, 1, 2, \ldots \qquad (4.72)$$

Conditions under which the representation (4.71) and (4.72) is valid will be taken up in the next section. For now it suffices to say that for certain functions the series (4.71) will converge throughout the interval $-1 \leq x \leq 1$, even at points of finite discontinuities of the given function. Series of this type are called *Legendre series*, and because they belong to the larger class of generalized Fourier series, the coefficients (4.72) are commonly called the *Fourier coefficients* of the series.

In practice, the evaluation of integrals like (4.72) must be performed numerically. However, if the function f is not too complicated, we can sometimes use various properties of the Legendre polynomials to evaluate such integrals in closed form. The following example illustrates the point.

Remark: Because the interval of convergence of (4.71) is confined to $-1 \leq x \leq 1$, it really doesn't matter if the function f is defined outside this interval. That is, even if f is defined for *all* x, the representation will not be valid beyond the interval $-1 \leq x \leq 1$ (unless f is a polynomial).

Example 2: Find the Legendre series for

$$f(x) = \begin{cases} -1, & -1 \leq x < 0 \\ 1, & 0 < x \leq 1 \end{cases}$$

Solution: The function f is an odd function. Hence, owing to the even-odd property of the Legendre polynomials depending upon the index n, we note that $f(x) P_n(x)$ is an odd function when n is even, and in this case it follows that (see problem 25)

$$c_n = \left(n + \tfrac{1}{2}\right) \int_{-1}^{1} f(x) P_n(x)\, dx = 0, \qquad n = 0, 2, 4, \ldots$$

For odd index n, the product $f(x)P_n(x)$ is even and therefore

$$c_n = (n + \tfrac{1}{2})\int_{-1}^{1} f(x)P_n(x)\,dx$$

$$= (2n + 1)\int_{0}^{1} P_n(x)\,dx, \qquad n = 1, 3, 5, \ldots$$

Let us use the identity [see Equation (4.40)]

$$P_n(x) = \frac{1}{2n+1}\left[P'_{n+1}(x) - P'_{n-1}(x)\right]$$

and set $n = 2k + 1$, thereby obtaining the result (for $k = 0, 1, 2, \ldots$)

$$c_{2k+1} = (4k+3)\int_0^1 P_{2k+1}(x)\,dx$$

$$= \int_0^1 \left[P'_{2k+2}(x) - P'_{2k}(x)\right]dx$$

$$= \left[P_{2k+2}(x) - P_{2k}(x)\right]\Big|_0^1$$

$$= P_{2k}(0) - P_{2k+2}(0)$$

where we have used the property $P_n(1) = 1$ for all n. Referring to Equation (4.27), we have

$$c_{2k+1} = \frac{(-1)^k (2k)!}{2^{2k}(k!)^2} - \frac{(-1)^{k+1}(2k+2)!}{2^{2k+2}[(k+1)!]^2}$$

$$= \frac{(-1)^k (2k)!}{2^{2k}(k!)^2}\left[1 + \frac{(2k+2)(2k+1)}{2^2(k+1)^2}\right]$$

$$= \frac{(-1)^k (2k)!}{2^{2k}(k!)^2}\left[1 + \frac{2k+1}{2k+2}\right]$$

$$= \frac{(-1)^k (2k)!(4k+3)}{2^{2k+1}k!(k+1)!}$$

and thus

$$f(x) = \sum_{k=0}^{\infty} \frac{(-1)^k (2k)!(4k+3)}{2^{2k+1}k!(k+1)!} P_{2k+1}(x), \qquad -1 \le x \le 1$$

EXERCISES 4.4

In problems 1–15, use the orthogonality property and/or any other relations to derive the integral formula.

1. $\int_{-1}^{1} xP_n(x)\,dx = \begin{cases} 0, & n \neq 1, \\ \frac{2}{3}, & n = 1. \end{cases}$

2. $\int_{-1}^{1} xP_n(x)P_{n-1}(x)\,dx = \dfrac{2n}{4n^2 - 1},\ n = 1, 2, 3, \ldots$.

3. $\int_{-1}^{1} P_n(x)P'_{n+1}(x)\,dx = 2,\ n = 0, 1, 2, \ldots$.

4. $\int_{-1}^{1} xP'_n(x)P_n(x)\,dx = \dfrac{2n}{2n + 1},\ n = 0, 1, 2, \ldots$.

5. $\int_{-1}^{1} (1 - x^2)P'_n(x)P'_k(x)\,dx = 0,\ k \neq n$.

6. $\int_{-1}^{1} (1 - 2xt + t^2)^{-1/2}P_n(x)\,dx = \dfrac{2t^n}{2n + 1},\ n = 0, 1, 2, \ldots$.

7. $\int_{-1}^{1} (1 - x)^{-1/2}P_n(x)\,dx = \dfrac{2\sqrt{2}}{2n + 1},\ n = 0, 1, 2, \ldots$.

   Hint: Let $t \to 1$ in problem 6.

8. $\int_{-1}^{1} x^2 P_{n+1}(x)P_{n-1}(x)\,dx = \dfrac{2n(n + 1)}{(4n^2 - 1)(2n + 3)},\ n = 1, 2, 3, \ldots$.

9. $\int_{-1}^{1} (x^2 - 1)P_{n+1}(x)P'_n(x)\,dx = \dfrac{2n(n + 1)}{(2n + 1)(2n + 3)},\ n = 1, 2, 3, \ldots$.

10. $\int_{-1}^{1} x^n P_n(x)\,dx = \dfrac{2^{n+1}(n!)^2}{(2n + 1)!},\ n = 0, 1, 2, \ldots$.

    Hint: Use problem 31.

11. If $k \le n$, $\int_{-1}^{1} P_n(x)P'_k(x)\,dx = \begin{cases} 0, & n + k \text{ even}, \\ k(k + 1), & n + k \text{ odd}. \end{cases}$

12. $\int_{-1}^{1} P_n(x)P'_k(x)\,dx = \begin{cases} 0, & k \le n, \\ 0, & k > n,\ k + n \text{ even}, \\ 2, & k > n,\ k + n \text{ odd}. \end{cases}$

13. $\int_0^1 P_{2n}(x)\,dx = 0,\ n = 1, 2, 3, \ldots$.

14. $\int_{-1}^{1} (1 - x^2)[P'_n(x)]^2\,dx = \dfrac{2n(n + 1)}{2n + 1},\ n = 0, 1, 2, \ldots$.

15. $\int_{-1}^{1} x^2 [P_n(x)]^2 \, dx = \dfrac{2}{(2n+1)^2} \left[\dfrac{(n+1)^2}{2n+3} + \dfrac{n^2}{2n-1} \right]$,

$n = 0, 1, 2, \ldots$.

16. Show that the orthogonality relation (4.57) for the functions $P_n(\cos\phi)$ is

$$\int_0^{\pi} P_n(\cos\phi) P_k(\cos\phi) \sin\phi \, d\phi = 0, \qquad k \neq n$$

In problems 17–21, derive the given integral formula.

17. $\int_0^{\pi} P_{2n}(\cos\phi) \, d\phi = \dfrac{\pi}{2^{2n}} \binom{2n}{n}$, $n = 0, 1, 2, \ldots$.

18. $\int_0^{2\pi} P_{2n}(\cos\phi) \, d\phi = \dfrac{\pi}{2^{4n-1}} \binom{2n}{n}^2$, $n = 1, 2, 3, \ldots$.

19. $\int_0^{2\pi} P_{2n}(\cos\phi) \cos\phi \, d\phi = \dfrac{1}{2^{4n+1}} \binom{2n}{n} \binom{2n+2}{n+1}$, $n = 1, 2, 3, \ldots$.

20. $\int_0^{\pi/2} P_{2n}(\cos\phi) \sin\phi \, d\phi = 0$, $n = 1, 2, 3, \ldots$.

21. $\int_0^{\pi} P_n(\cos\phi) \cos n\phi \, d\phi = B(n+\tfrac{1}{2}, \tfrac{1}{2})$, $n = 0, 1, 2, \ldots$.

22. Using Rodrigues's formula (4.47) for $P_n(x)$,

   (a) show that integration by parts leads to

   $$\int_{-1}^{1} P_n(x) P_k(x) \, dx = -\dfrac{1}{2^n n!} \int_{-1}^{1} P_k'(x) \dfrac{d^{n-1}}{dx^{n-1}} \left[(x^2-1)^n \right] dx$$

   (b) Show, by continued integration by parts, that

   $$\int_{-1}^{1} P_n(x) P_k(x) \, dx = \dfrac{(-1)^n}{2^n n!} \int_{-1}^{1} \dfrac{d^n}{dx^n} [P_k(x)] (x^2-1)^n \, dx$$

   (c) For $k \neq n$, show that the integral on the right in (b) is zero.

23. For $k = n$, show that problem 22(b) leads to ($n = 0, 1, 2, \ldots$)

   (a) $\int_{-1}^{1} [P_n(x)]^2 \, dx = \dfrac{(2n)!}{2^{2n}(n!)^2} \int_{-1}^{1} (1-x^2)^n \, dx$.

   (b) By making an appropriate change of variable, evaluate the integral in (a) through use of the gamma function and hence derive Equation (4.66).

24. Starting with the expression

$$(1 - 2xt + t^2)^{-1} = \sum_{n=0}^{\infty} \sum_{k=0}^{\infty} P_n(x) P_k(x) t^{n+k}$$

use the orthogonality property (4.57) to deduce Equation (4.66).

Hint: $\log \dfrac{1+t}{1-t} = 2 \sum_{n=0}^{\infty} \dfrac{t^{2n+1}}{2n+1}$.

25. Show that if f is an odd function

(a) $\int_{-1}^{1} f(x)P_n(x)\,dx = 0, \quad n = 0, 2, 4, \ldots,$

(b) $\int_{-1}^{1} f(x)P_n(x)\,dx = 2\int_{0}^{1} f(x)P_n(x)\,dx, \quad n = 1, 3, 5, \ldots.$

26. Show that if f is an even function

(a) $\int_{-1}^{1} f(x)P_n(x)\,dx = 2\int_{0}^{1} f(x)P_n(x)\,dx, \quad n = 0, 2, 4, \ldots,$

(b) $\int_{-1}^{1} f(x)P_n(x)\,dx = 0, \quad n = 1, 3, 5, \ldots.$

In problems 27–30, find the Legendre series for the given polynomial.

27. $q(x) = x^3$.

28. $q(x) = 9x^3 - 8x^2 + 7x - 6$.

29. $q(x) = 12x^4 - 8x^2 + 7$.

30. $q(x) = 1 + x + \dfrac{x^2}{2!} + \dfrac{x^3}{3!} + \dfrac{x^4}{4!}$.

31. Using Rodrigues's formula (4.47) and integration by parts, show that

$$\int_{-1}^{1} f(x)P_n(x)\,dx = \frac{(-1)^n}{2^n n!}\int_{-1}^{1} f^{(n)}(x)(x^2 - 1)^n\,dx$$

Hint: See problem 22.

32. From the result of problem 31, deduce that

$$\int_{-1}^{1} x^m P_n(x)\,dx = 0 \quad \text{if} \quad m < n$$

33. From the result of problem 31, deduce that

$$\int_{-1}^{1} x^{n+2k} P_n(x)\,dx = \frac{(n+2k)!\,\Gamma(k+\tfrac{1}{2})}{2^n(2k)!\,\Gamma(n+k+\tfrac{3}{2})}, \quad k = 0, 1, 2, \ldots$$

34. Show that

(a) $x^{2m} = \displaystyle\sum_{n=0}^{m} \dfrac{2^{2n}(4n+1)(2m)!\,(m+n)!}{(2m+2n+1)(m-n)!}\,P_{2n}(x),$

(b) $x^{2m+1} = \displaystyle\sum_{n=0}^{m} \dfrac{2^{2n+1}(4n+3)(2m+1)!\,(m+n+1)!}{(2m+2n+3)!\,(m-n)!}\,P_{2n+1}(x).$

Hint: Use problem 33.

In problems 35–40, develop the Legendre series for the given function.

35. $f(x) = P_6(x)$.

36. $f(x) = |x|$, $-1 \le x \le 1$.

37. $f(x) = \begin{cases} 0, & -1 \le x < 0, \\ 1, & 0 < x \le 1. \end{cases}$

38. $f(x) = \begin{cases} 1, & -1 \le x < 0, \\ 0, & 0 < x \le 1. \end{cases}$

39. $f(x) = \begin{cases} 0, & -1 \le x < 0, \\ x, & 0 < x \le 1. \end{cases}$

40. $f(x) = \begin{cases} 1, & -1 \le x < 0, \\ x, & 0 < x \le 1. \end{cases}$

41. Show that the Legendre series of a function f defined in the interval $-a \le x \le a$ is given by

$$f(x) = \sum_{n=0}^{\infty} c_n P_n(x/a), \quad -a \le x \le a$$

where

$$c_n = \frac{2n+1}{2a} \int_{-a}^{a} f(x) P_n(x/a)\, dx, \quad n = 0, 1, 2, \ldots$$

42. Making the change of variable $x = \cos\phi$, show that the Legendre series for a function $f(\phi)$ is given by

$$f(\phi) = \sum_{n=0}^{\infty} c_n P_n(\cos\phi), \quad 0 \le \phi \le \pi$$

where

$$c_n = (n + \tfrac{1}{2}) \int_0^{\pi} f(\phi) P_n(\cos\phi) \sin\phi\, d\phi, \quad n = 0, 1, 2, \ldots$$

Hint: See problem 16.

43. Using the result of problem 42, find the Legendre series for

(a) $f(\phi) = \begin{cases} 0, & 0 \le \phi < \pi/2 \\ 1, & \pi/2 < \phi \le \pi. \end{cases}$

(b) $f(\phi) = \cos^2\phi$, $0 \le \phi \le \pi$.

44. Show that

(a) $(1-x)^n P_n\left(\dfrac{1+x}{1-x}\right) = \sum_{k=0}^{n} \binom{n}{k}^2 x^k$.

(b) Letting $x \to 1$, use part (a) to derive the identity

$$\sum_{k=0}^{n} \binom{n}{k}^2 = \binom{2n}{n}$$

4.5 Convergence of the Series

Given the Legendre series of some function f, we now wish to discuss the validity of such a representation. What we mean is—if a value of x is selected in the chosen interval and each term of the series is evaluated for this value of x, will the sum of the series be $f(x)$? If so, we say the series *converges pointwise* to $f(x)$.* In order to establish pointwise convergence of the series, we need to obtain an expression for the partial sum†

$$S_n(x) = \sum_{k=0}^{n} c_k P_k(x) \qquad (4.73)$$

and then for a fixed value of x, show that

$$\lim_{n \to \infty} S_n(x) = f(x) \qquad (4.74)$$

4.5.1 Piecewise Continuous and Piecewise Smooth Functions

To be sure the Legendre series converges to the function which generates the series, it is essential to place certain restrictions on the function f. From a practical point of view, such conditions should be broad enough to cover most situations of concern and still simple enough to be easily checked for the given function.

Definition 4.1. A function f is said to be *piecewise continuous* in the interval $a \le x \le b$, provided that

(1) $f(x)$ is defined and continuous at all but a finite number of points in the interval, and
(2) the left-hand and right-hand limits exist at each point in the interval.

Remark: The left-hand and right-hand limits are defined, respectively, by

$$\lim_{\varepsilon \to 0^+} f(x - \varepsilon) = f(x^-), \qquad \lim_{\varepsilon \to 0^+} f(x + \varepsilon) = f(x^+)$$

Furthermore, when x is a point of continuity, $f(x^-) = f(x^+) = f(x)$.

It is not essential that a piecewise continuous function f be defined at every point in the interval of interest. In particular, it is often not defined at a point of discontinuity, and even when it is, it really doesn't matter what functional value is assigned at such a point. Also, the interval of interest may be open or closed, or open at one end and closed at the other (see Fig. 4.4).

---

*See also the discussion in Section 1.3.
†Although (4.73) has $n + 1$ terms, we still designate it by the symbol $S_n(x)$.

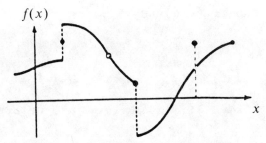

Figure 4.4 A Piecewise Continuous Function

Definition 4.2. A function f is said to be *smooth* in the interval $a \leq x \leq b$ if it has a continuous derivative there. We say the function is *piecewise smooth* if f and/or its derivative f' are only piecewise continuous in $a \leq x \leq b$.

Example 3: Classify the following functions as smooth, piecewise smooth, or neither in $-1 \leq x \leq 1$: (a) $f(x) = x$, (b) $f(x) = |x|$, (c) $f(x) = |x|^{1/2}$.

Solution: In (a), the function $f(x) = x$ and its derivative $f'(x) = 1$ are both continuous, and thus f is *smooth*. The function in (b) is also continuous, but because the derivative is discontinuous, i.e.,

$$f'(x) = \begin{cases} -1, & -1 < x < 0 \\ 1, & 0 < x < 1 \end{cases}$$

it is not smooth but only *piecewise smooth*. In (c), the function is once again continuous, but $|f'(x)| \to \infty$ as $x \to 0$, so it is *neither* smooth nor piecewise smooth.

4.5.2 A Theorem on Pointwise Convergence

Before stating and proving our main theorem on convergence, we must first establish two lemmas.

Lemma 4.1 (*Riemann*). If the function f is piecewise continuous in the closed interval $-1 \leq x \leq 1$, then

$$\lim_{n \to \infty} (n + \tfrac{1}{2})^{1/2} \int_{-1}^{1} f(x) P_n(x)\, dx = 0$$

Proof: Let the nth partial sum be denoted by

$$S_n(x) = \sum_{k=0}^{n} c_k P_k(x)$$

and consider the nonnegative quantity

$$\int_{-1}^{1} [f(x) - S_n(x)]^2 \, dx \geq 0$$

or

$$\int_{-1}^{1} f^2(x) \, dx - 2 \int_{-1}^{1} f(x) S_n(x) \, dx + \int_{-1}^{1} S_n^2(x) \, dx \geq 0$$

Now

$$\int_{-1}^{1} f(x) S_n(x) \, dx = \sum_{k=0}^{n} c_k \int_{-1}^{1} f(x) P_k(x) \, dx$$

$$= \sum_{k=0}^{n} \frac{c_k^2}{k + \frac{1}{2}}$$

and

$$\int_{-1}^{1} S_n^2(x) \, dx = \sum_{j=0}^{n} \sum_{k=0}^{n} c_j c_k \int_{-1}^{1} P_j(x) P_k(x) \, dx \quad \overset{0 \, (j \neq k)}{\nearrow}$$

$$= \sum_{k=0}^{n} c_k^2 \int_{-1}^{1} [P_k(x)]^2 \, dx$$

$$= \sum_{k=0}^{n} \frac{c_k^2}{k + \frac{1}{2}}$$

Accordingly, we have

$$\int_{-1}^{1} f^2(x) \, dx - 2 \sum_{k=0}^{n} \frac{c_k^2}{k + \frac{1}{2}} + \sum_{k=0}^{n} \frac{c_k^2}{k + \frac{1}{2}} \geq 0$$

from which we deduce

$$\sum_{k=0}^{n} \frac{c_k^2}{k + \frac{1}{2}} \leq \int_{-1}^{1} f^2(x) \, dx$$

Because this last inequality is valid for all n, we simply pass to the limit to get

$$\sum_{k=0}^{\infty} \frac{c_k^2}{k + \frac{1}{2}} \leq \int_{-1}^{1} f^2(x) \, dx$$

The integral on the right is necessarily bounded, since f is assumed to be piecewise continuous in the closed interval of integration. Hence, the series on the left is a convergent series (because its sum is finite), and therefore it follows that

$$\lim_{k \to \infty} \frac{c_k^2}{k + \frac{1}{2}} = 0$$

or equivalently (changing the index back to n),

$$\lim_{n \to \infty} (n + \tfrac{1}{2})^{1/2} \int_{-1}^{1} f(x) P_n(x) \, dx = 0 \qquad \blacksquare$$

Lemma 4.2 (*Christoffel-Darboux*). The Legendre polynomials satisfy the identity

$$\sum_{k=0}^{n} (2k + 1) P_k(t) P_k(x) = \frac{n + 1}{t - x} [P_{n+1}(t) P_n(x) - P_n(t) P_{n+1}(x)]$$

Proof: We begin by multiplying the recurrence relation (4.31) by $P_k(t)$ to get

$$(2k + 1) x P_k(t) P_k(x) = (k + 1) P_k(t) P_{k+1}(x) + k P_k(t) P_{k-1}(x)$$

If we now interchange the roles of x and t in this expression and subtract the two results, we obtain

$$(2k + 1)(t - x) P_k(t) P_k(x) = (k + 1)[P_{k+1}(t) P_k(x) - P_k(t) P_{k+1}(x)]$$
$$- k[P_k(t) P_{k-1}(x) - P_{k-1}(t) P_k(x)]$$

Finally, summing both sides of this identity as k runs from 0 to n and setting $P_{-1}(x) = 0$, we find

$$(t - x) \sum_{k=0}^{n} (2k + 1) P_k(t) P_k(x)$$
$$= (n + 1)[P_{n+1}(t) P_n(x) - P_n(t) P_{n+1}(x)]$$

and the lemma is proved. $\blacksquare$

We note that integration of the Christoffel-Darboux formula leads to

$$\sum_{k=0}^{n} (2k + 1) P_k(x) \int_{-1}^{1} P_k(t) \, dt$$
$$= (n + 1) \int_{-1}^{1} \frac{P_{n+1}(t) P_n(x) - P_n(t) P_{n+1}(x)}{t - x} \, dt$$

from which we deduce

$$(n + 1) \int_{-1}^{1} \frac{P_{n+1}(t) P_n(x) - P_n(t) P_{n+1}(x)}{t - x} \, dt = 2 \qquad (4.75)$$

152 • Special Functions for Engineers and Applied Mathematicians

where we are using the orthogonality property

$$\int_{-1}^{1} P_k(t)\, dt = \begin{cases} 0, & k \neq 0 \\ 2, & k = 0 \end{cases} \qquad (4.76)$$

We are now prepared to state and prove our main result.

Theorem 4.2 If the function f is piecewise smooth in the closed interval $-1 \le x \le 1$, then the Legendre series

$$f(x) = \sum_{n=0}^{\infty} c_n P_n(x)$$

where

$$c_n = \left(n + \tfrac{1}{2}\right) \int_{-1}^{1} f(x) P_n(x)\, dx, \qquad n = 0, 1, 2, \ldots$$

converges pointwise to $f(x)$ at every continuity point of the function f in the interval $-1 < x < 1$. At points of discontinuity of f in the interval $-1 < x < 1$, the series converges to the average value $\tfrac{1}{2}[f(x^+) + f(x^-)]$. Finally, at $x = -1$ the series converges to $f(-1^+)$, and at $x = 1$ it converges to $f(1^-)$.

Proof (for a point of continuity): Let us assume that x is a point of continuity of the function f, and consider the nth partial sum $(-1 < x < 1)$

$$S_n(x) = \sum_{k=0}^{n} c_k P_k(x)$$

$$= \sum_{k=0}^{n} \left[\left(k + \tfrac{1}{2}\right) \int_{-1}^{1} f(t) P_k(t)\, dt\right] P_k(x)$$

where we have replaced the constants c_k by their integral representation. Interchanging the order of summation and integration, and recalling the Christoffel-Darboux formula (Lemma 4.2), we obtain

$$S_n(x) = \frac{1}{2} \int_{-1}^{1} f(t) \sum_{k=0}^{n} (2k+1) P_k(t) P_k(x)\, dt$$

$$= \tfrac{1}{2}(n+1) \int_{-1}^{1} f(t) \frac{P_{n+1}(t) P_n(x) - P_n(t) P_{n+1}(x)}{t - x}\, dt$$

If we add and subtract the function $f(x)$ (which is independent of the variable of integration), we get

$$S_n(x) = \tfrac{1}{2}(n+1) f(x) \int_{-1}^{1} \frac{P_{n+1}(t) P_n(x) - P_n(t) P_{n+1}(x)}{t - x}\, dt$$

$$+ \tfrac{1}{2}(n+1) \int_{-1}^{1} \frac{f(t) - f(x)}{t - x} \left[P_{n+1}(t) P_n(x) - P_n(t) P_{n+1}(x)\right] dt$$

For notational convenience we introduce the function
$$g(t) = \frac{f(t) - f(x)}{t - x}$$
and use (4.75) to obtain
$$S_n(x) = f(x) + \tfrac{1}{2}(n+1)P_n(x)\int_{-1}^{1} g(t)P_{n+1}(t)\,dt$$
$$- \tfrac{1}{2}(n+1)P_{n+1}(x)\int_{-1}^{1} g(t)P_n(t)\,dt$$

At this point we wish to show that g satisfies the conditions of Riemann's lemma, i.e., that g is at least piecewise continuous. Because f is at least piecewise smooth, it follows that g is also piecewise smooth for all $t \neq x$. To investigate the behavior of g at $t = x$, we consider the limit (remembering that x is a point of continuity of f)
$$g(x) = \lim_{t \to x} \frac{f(t) - f(x)}{t - x} = f'(x)$$

Since by hypothesis f' is at least piecewise continuous (why?), we see that g is indeed a piecewise continuous function.

Letting
$$b_n = (n + \tfrac{1}{2})^{1/2} \int_{-1}^{1} g(t)P_n(t)\,dt$$
we can express the nth partial sum in the form
$$S_n(x) = f(x) + \frac{(n+1)P_n(x)}{2(n+\tfrac{3}{2})^{1/2}} b_{n+1} - \frac{(n+1)P_{n+1}(x)}{2(n+\tfrac{1}{2})^{1/2}} b_n$$

By recognizing that the Legendre polynomials are bounded on the interval $-1 < x < 1$ [see Equation (4.56)], and applying Riemann's lemma, it can now be shown that the last two terms in the expression for $S_n(x)$ vanish in the limit as $n \to \infty$ (see problem 10), and hence we deduce our intended result
$$\lim_{n \to \infty} S_n(x) = f(x)$$
at a point of continuity of f. ∎

To prove that*
$$\lim_{n \to \infty} S_n(x) = \tfrac{1}{2}[f(x^+) + f(x^-)]$$

---

*For details, see D. Jackson, *Fourier Series and Orthogonal Polynomials*, Carus Math. Monogr. 6, LaSalle, Ill.: Math. Assoc. Amer., Open Court Publ. Co., 1941.

at a point of discontinuity of f requires only a slight modification of the above proof. Similar comments can be made about the points $x = \pm 1$.

EXERCISES 4.5

In problems 1–8, discuss whether the function is piecewise continuous, continuous, piecewise smooth, smooth, or none of these in the interval $-1 \le x \le 1$.

1. $f(x) = \tan 2x$.
2. $f(x) = \sin x$.
3. $f(x) = \dfrac{x^2 - 1}{x - 1}, \; x \ne 1$.
4. $f(x) = \begin{cases} 1 & \text{if } x \text{ is rational.} \\ 0 & \text{if } x \text{ is irrational.} \end{cases}$
5. $f(x) = \dfrac{\sin x}{x}, \; x \ne 0, \; f(0) = 1$.
6. $f(x) = \dfrac{\sin x}{x}, \; x \ne 0$.
7. $f(x) = \sin(1/x), \; x \ne 0$.
8. $f(x) = xe^{-1/x}, \; x \ne 0$.

9. Suppose that a piecewise smooth function f is to be approximated on the interval $-1 \le x \le 1$ by the finite sum

$$S_n(x) = \sum_{k=0}^{n} b_k P_k(x), \qquad -1 \le x \le 1$$

Determine the constants b_k so that the *mean square error* is minimized, i.e., minimize

$$E_n = \int_{-1}^{1} [f(x) - S_n(x)]^2 \, dx$$

Hint: Set $\partial E_n / \partial b_k = 0, \; k = 1, 2, \ldots, n$.

10. Given that

$$P_n(x) < \left[\frac{\pi}{2n(1 - x^2)} \right]^{1/2}, \qquad |x| < 1$$

and

$$b_n = (n + \tfrac{1}{2}) \int_{-1}^{1} g(t) P_n(t) \, dt$$

where $g(t)$ is piecewise continuous, deduce that

(a) $\displaystyle \lim_{n \to \infty} \frac{(n+1) P_n(x)}{2(n + \tfrac{3}{2})^{1/2}} b_{n+1} = 0$,

(b) $\displaystyle \lim_{n \to \infty} \frac{(n+1) P_{n+1}(x)}{2(n + \tfrac{1}{2})^{1/2}} b_n = 0$.

4.6 Legendre Functions of the Second Kind

The Legendre polynomial $P_n(x)$ represents only one solution of Legendre's equation

$$(1 - x^2)y'' - 2xy' + n(n + 1)y = 0 \qquad (4.77)$$

Because the equation is second-order, we know from the general theory of differential equations that there exists a second linearly independent solution $Q_n(x)$ such that the combination

$$y = C_1 P_n(x) + C_2 Q_n(x) \qquad (4.78)$$

where C_1 and C_2 are arbitrary constants, is a general solution of (4.77).

Also from the theory of second-order linear DEs it is well known that if $y_1(x)$ is a nontrivial solution of

$$y'' + a(x)y' + b(x)y = 0 \qquad (4.79)$$

then a second linearly independent solution can be defined by*

$$y_2(x) = y_1(x) \int \frac{\exp\left[-\int a(x)\,dx\right]}{y_1^2(x)}\,dx \qquad (4.80)$$

Hence, if we express (4.77) in the form

$$y'' - \frac{2x}{1 - x^2}y' + \frac{n(n + 1)}{1 - x^2}y = 0$$

and let $y_1(x) = P_n(x)$, it follows that

$$y_2(x) = P_n(x) \int \frac{dx}{(1 - x^2)[P_n(x)]^2} \qquad (4.81)$$

is a second solution, linearly independent of $P_n(x)$. Because any linear combination of solutions is also a solution of a homogeneous DE, it has become customary to define the second solution of (4.77), not by (4.81), but by

$$Q_n(x) = P_n(x)\left\{ A_n + B_n \int \frac{dx}{(1 - x^2)[P_n(x)]^2} \right\} \qquad (4.82)$$

where A_n and B_n are constants to be chosen for each n. We refer to $Q_n(x)$ as the *Legendre function of the second kind* of integral order.

Accordingly, when $n = 0$ we choose $A_0 = 0$ and $B_0 = 1$, and hence

$$Q_0(x) = \int \frac{dx}{1 - x^2}$$

---

*See Theorem 4.5 in L.C. Andrews, *Ordinary Differential Equations with Applications*, Glenview, Ill.: Scott, Foresman and Co., 1982.

which leads to
$$Q_0(x) = \tfrac{1}{2}\log\frac{1+x}{1-x}, \qquad |x| < 1 \tag{4.83}$$

For $n = 1$, we set $A_1 = 0$ and $B_1 = 1$, from which we obtain
$$Q_1(x) = x\int \frac{dx}{(1-x^2)x^2}$$
$$= x\int\left(\frac{1}{1-x^2} + \frac{1}{x^2}\right)dx$$
$$= \tfrac{1}{2}x\log\frac{1+x}{1-x} - 1 \tag{4.84}$$

or
$$Q_1(x) = xQ_0(x) - 1, \qquad |x| < 1 \tag{4.85}$$

Rather than continuing in this fashion, which leads to more difficult integrals to evaluate, we recall the Remark made at the end of Section 4.2.3 which stated that all (properly normalized) solutions of Legendre's equation automatically satisfy the recurrence formulas for $P_n(x)$. Hence, we will select the Legendre functions $Q_n(x)$ so that necessarily

$$Q_{n+1}(x) = \frac{2n+1}{n+1}xQ_n(x) - \frac{n}{n+1}Q_{n-1}(x) \tag{4.86}$$

for $n = 1, 2, 3, \ldots$. With $Q_0(x)$ and $Q_1(x)$ already defined, the substitution of $n = 1$ into (4.86) yields
$$Q_2(x) = \tfrac{3}{2}xQ_1(x) - \tfrac{1}{2}Q_0(x)$$
$$= \tfrac{1}{2}(3x^2 - 1)Q_0(x) - \tfrac{3}{2}x$$

which we recognize as
$$Q_2(x) = P_2(x)Q_0(x) - \tfrac{3}{2}x, \qquad |x| < 1 \tag{4.87}$$

For $n = 2$, we find
$$Q_3(x) = P_3(x)Q_0(x) - \tfrac{5}{2}x^2 + \tfrac{2}{3}, \qquad |x| < 1 \tag{4.88}$$

whereas in general it has been shown that*

$$Q_n(x) = P_n(x)Q_0(x) - \sum_{k=0}^{[\frac{1}{2}(n-1)]} \frac{(2n - 4k - 1)}{(2k+1)(n-k)} P_{n-2k-1}(x),$$
$$|x| < 1 \tag{4.89}$$

for $n = 1, 2, 3, \ldots$.

---

*See W.W. Bell, *Special Functions for Scientists and Engineers*, London: Van Nostrand, 1968, pp. 71–77.

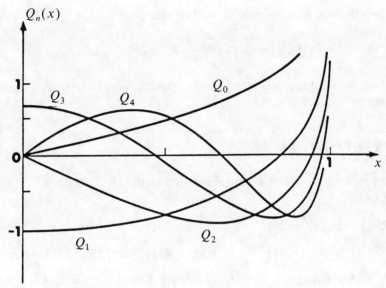

Figure 4.5 Graph of $Q_n(x)$, $n = 0, 1, 2, 3, 4$

Because of the logarithm term in $Q_0(x)$, it becomes clear that all $Q_n(x)$ have infinite discontinuities at $x = \pm 1$. However, within the interval $-1 < x < 1$ these functions are well defined. The first few Legendre functions of the second kind are sketched in Fig. 4.5 for the interval $0 \leq x < 1$.

In some applications it is important to consider $Q_n(x)$ defined on the interval $x > 1$. While Equation (4.89) is not valid for $x > 1$, the functions, $Q_n(x)$ can be expanded in a convergent asymptotic series (problem 16). Based on this series, it can then be shown that all $Q_n(x)$ approach zero as $x \to \infty$. Such behavior for large x is quite distinct from that of the Legendre polynomials $P_n(x)$, which become unbounded as $x \to \infty$ except for $P_0(x) = 1$.

4.6.1 Basic Properties

We have already mentioned that the Legendre functions $Q_n(x)$ satisfy all recurrence relations given in Section 4.2.2 for $P_n(x)$. In addition, there are several relations that directly involve both $P_n(x)$ and $Q_n(x)$. For example, if $|t| < |x|$, then*

$$\frac{1}{x - t} = \sum_{n=0}^{\infty} (2n + 1) P_n(t) Q_n(x) \tag{4.90}$$

---

*See E.T. Whittaker and G.N. Watson, *A Course of Modern Analysis*, Cambridge U.P., 1965, pp. 321–322.

From this result, it is easily shown that (see problem 13)

$$Q_n(x) = \frac{1}{2}\int_{-1}^{1} \frac{P_n(t)}{x-t}\,dt, \qquad n = 0,1,2,\ldots \tag{4.91}$$

which is called the *Neumann formula*. Other properties are taken up in the exercises.

EXERCISES 4.6

In problems 1–4, find a general solution of the DE in terms of $P_n(x)$ and $Q_n(x)$.

1. $(1-x^2)y'' - 2xy' = 0$.
2. $(1-x^2)y'' - 2xy' + 2y = 0$.
3. $(1-x^2)y'' - 2xy' + 12y = 0$.
4. $(1-x^2)y'' - 2xy' + 30y = 0$.

5. Given $P_0(x) = 1$ and $Q_0(x) = \frac{1}{2}\log[(1+x)/(1-x)]$, verify directly that their Wronskian* satisfies

$$W(P_0, Q_0)(x) = \frac{1}{1-x^2}$$

6. Use Equation (4.82) for $Q_n(x)$ to deduce that, in general, the Wronskian of $P_n(x)$ and $Q_n(x)$ is given by

$$W(P_n, Q_n)(x) = \frac{1}{1-x^2}, \qquad n = 0,1,2,\ldots$$

7. Show that $Q_n(x)$ satisfies the relations ($n = 1, 2, 3, \ldots$)
    (a) $Q'_{n+1}(x) - 2xQ'_n(x) + Q'_{n-1}(x) - Q_n(x) = 0$,
    (b) $Q'_{n+1}(x) - xQ'_n(x) - (n+1)Q_n(x) = 0$,
    (c) $xQ'_n(x) - Q'_{n-1}(x) - nQ_n(x) = 0$,
    (d) $Q'_{n+1}(x) - Q'_{n-1}(x) = (2n+1)Q_n(x)$,
    (e) $(1-x^2)Q'_n(x) = n[Q_{n-1}(x) - xQ_n(x)]$.

8. Show that
    (a) $Q_0(-x) = -Q_0(x)$,
    (b) $Q_n(-x) = (-1)^{n+1}Q_n(x)$, $n = 1, 2, 3, \ldots$.

9. Show that (for $n = 1, 2, 3, \ldots$)

$$n[Q_n(x)P_{n-1}(x) - Q_{n-1}(x)P_n(x)]$$
$$= (n-1)[Q_{n-1}(x)P_{n-2}(x) - Q_{n-2}(x)P_{n-1}(x)]$$

---
*Recall that the Wronskian is defined by $W(y_1, y_2) = y_1 y'_2 - y'_1 y_2$.

10. From the result of problem 9, deduce that ($n = 1, 2, 3, \ldots$)

$$Q_n(x)P_{n-1}(x) - Q_{n-1}(x)P_n(x) = -\frac{1}{n}$$

11. Deduce the result of problem 10 by using the Wronskian relation in problem 6 and appropriate recurrence relations.

12. Show that $Q_n(x)$ satisfies the Christoffel-Darboux formula

$$\sum_{k=0}^{n} (2k+1)Q_k(t)Q_k(x) = \frac{n+1}{t-x}[Q_{n+1}(t)Q_n(x) - Q_n(t)Q_{n+1}(x)]$$

13. Use the result of Equation (4.90) to deduce the Neumann formula

$$Q_n(x) = \frac{1}{2}\int_{-1}^{1} \frac{P_n(t)}{x-t}\,dt, \qquad |x| > 1$$

14. For $x > 1$, use the Neumann formula in problem 13 to show that

$$Q_n(x) = \frac{1}{2^{n+1}}\int_{-1}^{1} \frac{(1-t^2)^n}{(x-t)^{n+1}}\,dt$$

15. Using the result of problem 14, deduce that ($x > 1$)

(a) $Q_n(x) = \int_0^{\infty} \frac{d\theta}{[x+(x^2-1)^{1/2}\cosh\theta]^{n+1}}$.

Hint: Set $t = \dfrac{e^\theta(x+1)^{1/2} - (x-1)^{1/2}}{e^\theta(x+1)^{1/2} + (x-1)^{1/2}}$.

(b) $Q_n(x) \sim \dfrac{2^n}{x^{n+1}} \sum_{k=0}^{\infty} \dfrac{(n+k)!(n+2k)!}{k!(2n+2k+1)!} \dfrac{1}{x^{2k}}, \quad x \to \infty$.

16. Solve Legendre's equation

$$(1-x^2)y'' - 2xy' + n(n+1)y = 0$$

by assuming a power series solution of the form $y = \sum_{m=0}^{\infty} c_m x^m$.

(a) Show that the general solution is

$$y = Ay_1(x) + By_2(x)$$

where A and B are any constants and

$$y_1(x) = 1 - \frac{n(n+1)}{2!}x^2 + \frac{(n-2)n(n+1)(n+3)}{4!}x^4 - \cdots$$

and
$$y_2(x) = x - \frac{(n-1)(n+2)}{3!}x^3$$
$$+ \frac{(n-3)(n-1)(n+2)(n+4)}{5!}x^5 - \cdots$$

(b) For $n = 0$, show that
$$P_0(x) = \frac{y_1(x)}{y_1(1)}, \qquad Q_0(x) = y_1(1)y_2(x)$$

(c) For $n = 1$, show that
$$P_1(x) = \frac{y_2(x)}{y_2(1)}, \qquad Q_1(x) = -y_2(1)y_1(x)$$

4.7 Associated Legendre Functions

In applications involving either the Laplace or the Helmholtz equation in spherical, oblate spheroidal, or prolate spheroidal coordinates, it is not Legendre's equation (4.44) that ordinarily arises but rather the *associated Legendre equation*

$$(1 - x^2)y'' - 2xy' + \left[n(n+1) - \frac{m^2}{1 - x^2}\right]y = 0 \qquad (4.92)$$

Observe that for $m = 0$, (4.92) reduces to Legendre's equation (4.44). The DE (4.92) and its solutions, called *associated Legendre functions*, can be developed directly from Legendre's equation and its solutions. To show this we will need the *Leibniz formula* for the mth derivative of a product,

$$\frac{d^m}{dx^m}(fg) = \sum_{k=0}^{m} \binom{m}{k} \frac{d^{m-k}f}{dx^{m-k}} \frac{d^k g}{dx^k}, \qquad m = 1, 2, 3, \ldots \qquad (4.93)$$

If z is a solution of Legendre's equation, i.e., if
$$(1 - x^2)z'' - 2xz' + n(n+1)z = 0 \qquad (4.94)$$
we wish to show that
$$y = (1 - x^2)^{m/2} \frac{d^m z}{dx^m} \qquad (4.95)$$
is then a solution of (4.92). By taking m derivatives of (4.94), we get
$$\frac{d^m}{dx^m}\left[(1 - x^2)z''\right] - 2\frac{d^m}{dx^m}(xz') + n(n+1)\frac{d^m z}{dx^m} = 0$$

which, applying the Leibniz formula (4.93), becomes

$$(1 - x^2)\frac{d^{m+2}z}{dx^{m+2}} - 2mx\frac{d^{m+1}z}{dx^{m+1}} - m(m - 1)\frac{d^m z}{dx^m}$$

$$- 2\left[x\frac{d^{m+1}z}{dx^{m+1}} + m\frac{d^m z}{dx^m}\right] + n(n + 1)\frac{d^m z}{dx^m} = 0$$

Collecting like terms gives us

$$(1 - x^2)\frac{d^2 u}{dx^2} - 2(m + 1)x\frac{du}{dx} + [n(n + 1) - m(m + 1)]u = 0 \quad (4.96)$$

where, for notational convenience, we have set $u = d^m z/dx^m$. Next, by introducing the new variable $y = (1 - x^2)^{m/2} u$, or equivalently,

$$u = y(1 - x^2)^{-m/2}$$

we find that (4.96) takes the form

$$(1 - x^2)\frac{d^2}{dx^2}\left[y(1 - x^2)^{-m/2}\right] - 2(m + 1)x\frac{d}{dx}\left[y(1 - x^2)^{-m/2}\right]$$

$$+ [n(n + 1) - m(m + 1)]y(1 - x^2)^{-m/2} = 0 \quad (4.97)$$

Carrying out the indicated derivatives in (4.97) leads to

$$\frac{d}{dx}\left[y(1 - x^2)^{-m/2}\right] = y'(1 - x^2)^{-m/2} + mxy(1 - x^2)^{-1-m/2}$$

$$= \left[y' + \frac{mxy}{1 - x^2}\right](1 - x^2)^{-m/2} \quad (4.98)$$

and similarly

$$\frac{d^2}{dx^2}\left[y(1 - x^2)^{-m/2}\right]$$

$$= \left[y'' + \frac{m(2xy' + y)}{1 - x^2} + \frac{m(m + 2)x^2 y}{(1 - x^2)^2}\right](1 - x^2)^{-m/2} \quad (4.99)$$

Finally, the substitution of (4.98) and (4.99) into (4.97), and cancellation of the common factor $(1 - x^2)^{-m/2}$, then yields

$$(1 - x^2)\left[y'' + \frac{m(2xy' + y)}{1 - x^2} + \frac{m(m + 2)x^2 y}{(1 - x^2)^2}\right]$$

$$- 2(m + 1)x\left[y' + \frac{mxy}{1 - x^2}\right] + [n(n + 1) - m(m + 1)]y = 0$$

which reduces to (4.92) upon algebraic simplification.

We define the *associated Legendre functions of the first and second kinds*, respectively, by ($m = 0, 1, 2, \ldots, n$)

$$P_n^m(x) = (1 - x^2)^{m/2} \frac{d^m}{dx^m} P_n(x) \qquad (4.100)$$

and

$$Q_n^m(x) = (1 - x^2)^{m/2} \frac{d^m}{dx^m} Q_n(x) \qquad (4.101)$$

Since $P_n(x)$ and $Q_n(x)$ are solutions of Legendre's equation, it follows from (4.95) that $P_n^m(x)$ and $Q_n^m(x)$ are solutions of the associated Legendre equation (4.92).

The associated Legendre functions have many properties in common with the simpler Legendre polynomials $P_n(x)$ and Legendre functions of the second kind $Q_n(x)$. Many of these properties can be developed directly from the corresponding relation involving either $P_n(x)$ or $Q_n(x)$ by taking derivatives and applying the definitions (4.100) and (4.101).

4.7.1 Basic Properties of $P_n^m(x)$

Using the Rodrigues formula (4.47), it is possible to write (4.100) in the form

$$P_n^m(x) = \frac{1}{2^n n!} (1 - x^2)^{m/2} \frac{d^{n+m}}{dx^{n+m}} \left[(x^2 - 1)^n \right] \qquad (4.102)$$

Here we make the interesting observation that the right-hand side of (4.102) is well defined for all values of m such that $n + m \geq 0$, i.e., for $m \geq -n$, whereas (4.100) is valid only for $m \geq 0$. Thus, (4.102) may be used to extend the definition of $P_n^m(x)$ to include all integer values of m such that $-n \leq m \leq n$. (If $m > n$, then necessarily $P_n^m(x) \equiv 0$, which we leave to the reader to prove.) Moreover, using the Leibniz formula (4.93) once again, it can be shown that (see problem 5)

$$P_n^{-m}(x) = (-1)^m \frac{(n - m)!}{(n + m)!} P_n^m(x) \qquad (4.103)$$

Lastly, we note that for $m = 0$ we get the special case

$$P_n^0(x) = P_n(x) \qquad (4.104)$$

The associated Legendre functions $P_n^m(x)$ satisfy many recurrence relations, several of which are generalizations of the recurrence formulas for $P_n(x)$. But because $P_n^m(x)$ has two indices instead of just one, there exists a wider variety of possible relations than for $P_n(x)$.

To derive the *three-term recurrence formula* for $P_n^m(x)$, we start with the known relation [see Equation (4.31)]

$$(n + 1)P_{n+1}(x) - (2n + 1)xP_n(x) + nP_{n-1}(x) = 0 \quad (4.105)$$

and differentiate it m times to obtain

$$(n + 1)\frac{d^m}{dx^m}P_{n+1}(x) - (2n + 1)x\frac{d^m}{dx^m}P_n(x)$$
$$- m(2n + 1)\frac{d^{m-1}}{dx^{m-1}}P_n(x) + n\frac{d^m}{dx^m}P_{n-1}(x) = 0 \quad (4.106)$$

Now recalling [Equation (4.40)]

$$(2n + 1)P_n(x) = P'_{n+1}(x) - P'_{n-1}(x)$$

we find that taking $m - 1$ derivatives leads to

$$m(2n + 1)\frac{d^{m-1}}{dx^{m-1}}P_n(x) = m\frac{d^m}{dx^m}P_{n+1}(x) - m\frac{d^m}{dx^m}P_{n-1}(x)$$
$$(4.107)$$

and using this result, (4.106) becomes

$$(n - m + 1)\frac{d^m}{dx^m}P_{n+1}(x) - (2n + 1)x\frac{d^m}{dx^m}P_n(x)$$
$$+ (n + m)\frac{d^m}{dx^m}P_{n-1}(x) = 0$$

Finally, multiplication of this last result by $(1 - x^2)^{m/2}$ yields the desired recurrence formula

$$(n - m + 1)P_{n+1}^m(x) - (2n + 1)xP_n^m(x) + (n + m)P_{n-1}^m(x) = 0$$
$$(4.108)$$

Additional recurrence relations, which are left to the exercises for verification, include the following:

$$(1 - x^2)^{1/2}P_n^m(x) = \frac{1}{2n + 1}\left[P_{n+1}^{m+1}(x) - P_{n-1}^{m+1}(x)\right] \quad (4.109)$$

$$(1 - x^2)^{1/2}P_n^m(x) = \frac{1}{2n + 1}\big[(n + m)(n + m - 1)P_{n-1}^{m-1}(x)$$
$$- (n - m + 1)(n - m + 2)P_{n+1}^{m-1}(x)\big] \quad (4.110)$$

$$P_n^{m+1}(x) = 2mx(1 - x^2)^{-1/2}P_n^m(x)$$
$$- [n(n + 1) - m(m - 1)]P_n^{m-1}(x) \quad (4.111)$$

By constructing a proof exactly analogous to the proof of orthogonality of the Legendre polynomials, it can be shown that

$$\int_{-1}^{1} P_n^m(x) P_k^m(x)\, dx = 0, \qquad k \neq n \tag{4.112}$$

Also, the evaluation of

$$\int_{-1}^{1} [P_n^m(x)]^2\, dx = \frac{2(n+m)!}{(2n+1)(n-m)!} \tag{4.113}$$

follows exactly our derivation of (4.66) given in Section 4.4.1. The details of proving (4.112) and (4.113) are left for the exercises.

As a final comment we mention that, although it is essentially only a mathematical curiosity, there is another orthogonality relation for the associated Legendre functions given by

$$\int_{-1}^{1} P_n^m(x) P_n^k(x)(1-x^2)^{-1}\, dx = \begin{cases} 0, & k \neq m \\ \dfrac{(n+m)!}{m(n-m)!}, & k = m \end{cases} \tag{4.114}$$

EXERCISES 4.7

1. Directly from Equation (4.100), show that
   (a) $P_1^1(x) = (1-x^2)^{1/2}$,
   (b) $P_2^1(x) = 3x(1-x^2)^{1/2}$,
   (c) $P_2^2(x) = 3(1-x^2)$.
   (d) $P_3^1(x) = \frac{3}{2}(5x^2-1)(1-x^2)^{1/2}$,
   (e) $P_3^2(x) = 15x(1-x^2)$,

2. Show that
   (a) $P_n^m(-x) = (-1)^{n+m} P_n^m(x)$,
   (b) $P_n^m(\pm 1) = 0$, $m > 0$.

3. Show that (for $n = 0, 1, 2, \ldots$)
   (a) $P_{2n}^1(0) = 0$,
   (b) $P_{2n+1}^1(0) = \dfrac{(-1)^n (2n+1)!}{2^{2n}(n!)^2}$.

4. Show that
   (a) $P_n^m(0) = 0$, $n+m$ odd,
   (b) $P_n^m(0) = (-1)^{(n-m)/2} \dfrac{(n+m)!}{2^n[(n-m)/2]![(n+m)/2]!}$, $n+m$ even.

5. By applying the Leibniz formula (4.93) to the product $(x+1)^n(x-1)^n$ and using (4.102), verify that

$$P_n^{-m}(x) = (-1)^m \frac{(n-m)!}{(n+m)!} P_n^m(x)$$

6. Derive the generating function

$$\frac{(2m)!(1-x^2)^{m/2}}{2^m m!(1-2xt+t^2)^{m+\frac{1}{2}}} = \sum_{n=0}^{\infty} P_{n+m}^m(x) t^n$$

In problems 7–11, derive the given recurrence formula.

7. $(1-x^2)P_n^{m\prime}(x) = (n+m)P_{n-1}^m(x) - nxP_n^m(x)$.

8. $(1-x^2)P_n^{m\prime}(x) = (n+1)xP_n^m(x) - (n-m+1)P_{n+1}^m(x)$.

9. $(1-x^2)^{1/2}P_n^m(x) = \dfrac{1}{2n+1}[P_{n+1}^{m+1}(x) - P_{n-1}^{m+1}(x)]$.

10. $(1-x^2)^{1/2}P_n^m(x) = \dfrac{1}{2n+1}[(n+m)(n+m-1)P_{n-1}^{m-1}(x)$
    $-(n-m+1)(n-m+2)P_{n+1}^{m-1}(x)]$.

11. $P_n^{m+1}(x) = 2mx(1-x^2)^{-1/2}P_n^m(x) - [n(n+1) - m(m-1)]P_n^{m-1}(x)$.

12. Prove the orthogonality relation

$$\int_{-1}^{1} P_n^m(x)P_k^m(x)\,dx = 0, \qquad k \neq n$$

13. Prove the orthogonality relation

$$\int_{-1}^{1} P_n^m(x)P_n^k(x)(1-x^2)^{-1}\,dx = 0, \qquad k \neq m$$

14. By defining

$$A_n = \int_{-1}^{1} [P_n^m(x)]^2\,dx, \qquad n = 0, 1, 2, \ldots$$

show that

(a) $A_n = \dfrac{(2n-1)(n+m)}{(2n+1)(n-m)} A_{n-1}$, $n = 2, 3, 4, \ldots$.

(b) Evaluate A_0 and A_1 directly and use (a) to deduce that

$$A_n = \frac{2(n+m)!}{(2n+1)(n-m)!}, \qquad n = 0, 1, 2, \ldots$$

15. Show that

$$\int_{-1}^{1} [P_n^m(x)]^2 (1-x^2)^{-1}\,dx = \frac{(n+m)!}{m(n-m)!}$$

5

Other Orthogonal Polynomials

5.1 Introduction

A set of functions $\{\phi_n(x)\}$, $n = 0, 1, 2, \ldots$, is said to be *orthogonal* on the interval $a < x < b$, with respect to a weight function $r(x) > 0$, if*

$$\int_a^b r(x)\phi_n(x)\phi_k(x)\,dx = 0, \qquad k \neq n$$

Sets of orthogonal functions play an extremely important role in analysis, primarily because functions belonging to a very general class can be represented by series of orthogonal functions, called *generalized Fourier series*.

A special case of orthogonal functions consists of the sets of *orthogonal polynomials* $\{p_n(x)\}$, where n denotes the degree of the polynomial $p_n(x)$. The Legendre polynomials discussed in Chapter 4 are probably the simplest set of polynomials belonging to this class. Other polynomial sets which commonly occur in applications are the *Hermite, Laguerre,* and *Chebyshev polynomials*. More general polynomial sets are defined by the *Gegenbauer* and *Jacobi polynomials*, which include the others as special cases.

The study of general polynomial sets like the Jacobi polynomials facilitates the study of each polynomial set by focusing upon those properties

---

*In some cases the interval of orthogonality may be of infinite extent.

that are characteristic of all the individual sets. For example, the sets $\{p_n(x)\}$ that we will study all satisfy a second-order linear DE and Rodrigues formula, and the related set $\{(d^m/dx^m)p_n(x)\}$ (e.g., the associated Legendre functions) is also orthogonal. Moreover, it can be shown that any orthogonal polynomial set satisfying these three conditions is necessarily a member of the Jacobi polynomial set, or a limiting case such as the Hermite and Laguerre polynomials.

5.2 Hermite Polynomials

The *Hermite polynomials* play an important role in problems involving Laplace's equation in parabolic coordinates, in various problems in quantum mechanics, and in probability theory.

We define the Hermite polynomials $H_n(x)$ by means of the generating function*

$$\exp(2xt - t^2) = \sum_{n=0}^{\infty} H_n(x) \frac{t^n}{n!}, \qquad |t| < \infty, \quad |x| < \infty \qquad (5.1)$$

By writing

$$\exp(2xt - t^2) = e^{2xt} \cdot e^{-t^2}$$

$$= \left(\sum_{m=0}^{\infty} \frac{(2xt)^m}{m!} \right) \left(\sum_{k=0}^{\infty} \frac{(-t^2)^k}{k!} \right)$$

$$= \sum_{n=0}^{\infty} \sum_{k=0}^{[n/2]} \frac{(-1)^k (2x)^{n-2k}}{k!(n-2k)!} t^n \qquad (5.2)$$

where the last step follows from the index change $m = n - 2k$ [see Equation (1.17) in Section 1.2.3], we identify

$$H_n(x) = \sum_{k=0}^{[n/2]} \frac{(-1)^k n!}{k!(n-2k)!} (2x)^{n-2k} \qquad (5.3)$$

Examination of the series (5.3) reveals that $H_n(x)$ is a polynomial of degree n, and further, is an even function of x for even n and an odd function of x for odd n. Thus, it follows that

$$H_n(-x) = (-1)^n H_n(x) \qquad (5.4)$$

The first few Hermite polynomials are listed in Table 5.1 for easy reference.

---

*There is another definition of the Hermite polynomials that uses the generating function $\exp(xt - \frac{1}{2}t^2)$. This definition occurs most often in statistical applications.

Table 5.1 Hermite polynomials

$$H_0(x) = 1$$
$$H_1(x) = 2x$$
$$H_2(x) = 4x^2 - 2$$
$$H_3(x) = 8x^3 - 12x$$
$$H_4(x) = 16x^4 - 48x^2 + 12$$
$$H_5(x) = 32x^5 - 160x^3 + 120x$$

In addition to the series (5.3), the Hermite polynomials can be defined by the *Rodrigues formula* (see problem 3)

$$H_n(x) = (-1)^n e^{x^2} \frac{d^n}{dx^n}(e^{-x^2}), \qquad n = 0, 1, 2, \ldots \quad (5.5)$$

and the *integral representation* (see problem 5)

$$H_n(x) = \frac{(-i)^n 2^n e^{x^2}}{\sqrt{\pi}} \int_{-\infty}^{\infty} e^{-t^2 + 2ixt} t^n \, dt, \qquad n = 0, 1, 2, \ldots \quad (5.6)$$

The Hermite polynomials have many properties in common with the Legendre polynomials, and in fact, there are many relations connecting the two sets of polynomials. For example, two of the simplest relations are given by ($n = 0, 1, 2, \ldots$)

$$\frac{2}{n!\sqrt{\pi}} \int_0^{\infty} e^{-t^2} t^n H_n(xt) \, dt = P_n(x) \quad (5.7)$$

and

$$2^{n+1} e^{x^2} \int_x^{\infty} e^{-t^2} t^{n+1} P_n(x/t) \, dt = H_n(x) \quad (5.8)$$

the verifications of which are left for the exercises.

Example 1: Use the generating function to derive the relation

$$x^n = \sum_{k=0}^{[n/2]} \frac{n! H_{n-2k}(x)}{2^n k!(n-2k)!}$$

Solution: From (5.1) we have

$$\exp(2xt - t^2) = \sum_{k=0}^{\infty} H_k(x) \frac{t^k}{k!}$$

or

$$e^{2xt} = e^{t^2} \sum_{k=0}^{\infty} H_k(x) \frac{t^k}{k!}$$

Expressing both exponentials in power series leads to

$$\sum_{n=0}^{\infty} \frac{(2x)^n t^n}{n!} = \sum_{m=0}^{\infty} \frac{t^{2m}}{m!} \cdot \sum_{k=0}^{\infty} H_k(x) \frac{t^k}{k!}$$

$$= \sum_{n=0}^{\infty} \sum_{k=0}^{[n/2]} \frac{H_{n-2k}(x) t^n}{k!(n-2k)!}$$

where the last step results from the change of index $m = n - 2k$. Finally, by comparing the coefficients of t^n in the two series, we deduce that

$$x^n = \sum_{k=0}^{[n/2]} \frac{n! H_{n-2k}(x)}{2^n k!(n-2k)!}$$

5.2.1 Recurrence Relations

By substituting the series for $w(x, t) = \exp(2xt - t^2)$ into the identity

$$\frac{\partial w}{\partial t} - 2(x - t)w = 0 \qquad (5.9)$$

we obtain (after some manipulation)

$$\sum_{n=1}^{\infty} [H_{n+1}(x) - 2xH_n(x) + 2nH_{n-1}(x)] \frac{t^n}{n!} + H_1(x) - 2xH_0(x) = 0$$
$$(5.10)$$

But $H_1(x) - 2xH_0(x) = 0$, and thus we deduce the recurrence formula

$$H_{n+1}(x) - 2xH_n(x) + 2nH_{n-1}(x) = 0 \qquad (5.11)$$

for $n = 1, 2, 3, \ldots$.

Another recurrence relation satisfied by the Hermite polynomials follows the substitution of the series for $w(x, t)$ into

$$\frac{\partial w}{\partial t} - 2tw = 0 \qquad (5.12)$$

This time we find

$$\sum_{n=1}^{\infty} [H_n'(x) - 2nH_{n-1}(x)] \frac{t^n}{n!} = 0$$

which leads to

$$H_n'(x) = 2nH_{n-1}(x), \qquad n = 1, 2, 3, \ldots \qquad (5.13)$$

The elimination of $H_{n-1}(x)$ from (5.11) and (5.13) yields

$$H_{n+1}(x) - 2xH_n(x) + H_n'(x) = 0 \qquad (5.14)$$

and by differentiating this expression and using (5.13) once again, we find

$$H_n''(x) - 2xH_n'(x) + 2nH_n(x) = 0 \qquad (5.15)$$

for $n = 0, 1, 2, \ldots$. Therefore we see that $y = H_n(x)$ ($n = 0, 1, 2, \ldots$) is a solution of the linear second-order DE

$$y'' - 2xy' + 2ny = 0 \qquad (5.16)$$

called *Hermite's equation*.

5.2.2 Hermite Series

The *orthogonality property* of the Hermite polynomials is given by*

$$\int_{-\infty}^{\infty} e^{-x^2} H_n(x) H_k(x) \, dx = 0, \qquad k \neq n \qquad (5.17)$$

We could construct a proof of (5.17) analogous to that given in Section 4.4.1 for the Legendre polynomials, but for the Hermite polynomials an interesting alternative proof exists.

Let us start with the generating-function relations

$$\sum_{n=0}^{\infty} \frac{t^n}{n!} H_n(x) = e^{2xt - t^2} \qquad (5.18a)$$

$$\sum_{k=0}^{\infty} \frac{s^k}{k!} H_k(x) = e^{2xs - s^2} \qquad (5.18b)$$

and multiply these two series to obtain

$$\sum_{n=0}^{\infty} \sum_{k=0}^{\infty} \frac{t^n}{n!} \frac{s^k}{k!} H_n(x) H_k(x) = \exp[-(t^2 + s^2) + 2x(t + s)] \qquad (5.19)$$

Next, we multiply both sides of (5.19) by the weight function e^{-x^2} and integrate (assuming that termwise integration is permitted), to find

$$\sum_{n=0}^{\infty} \sum_{k=0}^{\infty} \frac{t^n}{n!} \frac{s^k}{k!} \int_{-\infty}^{\infty} e^{-x^2} H_n(x) H_k(x) \, dx = e^{-(t^2 + s^2)} \int_{-\infty}^{\infty} e^{-x^2 + 2x(t+s)} \, dx$$

$$= \sqrt{\pi} \, e^{2ts}$$

where we have made the observation (see Example 2 below)

$$\int_{-\infty}^{\infty} e^{-x^2 + 2bx} \, dx = \sqrt{\pi} \, e^{b^2} \qquad (5.20)$$

Finally, expanding e^{2ts} in a power series, we have

$$\sum_{n=0}^{\infty} \sum_{k=0}^{\infty} \frac{t^n}{n!} \frac{s^k}{k!} \int_{-\infty}^{\infty} e^{-x^2} H_n(x) H_k(x) \, dx = \sqrt{\pi} \sum_{n=0}^{\infty} \frac{2^n t^n s^n}{n!} \qquad (5.21)$$

---

*The function e^{-x^2} in (5.17) is called a *weight function*. In the case of the Legendre polynomials, the weight function is unity.

and by comparing like coefficients of $t^n s^k$, we deduce that

$$\int_{-\infty}^{\infty} e^{-x^2} H_n(x) H_k(x)\, dx = 0, \quad k \neq n$$

As a bonus, we find that when $k = n$ in (5.21), we get the additional important result (for $n = 0, 1, 2, \ldots$)

$$\int_{-\infty}^{\infty} e^{-x^2} [H_n(x)]^2\, dx = 2^n n! \sqrt{\pi} \tag{5.22}$$

Based upon the relations (5.17) and (5.22), we can generate a theory concerning the expansion of arbitrary polynomials, or functions in general, in a series of Hermite polynomials. Specifically, if f is a suitable function defined for all x, we look for expansions of the general form

$$f(x) = \sum_{n=0}^{\infty} c_n H_n(x), \quad -\infty < x < \infty \tag{5.23}$$

where the (Fourier) coefficients are given by*

$$c_n = \frac{1}{2^n n! \sqrt{\pi}} \int_{-\infty}^{\infty} e^{-x^2} f(x) H_n(x)\, dx, \quad n = 0, 1, 2, \ldots \tag{5.24}$$

Series of this type are called *Hermite series*. We have the following theorem for them.

Theorem 5.1. If f is piecewise smooth in every finite interval and

$$\int_{-\infty}^{\infty} e^{-x^2} f^2(x)\, dx < \infty$$

then the Hermite series (5.23) with constants defined by (5.24) converges pointwise to $f(x)$ at every continuity point of f. At points of discontinuity, the series converges to the average value $\frac{1}{2}[f(x^+) + f(x^-)]$.

The proof of Theorem 5.1 closely follows that of Theorem 4.1 [see N.N. Lebedev, *Special Functions and Their Applications*, New York: Dover, 1972, pp. 71–73].

Example 2: Express $f(x) = e^{2bx}$ in a Hermite series and use this result to deduce the value of the integral

$$\int_{-\infty}^{\infty} e^{-x^2 + 2bx} H_n(x)\, dx$$

---

*The constants c_n can be formally derived through use of the orthogonality property analogous to the technique used in Section 4.4.2.

Solution: In this case we can obtain the series in an indirect way. We simply set $t = b$ in the generating function (5.1) to obtain

$$\exp(2bx - b^2) = \sum_{n=0}^{\infty} \frac{b^n}{n!} H_n(x)$$

and hence we have our intended series

$$e^{2bx} = e^{b^2} \sum_{n=0}^{\infty} \frac{b^n}{n!} H_n(x)$$

The direct derivation of this result from (5.24) leads to

$$c_n = \frac{1}{2^n n! \sqrt{\pi}} \int_{-\infty}^{\infty} e^{-x^2 + 2bx} H_n(x)\, dx, \qquad n = 0, 1, 2, \ldots$$

However, we have already shown that

$$c_n = \frac{b^n}{n!} e^{b^2}$$

and thus it follows that

$$\int_{-\infty}^{\infty} e^{-x^2 + 2bx} H_n(x)\, dx = \sqrt{\pi}\, (2b)^n e^{b^2}, \qquad n = 0, 1, 2, \ldots$$

In particular, for $n = 0$ we get the result of Equation (5.20).

5.2.3 Simple Harmonic Oscillator

A fundamental problem in quantum mechanics involving Schrödinger's equation concerns the one-dimensional motion of a particle bound in a potential well. It has been established that bounded solutions of Schrödinger's equation for such problems are obtainable only for certain discrete energy levels of the particle within the well. A particular example of this important class of problems is the *harmonic oscillator problem*, the solutions of which lead to Hermite polynomials.

In terms of dimensionless parameters, Schrödinger's equation for the harmonic-oscillator problem takes the form

$$\psi'' + (\lambda - x^2)\psi = 0, \qquad -\infty < x < \infty \qquad (5.25)$$

The parameter λ is proportional to the possible energy levels of the oscillator and ψ is related to the corresponding wave function. In addition to (5.25), the solution ψ must satisfy the boundary condition

$$\lim_{|x| \to \infty} \psi(x) = 0 \qquad (5.26)$$

In looking for bounded solutions of (5.25), we start with the observation that λ becomes negligible compared with x^2 for large values of x. Thus, asymptotically we expect the solution of (5.25) to behave like

$$\psi(x) \sim e^{\pm x^2/2}, \quad |x| \to \infty \tag{5.27}$$

where only the negative sign in the exponent is appropriate in order that (5.26) be satisfied. Based upon this observation, we make the assumption that (5.25) has solutions of the form

$$\psi(x) = y(x) e^{-x^2/2} \tag{5.28}$$

for suitable y. The substitution of (5.28) into (5.25) yields the DE

$$y'' - 2xy' + (\lambda - 1) y = 0 \tag{5.29}$$

The boundary condition (5.26) suggests that whatever functional form y assumes, it must either be finite for all x or approach infinity at a rate slower than $e^{-x^2/2}$ approaches zero. It has been shown* that the only solutions of (5.29) satisfying this condition are those for which $\lambda - 1 = 2n$, or

$$\lambda \equiv \lambda_n = 2n + 1, \quad n = 0, 1, 2, \ldots \tag{5.30}$$

These allowed values of λ are called *eigenvalues*, or energy levels, of the oscillator. With λ so restricted, we see that (5.29) becomes

$$y'' - 2xy' + 2ny = 0 \tag{5.31}$$

which is Hermite's equation with solutions $y = H_n(x)$. (The other solutions of Hermite's equation are not appropriate in this problem.) Hence, we conclude that to each eigenvalue λ_n given by (5.30), there corresponds the solution of (5.25) (called an *eigenfunction* or *eigenstate*) given by

$$\psi_n(x) = H_n(x) e^{-x^2/2}, \quad n = 0, 1, 2, \ldots \tag{5.32}$$

EXERCISES 5.2

1. Show that (for $n = 0, 1, 2, \ldots$)

    (a) $H_{2n}(0) = (-1)^n \dfrac{(2n)!}{n!}$, (b) $H_{2n+1}(0) = 0$,

    (c) $H'_{2n}(0) = 0$, (d) $H'_{2n+1}(0) = (-1)^n \dfrac{(2n+2)!}{(n+1)!}$.

---

*See E.C. Kemble, *The Fundamental Principles of Quantum Mechanics with Elementary Applications*, New York: Dover, 1958, p. 87.

2. Derive the generating-function relations

(a) $e^{t^2}\cos 2xt = \sum_{n=0}^{\infty}(-1)^n H_{2n}(x)\dfrac{t^{2n}}{(2n)!}$, $|t|<\infty$,

(b) $e^{t^2}\sin 2xt = \sum_{n=0}^{\infty}(-1)^n H_{2n+1}(x)\dfrac{t^{2n+1}}{(2n+1)!}$, $|t|<\infty$.

3. Derive the Rodrigues formula (for $n = 0, 1, 2, \ldots$)

$$H_n(x) = (-1)^n e^{x^2}\dfrac{d^n}{dx^n}(e^{-x^2})$$

4. Starting with the integral formula

$$\int_{-\infty}^{\infty} e^{-t^2+2bt}\,dt = \sqrt{\pi}\,e^{b^2}$$

(a) show that differentiating both sides with respect to b leads to

$$\int_{-\infty}^{\infty} te^{-t^2+2bt}\,dt = \sqrt{\pi}\,be^{b^2}$$

(b) For $n = 1, 2, 3, \ldots$, show that

$$\int_{-\infty}^{\infty} t^n e^{-t^2+2bt}\,dt = \dfrac{\sqrt{\pi}}{2^n}\dfrac{d^n}{db^n}(e^{b^2})$$

5. Set $b = ix$ in the result of problem 4(b) to deduce that (for $n = 0, 1, 2, \ldots$)

$$H_n(x) = \dfrac{(-i)^n 2^n e^{x^2}}{\sqrt{\pi}}\int_{-\infty}^{\infty} e^{-t^2+2ixt}t^n\,dt$$

6. Using the result of problem 5, show that (for $n = 0, 1, 2, \ldots$)

(a) $H_{2n}(x) = \dfrac{(-1)^n 2^{2n+1}}{\sqrt{\pi}}e^{x^2}\int_0^{\infty} e^{-t^2}t^{2n}\cos 2xt\,dt$,

(b) $H_{2n+1}(x) = \dfrac{(-1)^n 2^{2n+2}}{\sqrt{\pi}}e^{x^2}\int_0^{\infty} e^{-t^2}t^{2n+1}\sin 2xt\,dt$.

7. Derive the Fourier transform relations

(a) $\dfrac{1}{\sqrt{2\pi}}\int_{-\infty}^{\infty} e^{-\frac{1}{2}t^2+ixt}H_n(t)\,dt = i^n e^{-x^2/2}H_n(x)$,

(b) $\sqrt{\dfrac{2}{\pi}}\int_0^{\infty} e^{-\frac{1}{2}t^2}H_{2n}(t)\cos xt\,dt = (-1)^n e^{-x^2/2}H_{2n}(x)$,

(c) $\sqrt{\dfrac{2}{\pi}}\int_0^{\infty} e^{-\frac{1}{2}t^2}H_{2n+1}(t)\sin xt\,dt = (-1)^n e^{-x^2/2}H_{2n+1}(x)$.

Other Orthogonal Polynomials • 175

In problems 8–11, verify the integral relation.

8. $\int_{-\infty}^{\infty} x^k e^{-x^2} H_n(x)\,dx = 0, \ k = 0, 1, \ldots, n-1.$

9. $\int_{-\infty}^{\infty} x^2 e^{-x^2} [H_n(x)]^2\,dx = \sqrt{\pi}\, 2^n n!(n + \tfrac{1}{2}).$

10. $\int_{0}^{\infty} t^n e^{-t^2} H_n(xt)\,dt = \dfrac{\sqrt{\pi}\, n!}{2} P_n(x).$

11. $\int_{x}^{\infty} e^{-t^2} t^{n+1} P_n(x/t)\,dt = \dfrac{1}{2^{n+1}} e^{-x^2} H_n(x).$

12. Use the result of problem 5 to deduce that

(a) $(1-t^2)^{-1/2} \exp\left[\dfrac{2xyt - (x^2 + y^2)t^2}{1 - t^2}\right] = \sum_{n=0}^{\infty} H_n(x) H_n(y) \dfrac{(t/2)^n}{n!},$

(b) $(1-t^2)^{-1/2} \exp\left(\dfrac{2x^2 t}{1+t}\right) = \sum_{n=0}^{\infty} [H_n(x)]^2 \dfrac{(t/2)^n}{n!}.$

13. Use problem 12 to show that $(n = 0, 1, 2, \ldots)$

$$\int_{-\infty}^{\infty} e^{-2x^2} [H_n(x)]^2\,dx = 2^{n-\tfrac{1}{2}} \Gamma(n + \tfrac{1}{2})$$

14. Derive the Hermite series relations

(a) $x^{2k} = \dfrac{(2k)!}{2^{2k}} \sum_{n=0}^{k} \dfrac{H_{2n}(x)}{(2n)!(k-n)!},$

(b) $x^{2k+1} = \dfrac{(2k+1)!}{2^{2k+1}} \sum_{n=0}^{k} \dfrac{H_{2n+1}(x)}{(2n+1)!(k-n)!}.$

15. Show that the functions $\psi_n(x) = H_n(x) e^{-x^2/2}$ satisfy the relations
  (a) $2n\psi_{n-1}(x) = x\psi_n(x) + \psi'_n(x),$
  (b) $2x\psi_n(x) - 2n\psi_{n-1}(x) = \psi_{n+1}(x),$
  (c) $\psi'_n(x) = x\psi_n(x) - \psi_{n+1}(x).$

16. For $m < n$, prove that

$$\dfrac{d^m}{dx^m} H_n(x) = \dfrac{2^m n!}{(n-m)!} H_{n-m}(x)$$

In problems 17–20, derive the series relationship.

17. $[H_n(x)]^2 = 2^n (n!)^2 \sum_{k=0}^{n} \dfrac{H_{2k}(x)}{2^k (k!)^2 (n-k)!}.$

18. $\sum_{k=0}^{n} \dfrac{H_k(x) H_k(y)}{2^k k!} = \dfrac{H_n(x) H_{n+1}(y) - H_{n+1}(x) H_n(y)}{2^{n+1} n!(y - x)}.$

176 • Special Functions for Engineers and Applied Mathematicians

19. $H_n(x+y) = \dfrac{1}{2^{n/2}} \displaystyle\sum_{k=0}^{n} \binom{n}{k} H_{n-k}(x\sqrt{2}) H_k(y\sqrt{2}).$

20. $H_n(x)H_{n+p}(x) = 2^n n!(n+p)! \displaystyle\sum_{k=0}^{n} \dfrac{H_{2k+p}(x)}{2^k k!(k+p)!(n-k)!}.$

5.3 Laguerre Polynomials

The generating function

$$(1-t)^{-1}\exp\left[-\dfrac{xt}{1-t}\right] = \sum_{n=0}^{\infty} L_n(x)t^n, \qquad |t|<1,\ 0 \le x < \infty \tag{5.33}$$

leads to yet another important class of polynomials, called *Laguerre polynomials*. By expressing the exponential function in a series, we have

$$\begin{aligned}(1-t)^{-1}\exp\left[-\dfrac{xt}{1-t}\right] &= \sum_{k=0}^{\infty} \dfrac{(-1)^k}{k!}(xt)^k (1-t)^{-k-1} \\ &= \sum_{k=0}^{\infty} \dfrac{(-1)^k}{k!}(xt)^k \sum_{m=0}^{\infty} \binom{-k-1}{m}(-1)^m t^m \end{aligned} \tag{5.34}$$

but since [see Equation (1.27) in Section 1.2.4]

$$\binom{-k-1}{m} = (-1)^m \binom{k+m}{m}$$

it follows that (5.34) becomes

$$(1-t)^{-1}\exp\left[-\dfrac{xt}{1-t}\right] = \sum_{m=0}^{\infty} \sum_{k=0}^{\infty} \dfrac{(-1)^k (k+m)! x^k}{(k!)^2 m!} t^{k+m} \tag{5.35}$$

where we have reversed the order of summation. Finally, the change of index $m = n - k$ leads to (5.33) where

$$L_n(x) = \sum_{k=0}^{n} \dfrac{(-1)^k n! x^k}{(k!)^2 (n-k)!} \tag{5.36}$$

In Table 5.2 we have listed the first few Laguerre polynomials $L_n(x)$.
The *Rodrigues formula* for the polynomials $L_n(x)$ is given by

$$L_n(x) = \dfrac{e^x}{n!} \dfrac{d^n}{dx^n}(x^n e^{-x}), \qquad n = 0,1,2,\ldots \tag{5.37}$$

Table 5.2 Laguerre polynomials

$L_0(x) = 1$
$L_1(x) = -x + 1$
$L_2(x) = \frac{1}{2!}(x^2 - 4x + 2)$
$L_3(x) = \frac{1}{3!}(-x^3 + 9x^2 - 18x + 6)$
$L_4(x) = \frac{1}{4!}(x^4 - 16x^3 + 72x^2 - 96x + 24)$

which can be verified by application of the Leibniz formula

$$\frac{d^n}{dx^n}(fg) = \sum_{k=0}^{n} \binom{n}{k} \frac{d^{n-k}f}{dx^{n-k}} \frac{d^k g}{dx^k}, \qquad n = 1, 2, 3, \ldots \qquad (5.38)$$

5.3.1 Recurrence Relations

It is easily verified that the generating function

$$w(x, t) = (1 - t)^{-1} \exp\left[-\frac{xt}{1-t}\right]$$

satisfies the identity

$$(1 - t)^2 \frac{\partial w}{\partial t} + (x - 1 + t)w = 0 \qquad (5.39)$$

By substituting the series (5.33) for $w(x, t)$ into (5.39), we find upon simplification that

$$\sum_{n=1}^{\infty} [(n + 1)L_{n+1}(x) + (x - 1 - 2n)L_n(x) + nL_{n-1}(x)]t^n = 0 \qquad (5.40)$$

Hence, equating the coefficient of t^n to zero, we obtain the recurrence formula

$$(n + 1)L_{n+1}(x) + (x - 1 - 2n)L_n(x) + nL_{n-1}(x) = 0 \qquad (5.41)$$

for $n = 1, 2, 3, \ldots$.

Similarly, substituting (5.33) into the identity

$$(1 - t)\frac{\partial w}{\partial x} + tw = 0 \qquad (5.42)$$

leads to the derivative relation

$$L'_n(x) - L'_{n-1}(x) + L_{n-1}(x) = 0 \qquad (5.43)$$

where $n = 1, 2, 3, \ldots$.

If we now differentiate (5.41), we obtain

$$(n+1)L'_{n+1}(x) + (x-1-2n)L'_n(x) + L_n(x) + nL'_{n-1}(x) = 0 \tag{5.44}$$

and by writing (5.43) in the equivalent forms

$$L'_{n+1}(x) = L'_n(x) - L_n(x) \tag{5.45a}$$
$$L'_{n-1}(x) = L'_n(x) + L_{n-1}(x) \tag{5.45b}$$

we can eliminate $L'_{n+1}(x)$ and $L'_{n-1}(x)$ from (5.44), which yields

$$xL'_n(x) = nL_n(x) - nL_{n-1}(x) \tag{5.46}$$

This last relation allows us to express the derivative of a Laguerre polynomial in terms of Laguerre polynomials.

To obtain the governing DE for the Laguerre polynomials, we begin by differentiating (5.46) and using (5.43) to get

$$xL''_n(x) + L'_n(x) = nL'_n(x) - nL'_{n-1}(x)$$
$$= -nL_{n-1}(x)$$

We can eliminate $L_{n-1}(x)$ by use of (5.46), which leads to

$$xL''_n(x) + (1-x)L'_n(x) + nL_n(x) = 0 \tag{5.47}$$

Hence we conclude that $y = L_n(x)$ $(n = 0, 1, 2, \ldots)$ is a solution of Laguerre's equation

$$xy'' + (1-x)y' + ny = 0 \tag{5.48}$$

5.3.2 Laguerre Series

Like the Legendre polynomials and Hermite polynomials, various functions satisfying rather general conditions can be expanded in a series of Laguerre polynomials. Fundamental to the theory of such series is the orthogonality property

$$\int_0^\infty e^{-x} L_n(x) L_k(x)\, dx = 0, \qquad k \neq n \tag{5.49}$$

Our proof of (5.49) will parallel that given for the Hermite polynomials. We begin by multiplying the two series

$$\sum_{n=0}^\infty L_n(x) t^n = (1-t)^{-1} \exp\left[-\frac{xt}{1-t}\right] \tag{5.50a}$$

$$\sum_{k=0}^\infty L_k(x) s^k = (1-s)^{-1} \exp\left[-\frac{xs}{1-s}\right] \tag{5.50b}$$

to obtain

$$\sum_{n=0}^{\infty}\sum_{k=0}^{\infty} t^n s^k L_n(x)L_k(x) = \frac{\exp\left[-x\left(\frac{t}{1-t} + \frac{s}{1-s}\right)\right]}{(1-t)(1-s)} \quad (5.51)$$

Next, multiplication of both sides of (5.51) by the weight function e^{-x} and subsequent integration leads to (see problem 29)

$$\sum_{n=0}^{\infty}\sum_{k=0}^{\infty} t^n s^k \int_0^{\infty} e^{-x} L_n(x)L_k(x)\, dx = (1-ts)^{-1}$$

$$= \sum_{n=0}^{\infty} t^n s^n \quad (5.52)$$

By comparing the coefficient of $t^n s^k$ on both sides of (5.52) we deduce the result (5.49), while for $k = n$, we also see that (for $n = 0, 1, 2, \ldots$)

$$\int_0^{\infty} e^{-x}[L_n(x)]^2\, dx = 1 \quad (5.53)$$

By a *Laguerre series*, we mean a series of the form

$$f(x) = \sum_{n=0}^{\infty} c_n L_n(x), \quad 0 < x < \infty \quad (5.54)$$

where

$$c_n = \int_0^{\infty} e^{-x} f(x) L_n(x)\, dx, \quad n = 0, 1, 2, \ldots \quad (5.55)$$

Without proof, we state the following theorem.

Theorem 5.2. If f is piecewise smooth in every finite interval $x_1 \le x \le x_2$, $0 < x_1 < x_2 < \infty$, and

$$\int_0^{\infty} e^{-x} f^2(x)\, dx < \infty$$

then the Laguerre series (5.54) with constants defined by (5.55) converges pointwise to $f(x)$ at every continuity point of f. At points of discontinuity, the series converges to the average value $\tfrac{1}{2}[f(x^+) + f(x^-)]$.

5.3.3 Associated Laguerre Polynomials

In many applications, particularly in quantum-mechanical problems, we need a generalization of the Laguerre polynomials called the *associated Laguerre polynomials*, i.e.,

$$L_n^{(m)}(x) = (-1)^m \frac{d^m}{dx^m}[L_{n+m}(x)], \quad m = 0, 1, 2, \ldots \quad (5.56)$$

By repeated differentiation of the series representation (5.36), it readily follows that (see problem 4)

$$L_n^{(m)}(x) = \sum_{k=0}^{n} \frac{(-1)^k (n+m)! x^k}{(n-k)!(m+k)!k!}, \qquad m = 0, 1, 2, \ldots \quad (5.57)$$

A generating function for the Laguerre polynomials $L_n^{(m)}(x)$ can be derived from that for $L_n(x)$. We first replace n by $n + m$ in (5.33) to get

$$(1-t)^{-1} \exp\left[-\frac{xt}{1-t}\right] = \sum_{n=-m}^{\infty} L_{n+m}(x) t^{n+m}$$

and then differentiate both sides m times with respect to x, i.e.,

$$(-1)^m t^m (1-t)^{-1-m} \exp\left[-\frac{xt}{1-t}\right] = \sum_{n=-m}^{\infty} \frac{d^m}{dx^m} [L_{n+m}(x)] t^{n+m}$$

The terms of the series for which $n = -1, -2, \ldots, -m$ are all zero, since the mth derivative of a polynomial of degree less than m is zero, and hence we deduce that

$$(1-t)^{-1-m} \exp\left[-\frac{xt}{1-t}\right] = \sum_{n=0}^{\infty} L_n^{(m)}(x) t^n, \qquad |t| < 1 \quad (5.58)$$

The associated Laguerre polynomials have many properties that are simple generalizations of those for the Laguerre polynomials. Among these are the recurrence relations*

$$(n+1) L_{n+1}^{(m)}(x)$$
$$+ (x - 1 - 2n - m) L_n^{(m)}(x) + (n+m) L_{n-1}^{(m)}(x) = 0 \quad (5.59)$$
$$x L_n^{(m)\prime}(x) - n L_n^{(m)}(x) + (n+m) L_{n-1}^{(m)}(x) = 0 \quad (5.60)$$

and the Rodrigues formula

$$L_n^{(m)}(x) = \frac{1}{n!} e^x x^{-m} \frac{d^n}{dx^n} (e^{-x} x^{n+m}) \quad (5.61)$$

The polynomials $L_n^{(m)}(x)$ also satisfy numerous relations where the upper index does not remain constant. Two such relations are given by

$$L_{n-1}^{(m)}(x) + L_n^{(m-1)}(x) - L_n^{(m)}(x) = 0 \quad (5.62)$$

and

$$L_n^{(m)\prime}(x) = -L_{n-1}^{(m+1)}(x) \quad (5.63)$$

---
*Note that for $m = 0$, (5.59) reduces to (5.41).

The second-order DE satisfied by the polynomials $L_n^{(m)}(x)$ is the *associated Laguerre's equation*

$$xy'' + (m + 1 - x)y' + ny = 0 \qquad (5.64)$$

To show this, we first note that the polynomial $z = L_{n+m}(x)$ is a solution of Laguerre's equation

$$xz'' + (1 - x)z' + (n + m)z = 0 \qquad (5.65)$$

By differentiating (5.65) m times, using the Leibniz rule (5.38), we obtain

$$x\frac{d^{m+2}z}{dx^{m+2}} + m\frac{d^{m+1}z}{dx^{m+1}} + (1-x)\frac{d^{m+1}z}{dx^{m+1}} + n\frac{d^m z}{dx^m} = 0$$

or equivalently,

$$x\frac{d^2}{dx^2}\left(\frac{d^m z}{dx^m}\right) + (m+1-x)\frac{d}{dx}\left(\frac{d^m z}{dx^m}\right) + n\left(\frac{d^m z}{dx^m}\right) = 0 \qquad (5.66)$$

Comparing (5.64) and (5.66), we see that any function $y = C_1(d^m z/dx^m)$ is a solution of (5.64) where C_1 is arbitrary. In particular, $y = L_n^{(m)}(x)$ is a solution.

Example 3: Prove the *addition formula*

$$L_n^{(a+b+1)}(x+y) = \sum_{k=0}^n L_k^{(a)}(x) L_{n-k}^{(b)}(y), \qquad a,b > -1$$

Solution: From the generating function (5.58), we have

$$\sum_{n=0}^\infty L_n^{(a+b+1)}(x+y) t^n = \frac{\exp[-(x+y)t/(1-t)]}{(1-t)^{a+b+2}}$$

$$= \frac{\exp[-xt/(1-t)]}{(1-t)^{a+1}} \cdot \frac{\exp[-yt/(1-t)]}{(1-t)^{b+1}}$$

$$= \sum_{k=0}^\infty L_k^{(a)}(x) t^k \cdot \sum_{m=0}^\infty L_m^{(b)}(y) t^m$$

$$= \sum_{m=0}^\infty \sum_{k=0}^\infty L_k^{(a)}(x) L_m^{(b)}(y) t^{m+k}$$

Next, making the change of index $m = n - k$ leads to

$$\sum_{n=0}^\infty L_n^{(a+b+1)}(x+y) t^n = \sum_{n=0}^\infty \sum_{k=0}^n L_k^{(a)}(x) L_{n-k}^{(b)}(y) t^n$$

and by comparing the coefficient of t^n in each series, we get our intended result.

Remark: The associated Laguerre polynomial $L_n^{(m)}(x)$ can be generalized to the case where m is not restricted to integer values by writing

$$L_n^{(a)}(x) = \sum_{k=0}^{n} \frac{(-1)^k \Gamma(n+a+1) x^k}{(n-k)!\Gamma(k+a+1) k!}, \qquad a > -1$$

Most of the above relations are also valid for this more general polynomial.

EXERCISES 5.3

1. Show that (for $n = 0, 1, 2, \ldots$)
   (a) $L_n(0) = 1$,
   (b) $L_n'(0) = -n$,
   (c) $L_n''(0) = \tfrac{1}{2} n(n-1)$.

2. Derive the Rodrigues formula
   (a) $L_n(x) = \dfrac{e^x}{n!} \dfrac{d^n}{dx^n}(x^n e^{-x})$,
   (b) $L_n^{(m)}(x) = \dfrac{1}{n!} x^{-m} e^x \dfrac{d^n}{dx^n}(x^{n+m} e^{-x})$.

   Hint: Use the Leibniz formula (5.38).

3. Derive the recurrence formulas
   (a) $L_n'(x) - L_{n-1}'(x) + L_{n-1}(x) = 0$.
   (b) $L_n'(x) = -\sum_{k=0}^{n-1} L_k(x)$.

4. By repeated differentiation of the series (5.36), show that

$$L_n^{(m)}(x) = \sum_{k=0}^{n} \frac{(-1)^k (m+n)! x^k}{(n-k)!(m+k)! k!}, \qquad m = 0, 1, 2, \ldots$$

5. Show that

$$n! \frac{d^k}{dx^k}\left[e^{-x} x^m L_n^{(m)}(x) \right] = (n+k)! e^{-x} x^{m-k} L_{n+k}^{(m-k)}(x)$$

6. Show that

$$L_n^{(m)}(0) = \frac{(n+m)!}{n!\, m!}$$

In problems 7–10, verify the given recurrence relation.

7. $(n+1)L_{n+1}^{(m)}(x) + (x-1-2n-m)L_n^{(m)}(x) + (n+m)L_{n-1}^{(m)}(x) = 0$.
8. $xL_n^{(m)'}(x) - nL_n^{(m)}(x) + (n+m)L_{n-1}^{(m)}(x) = 0$.
9. $L_{n-1}^{(m)}(x) + L_n^{(m-1)}(x) - L_n^{(m)}(x) = 0$.
10. $L_n^{(m)'}(x) = -L_{n-1}^{(m+1)}(x)$.

In problems 11–18, verify the integral formula.

11. $\int_0^\infty e^{-x} x^k L_n(x)\, dx = \begin{cases} 0, & k < n, \\ (-1)^n n!, & k = n. \end{cases}$

12. $\int_0^x L_k(t) L_n(x-t)\, dt = \int_0^x L_{n+k}(t)\, dt = L_{n+k}(x) - L_{n+k+1}(x)$.

13. $\int_x^\infty e^{-t} L_n^{(m)}(t)\, dt = e^{-x}[L_n^{(m)}(x) - L_{n-1}^{(m)}(x)]$, $m = 0, 1, 2, \ldots$.

14. $\int_0^x (x-t)^m L_n(t)\, dt = \dfrac{m!\, n!}{(m+n+1)!} x^{m+1} L_n^{(m+1)}(x)$, $m = 0, 1, 2, \ldots$.

15. $\int_0^1 t^a (1-t)^{b-1} L_n^{(a)}(xt)\, dt = \dfrac{\Gamma(b)\Gamma(n+a+1)}{\Gamma(n+a+b+1)} L_n^{(a+b)}(x)$, $a > -1$, $b > 0$.

16. $\int_0^\infty e^{-x} x^a L_n^{(a)}(x) L_k^{(a)}(x)\, dx = 0$, $k \neq n$, $a > -1$.

17. $\int_0^\infty e^{-x} x^a [L_n^{(a)}(x)]^2\, dx = \dfrac{\Gamma(n+a+1)}{n!}$, $a > -1$.

18. $\int_0^\infty e^{-x} x^{a+1} [L_n^{(a)}(x)]^2\, dx = \dfrac{\Gamma(n+a+1)}{n!}(2n+a+1)$, $a > -1$.

In problems 19–23, derive the given relation between the Hermite and Laguerre polynomials.

19. $L_n^{(-1/2)}(x) = \dfrac{(-1)^n}{2^{2n} n!} H_{2n}(\sqrt{x})$.

20. $L_n^{(1/2)}(x) = \dfrac{(-1)^n}{2^{2n+1} n! \sqrt{x}} H_{2n+1}(\sqrt{x})$.

21. $\int_0^\infty e^{-t^2} [H_n(t)]^2 \cos(\sqrt{2x}\, t)\, dt = \sqrt{\pi}\, 2^{n-1} n!\, e^{-x/2} L_n(x)$.

22. $\int_{-1}^1 (1-t^2)^{a-\frac{1}{2}} H_{2n}(\sqrt{x}\, t)\, dt = (-1)^n \sqrt{\pi}\, \dfrac{\Gamma(a+\frac{1}{2})(2n)!}{\Gamma(n+a+1)} L_n^{(a)}(x)$, $a > -\frac{1}{2}$.

23. $L_n(x^2 + y^2) = \dfrac{(-1)^n}{2^{2n}} \sum_{k=0}^n \dfrac{H_{2k}(x) H_{2n-2k}(y)}{k!(n-k)!}$.

184 • Special Functions for Engineers and Applied Mathematicians

In problems 24 and 25, derive the Laguerre series.

24. $x^p = p! \sum_{n=0}^{p} \binom{p}{n}(-1)^n L_n(x)$.

25. $e^{-ax} = (a+1)^{-1} \sum_{n=0}^{\infty} \left(\dfrac{a}{a+1}\right)^n L_n(x)$, $a > -\tfrac{1}{2}$.

   Hint: Set $t = a/(a+1)$ in the generating function.

26. Show that $(x > 0)$

$$\int_0^\infty \frac{e^{-xt}}{t+1}\, dt = \sum_{n=0}^{\infty} \frac{L_n(x)}{n+1}$$

   Hint: Use problem 25.

27. Show that $(x > 0)$

$$e^t(xt)^{-m/2} J_m(2\sqrt{xt}) = \sum_{n=0}^{\infty} \frac{L_n^{(m)}(x)}{(n+m)!} t^n, \qquad m = 0, 1, 2, \ldots$$

where $J_m(x)$ is the *Bessel function* defined by (see Chapter 6)

$$J_m(x) = \sum_{k=0}^{\infty} \frac{(-1)^k (x/2)^{2k+m}}{k!(k+m)!}$$

28. Show that for $m > 1$,

$$\int_0^\infty t^{n+m/2} J_m(2\sqrt{xt}) e^{-t}\, dt = n!\, e^{-x} x^{m/2} L_n^{(m)}(x)$$

   Hint: See problem 27.

29. Show that

$$\frac{1}{(1-t)(1-s)} \int_0^\infty \exp\left[-x\left(1 + \frac{t}{1-t} + \frac{s}{1-s}\right)\right] dx = \frac{1}{1-ts}$$

30. Show that the Laplace transform of $L_n(t)$ leads to

$$\int_0^\infty e^{-st} L_n(t)\, dt = \frac{1}{s}\left(1 - \frac{1}{s}\right)^n, \qquad s > 0$$

5.4 Generalized Polynomial Sets

The many properties that are shared by the Legendre, Hermite, and Laguerre polynomials suggest that there may exist more general polynomial sets of which these are certain specializations. Indeed, the *Gegenbauer* and *Jacobi polynomials* are two such generalizations. The Gegenbauer polynomials are closely connected with axially symmetric potentials in n dimensions and contain the Legendre, Hermite, and Chebyshev polynomials as special cases. The Jacobi polynomials are more general yet, as they contain the Gegenbauer polynomials as a special case.

5.4.1 Gegenbauer Polynomials

The *Gegenbauer polynomials** $C_n^\lambda(x)$ are defined by the generating function

$$(1 - 2xt + t^2)^{-\lambda} = \sum_{n=0}^{\infty} C_n^\lambda(x) t^n, \qquad |t| < 1, \quad |x| \le 1 \qquad (5.67)$$

where $\lambda > -\frac{1}{2}$. By expanding the function $w(x,t) = (1 - 2xt + t^2)^{-\lambda}$ in a binomial series, and following our approach in Section 4.2.1, we find

$$\begin{aligned} w(x,t) &= \sum_{n=0}^{\infty} \binom{-\lambda}{n} (-1)^n t^n (2x - t)^n \\ &= \sum_{n=0}^{\infty} \sum_{k=0}^{\infty} \binom{-\lambda}{n} \binom{n}{k} (-1)^{n+k} (2x)^{n-k} t^{n+k} \\ &= \sum_{n=0}^{\infty} \sum_{k=0}^{[n/2]} \binom{-\lambda}{n-k} \binom{n-k}{k} (-1)^n (2x)^{n-2k} t^n \end{aligned} \qquad (5.68)$$

and thus deduce that

$$C_n^\lambda(x) = (-1)^n \sum_{k=0}^{[n/2]} \binom{-\lambda}{n-k} \binom{n-k}{k} (2x)^{n-2k} \qquad (5.69)$$

By substituting the series (5.67) into the identity

$$(1 - 2xt + t^2) \frac{\partial w}{\partial t} + 2\lambda (t - x) w = 0 \qquad (5.70)$$

where $w(x,t) = (1 - 2xt + t^2)^{-\lambda}$, we obtain the three-term recurrence formula ($n = 1, 2, 3, \ldots$)

$$(n+1) C_{n+1}^\lambda(x) - 2(\lambda + n) x C_n^\lambda(x) + (2\lambda + n - 1) C_{n-1}^\lambda(x) = 0 \qquad (5.71)$$

Other recurrence formulas satisfied by the Gegenbauer polynomials include

---

*The polynomials $C_n^\lambda(x)$ are also called *ultraspherical polynomials*.

the following:

$$(n+1)C_{n+1}^\lambda(x) - 2\lambda x C_n^{\lambda+1}(x) + 2\lambda C_{n-1}^{\lambda+1}(x) = 0 \quad (5.72)$$

$$(n+2\lambda)C_n^\lambda(x) - 2\lambda C_n^{\lambda+1}(x) + 2\lambda x C_{n-1}^{\lambda+1}(x) = 0 \quad (5.73)$$

$$C_n^{\lambda\prime}(x) = 2\lambda C_{n+1}^{\lambda+1}(x) \quad (5.74)$$

The orthogonality property is given by (see problem 13)

$$\int_{-1}^{1} (1-x^2)^{\lambda-\frac{1}{2}} C_n^\lambda(x) C_k^\lambda(x)\, dx = 0, \qquad k \ne n \quad (5.75)$$

and the governing DE is

$$(1-x^2)y'' - (2\lambda+1)xy' + n(n+2\lambda)y = 0 \quad (5.76)$$

which can be verified by substituting the series (5.69) directly into (5.76).

One of the main advantages of developing properties of the Gegenbauer polynomials is that each recurrence formula, etc., becomes a master formula for all the polynomial sets that are generated as special cases. For example, when $\lambda = \frac{1}{2}$ we see that (5.67) is the generating function for the Legendre polynomials, and thus

$$P_n(x) = C_n^{1/2}(x), \qquad n = 0,1,2,\ldots \quad (5.77)$$

By setting $\lambda = \frac{1}{2}$ in (5.71), (5.75), and (5.76), we immediately obtain the recurrence formula, orthogonality property, and governing DE, respectively, for the Legendre polynomials.

The Hermite polynomials can also be generated from the Gegenbauer polynomials through the limit relation

$$H_n(x) = n! \lim_{\lambda \to \infty} \lambda^{-n/2} C_n^\lambda(x/\sqrt{\lambda}), \qquad n = 0,1,2,\ldots \quad (5.78)$$

To show this, we start with the series representation

$$\lambda^{-n/2} C_n^\lambda(x/\sqrt{\lambda}) = (-1)^n \sum_{k=0}^{[n/2]} \binom{-\lambda}{n-k} \binom{n-k}{k} \frac{(2x)^{n-2k}}{\lambda^{n-k}} \quad (5.79)$$

From Equation (1.27) in Section 1.2.4, we obtain the relation

$$\frac{(-1)^n}{\lambda^{n-k}} \binom{-\lambda}{n-k} = \frac{(-1)^k}{\lambda^{n-k}} \binom{\lambda+n-k-1}{n-k}$$

$$= \frac{(-1)^k \Gamma(\lambda+n-k)}{\lambda^{n-k} \Gamma(\lambda)(n-k)!}$$

and thus establish that (see problem 3)

$$\lim_{\lambda \to \infty} \frac{(-1)^n}{\lambda^{n-k}} \binom{-\lambda}{n-k} = \frac{(-1)^k}{(n-k)!} \qquad (5.80)$$

Hence, from (5.79) we now deduce our intended result

$$n! \lim_{\lambda \to \infty} \lambda^{-n/2} C_n^\lambda(x/\sqrt{\lambda}) = \sum_{k=0}^{[n/2]} \frac{(-1)^k n!}{k!(n-2k)!} (2x)^{n-2k}$$
$$= H_n(x)$$

Properties of the Hermite polynomials can be obtained from properties of the Gegenbauer polynomials, although most such relations are more difficult to deduce than for the Legendre polynomials.

5.4.2 Chebyshev Polynomials*

An important subclass of Gegenbauer polynomials are the Chebyshev polynomials, of which there are two kinds. The *Chebyshev polynomials of the first kind* are defined by

$$T_0(x) = 1, \qquad T_n(x) = \frac{n}{2} \lim_{\lambda \to 0} \frac{C_n^\lambda(x)}{\lambda}, \qquad n = 1, 2, 3, \ldots \qquad (5.81)$$

Because the Gegenbauer polynomials vanish when $\lambda = 0$, we cannot just simply define the polynomials $T_n(x)$ by $C_n^0(x)$. The choice $T_0(x) = 1$ is made to preserve the recurrence relation (5.85) given below. By following a procedure similar to that used to verify the relation (5.78), it can be established that (see problem 15)

$$T_n(x) = \frac{n}{2} \sum_{k=0}^{[n/2]} \frac{(-1)^k (n-k-1)!}{k!(n-2k)!} (2x)^{n-2k} \qquad (5.82)$$

The *Chebyshev polynomials of the second kind* are simply†

$$U_n(x) = C_n^1(x), \qquad n = 0, 1, 2, \ldots \qquad (5.83)$$

and thus by setting $\lambda = 1$ in (5.69) we immediately deduce that

$$U_n(x) = \sum_{k=0}^{[n/2]} \binom{n-k}{k} (-1)^k (2x)^{n-2k} \qquad (5.84)$$

---

*There are numerous spellings of Chebyshev that occur throughout the literature, e.g., Tchebysheff, Tchebycheff, Tchebichef, and Chebysheff, among others.

†Some authors call $(1-x^2)^{1/2} U_n(x)$ the Chebyshev functions of the second kind.

By using properties previously cited for the Gegenbauer polynomials, we readily obtain the recurrence formulas

$$T_{n+1}(x) - 2xT_n(x) + T_{n-1}(x) = 0 \tag{5.85}$$
$$U_{n+1}(x) - 2xU_n(x) + U_{n-1}(x) = 0 \tag{5.86}$$

orthogonality properties

$$\int_{-1}^{1} (1-x^2)^{-1/2} T_n(x) T_k(x)\, dx = 0, \quad k \neq n \tag{5.87}$$
$$\int_{-1}^{1} (1-x^2)^{1/2} U_n(x) U_k(x)\, dx = 0, \quad k \neq n \tag{5.88}$$

and governing DE for $T_n(x)$,

$$(1-x^2)y'' - xy' + n^2 y = 0 \tag{5.89}$$

and for $U_n(x)$,

$$(1-x^2)y'' - 3xy' + n(n+2)y = 0 \tag{5.90}$$

There are also several recurrence-type formulas connecting the polynomials $T_n(x)$ and $U_n(x)$, such as

$$T_n(x) = U_n(x) - xU_{n-1}(x) \tag{5.91}$$

and

$$(1-x^2)U_n(x) = xT_n(x) - T_{n+1}(x) \tag{5.92}$$

the proofs of which are left for the exercises.

By making the substitution $x = \cos\phi$ in (5.89), we find it reduces to

$$\frac{d^2 y}{d\phi^2} + n^2 y = 0$$

with solutions $\cos n\phi$ and $\sin n\phi$. Thus we speculate that

$$T_n(\cos\phi) = c_n \cos n\phi$$

for some constant c_n. But since $T_n(1) = 1$ for all n (see problem 26), it follows that $c_n = 1$ for all n. It turns out that this speculation is correct, and in general we write

$$T_n(x) = \cos n\phi = \cos(n \cos^{-1} x) \tag{5.93}$$

Similarly, it can be shown that

$$U_n(x) = \frac{\sin\left[(n+1)\cos^{-1} x\right]}{\sqrt{1-x^2}} \tag{5.94}$$

The significance of these observations is that the properties of sines and cosines can be used to establish many of the properties of the Chebyshev polynomials.

The Chebyshev polynomials have acquired great practical importance in polynomial approximation methods. Specifically, it has been shown that a series of Chebyshev polynomials converges more rapidly than any other series of Gegenbauer polynomials, and converges much more rapidly than power series.*

5.4.3 Jacobi Polynomials

The *Jacobi polynomials*, which are generalizations of the Gegenbauer polynomials, are defined by the generating function

$$\frac{2^{a+b}}{R}(1 - t + R)^{-a}(1 + t + R)^{-b} = \sum_{n=0}^{\infty} P_n^{(a,b)}(x) t^n,$$

$$a > -1, \quad b > -1 \quad (5.95)$$

where

$$R = (1 - 2xt + t^2)^{1/2} \quad (5.96)$$

The Jacobi polynomials have the following three series representations (among others), which are somewhat involved to derive:

$$P_n^{(a,b)}(x) = \sum_{k=0}^{n} \binom{n+a}{n-k}\binom{n+b}{n-k}\left(\frac{x-1}{2}\right)^k \left(\frac{x+1}{2}\right)^{n-k} \quad (5.97)$$

$$P_n^{(a,b)}(x) = \sum_{k=0}^{n} \binom{n+a}{n-k}\binom{n+k+a+b}{k}\left(\frac{x-1}{2}\right)^k \quad (5.98)$$

$$P_n^{(a,b)}(x) = \sum_{k=0}^{n} (-1)^{n-k}\binom{n+b}{n-k}\binom{n+k+a+b}{k}\left(\frac{x+1}{2}\right)^k \quad (5.99)$$

By examination of the generating function (5.95), we observe that the Legendre polynomials are a specialization of the Jacobi polynomials for which $a = b = 0$, i.e.,

$$P_n(x) = P_n^{(0,0)}(x), \quad n = 0, 1, 2, \ldots \quad (5.100)$$

---

*For theory and applications involving the Chebyshev polynomials, see L. Fox and I.B. Parker, *Chebyshev Polynomials in Numerical Analysis*, London: Oxford U.P., 1968.

whereas the associated Laguerre polynomials arise as the limit (see problem 37)

$$L_n^{(a)}(x) = \lim_{b \to \infty} P_n^{(a,b)}(1 - 2x/b), \qquad n = 0, 1, 2, \ldots \qquad (5.101)$$

In addition to the Legendre and Laguerre polynomials, the Gegenbauer polynomials are also a special case of the Jacobi polynomials. To derive the relation between the Gegenbauer and Jacobi polynomials, we start with the identity

$$(1 - 2xt + t^2)^{-\lambda} = (1 - t)^{-2\lambda} \left[1 - \frac{2t(x - 1)}{(1 - t)^2} \right]^{-\lambda} \qquad (5.102)$$

and expand the right-hand side in a series. This action leads to

$$(1 - 2xt + t^2)^{-\lambda} = \sum_{k=0}^{\infty} \binom{-\lambda}{k} \frac{(-1)^k (2t)^k (x - 1)^k}{(1 - t)^{2(k + \lambda)}}$$

$$= \sum_{m=0}^{\infty} \sum_{k=0}^{\infty} \binom{-\lambda}{k} \binom{-2k - 2\lambda}{m} (-1)^{m+k} 2^k (x - 1)^k t^{m+k}$$

$$(5.103)$$

where we have expanded $(1 - t)^{-2(k+\lambda)}$ in another binomial series and interchanged the order of summation. Next, replacing the left-hand side of (5.103) by the series (5.67) and making the change of index $m = n - k$, we get

$$\sum_{n=0}^{\infty} C_n^{\lambda}(x) t^n = \sum_{n=0}^{\infty} \sum_{k=0}^{n} \binom{-\lambda}{k} \binom{-2k - 2\lambda}{n - k} (-1)^n 2^k (x - 1)^k t^n$$

from which we deduce

$$C_n^{\lambda}(x) = (-1)^n \sum_{k=0}^{n} \binom{-\lambda}{k} \binom{-2k - 2\lambda}{n - k} 2^k (x - 1)^k \qquad (5.104)$$

Recalling Equation (1.27) in Section 1.2.4 and the Legendre duplication formula, we see that

$$(-1)^n \binom{-\lambda}{k} \binom{-2k - 2\lambda}{n - k} = \binom{\lambda + k - 1}{k} \binom{n + k + 2\lambda - 1}{n - k}$$

$$= \frac{\Gamma(\lambda + k) \Gamma(n + k + 2\lambda)}{\Gamma(\lambda) k! \Gamma(2\lambda + 2k)(n - k)!}$$

$$= \frac{\Gamma(\lambda + \tfrac{1}{2}) \Gamma(n + k + 2\lambda)}{\Gamma(2\lambda) \Gamma(\lambda + k + \tfrac{1}{2}) k! (n - k)! 2^{2k}}$$

and hence (5.104) can be expressed in the form

$$C_n^\lambda(x) = \frac{\Gamma(\lambda + \tfrac{1}{2})\Gamma(n + 2\lambda)}{\Gamma(2\lambda)\Gamma(n + \lambda + \tfrac{1}{2})}$$
$$\times \sum_{k=0}^{n} \binom{n + \lambda - \tfrac{1}{2}}{n - k}\binom{n + k + 2\lambda - 1}{k}\left(\frac{x - 1}{2}\right)^k \quad (5.105)$$

or, by comparing with (5.98),

$$C_n^\lambda(x) = \frac{\Gamma(\lambda + \tfrac{1}{2})\Gamma(n + 2\lambda)}{\Gamma(2\lambda)\Gamma(n + \lambda + \tfrac{1}{2})} P_n^{(\lambda - \tfrac{1}{2}, \lambda - \tfrac{1}{2})}(x) \quad (5.106)$$

The basic recurrence formula for the polynomials $P_n^{(a,b)}(x)$ is

$$2(n + 1)(a + b + n + 1)(a + b + 2n)P_{n+1}^{(a,b)}(x) = (a + b + 2n + 1)$$
$$\times \left[a^2 - b^2 + x(a + b + 2n + 2)(a + b + 2n)\right]P_n^{(a,b)}(x)$$
$$- 2(a + n)(b + n)(a + b + 2n + 2)P_{n-1}^{(a,b)}(x) \quad (5.107)$$

for $n = 1, 2, 3, \ldots$. Also, the orthogonality property and governing DE are given respectively by

$$\int_{-1}^{1} (1 - x)^a (1 + x)^b P_n^{(a,b)}(x) P_k^{(a,b)}(x)\, dx = 0, \quad k \neq n \quad (5.108)$$

and

$$(1 - x^2)y'' + [b - a - (a + b + 2)x]y' + n(n + a + b + 1)y = 0 \quad (5.109)$$

Some additional properties concerning the Jacobi polynomials are taken up in the exercises.

EXERCISES 5.4

1. Show that (for $n = 0, 1, 2, \ldots$)

$$C_n^\lambda(-x) = (-1)^n C_n^\lambda(x)$$

2. Show that (for $n = 0, 1, 2, \ldots$)
   (a) $C_{2n}^\lambda(0) = \binom{-\lambda}{n}$,
   (b) $C_{2n+1}^\lambda(0) = 0$,
   (c) $C_n^\lambda(1) = (-1)^n \binom{-2\lambda}{n}$,
   (d) $C_n^\lambda(-1) = \binom{-2\lambda}{n}$.

3. Show that
$$\lim_{\lambda \to \infty} \frac{\Gamma(\lambda + n - k)}{\lambda^{n-k} \Gamma(\lambda)} = 1$$

In problems 4–8, derive the given recurrence relation.

4. $xC_n^{\lambda'}(x) = nC_n^{\lambda}(x) + C_{n-1}^{\lambda'}(x)$.
5. $2(\lambda + n)C_n^{\lambda}(x) = C_{n+1}^{\lambda'}(x) - C_{n-1}^{\lambda'}(x)$.
6. $xC_n^{\lambda'}(x) = C_{n+1}^{\lambda'}(x) - (2\lambda + n)C_n^{\lambda}(x)$.
7. $(x^2 - 1)C_n^{\lambda'}(x) = nxC_n^{\lambda}(x) - (2\lambda - 1 + n)C_{n-1}^{\lambda}(x)$.
8. $nC_n^{\lambda}(x) = 2x(\lambda + n - 1)C_{n-1}^{\lambda}(x) - (2\lambda + n - 2)C_{n-2}^{\lambda}(x)$.

9. Use any of the results of problems 4–8 and the recurrence formula (5.71) to show that $y = C_n^{\lambda}(x)$ is a solution of
$$(1 - x^2)y'' - (2\lambda + 1)xy' + n(n + 2\lambda)y = 0$$

10. Show that (for $k = 1, 2, 3, \ldots$)
$$\frac{d^k}{dx^k} C_n^{\lambda}(x) = 2^k \frac{\Gamma(\lambda + k)}{\Gamma(\lambda)} C_{n-k}^{\lambda+k}(x)$$

Hint: Use Equation (5.74).

11. Verify that (for $k = 1, 2, 3, \ldots$)*
$$C_{n-k}^{k+\frac{1}{2}}(x) = \frac{1}{(2k - 1)!!} \frac{d^k}{dx^k} P_n(x)$$

12. Derive the recurrence relation
$$\sum_{k=0}^{n} (n + \lambda)C_k^{\lambda}(x) = \frac{(n + 2\lambda)C_n^{\lambda}(x) - (n + 1)C_{n+1}^{\lambda}(x)}{2(1 - x)}$$

13. Verify the orthogonality property
$$\int_{-1}^{1} (1 - x^2)^{\lambda - \frac{1}{2}} C_n^{\lambda}(x) C_k^{\lambda}(x)\, dx = 0, \qquad k \neq n$$

14. Show that (for $n = 0, 1, 2, \ldots$)
$$\int_{-1}^{1} (1 - x^2)^{\lambda - \frac{1}{2}} [C_n^{\lambda}(x)]^2\, dx = \frac{2^{1-2\lambda}\pi}{(n + \lambda)} \frac{\Gamma(n + 2\lambda)}{[\Gamma(\lambda)]^2 n!}$$

---

*See problem 15 in Exercises 2.2 for definition of the symbol !!.

15. By using Equation (5.69) and the definition

$$T_n(x) = \frac{n}{2} \lim_{\lambda \to 0} \frac{C_n^\lambda(x)}{\lambda}, \qquad n = 1, 2, 3, \ldots$$

show that

$$T_n(x) = \frac{n}{2} \sum_{k=0}^{[n/2]} \frac{(-1)^k (n-k-1)!}{k!(n-2k)!} (2x)^{n-2k}$$

16. Using the recurrence formula (5.71), deduce the relations

(a) $T_{n+1}(x) - 2xT_n(x) + T_{n-1}(x) = 0$.
(b) $U_{n+1}(x) - 2xU_n(x) + U_{n-1}(x) = 0$.

In problems 17–22, derive the given relation for the Chebyshev polynomials.

17. $T_n(x) = U_n(x) - xU_{n-1}(x)$.

18. $(1 - x^2)U_n(x) = xT_n(x) - T_{n+1}(x)$.

19. $T_n'(x) = nU_{n-1}(x)$.

20. $2[T_n(x)]^2 = 1 + T_{2n}(x)$.

21. $[T_n(x)]^2 - T_{n+1}(x)T_{n-1}(x) = 1 - x^2$.

22. $[U_n(x)]^2 - U_{n+1}(x)U_{n-1}(x) = 1$.

23. By making the substitution $x = \cos\phi$ in the orthogonality relation (5.87), show that

$$\int_0^\pi \cos n\phi \cos k\phi \, d\phi = 0, \qquad k \neq n$$

In problems 24 and 25, derive the generating-function relation.

24. $\dfrac{1 - t^2}{1 - 2xt + t^2} = T_0(x) + 2 \sum_{n=1}^\infty T_n(x) t^n$.

25. $\dfrac{1 - xt}{1 - 2xt + t^2} = \sum_{n=0}^\infty T_n(x) t^n$.

26. Show that $T_n(1) = 1$, $n = 0, 1, 2, \ldots$, by using

(a) problem 24,
(b) problem 25.

27. Verify the special values (for $n = 0, 1, 2, \ldots$)

(a) $T_n(-1) = (-1)^n$,
(b) $T_{2n}(0) = (-1)^n$,
(c) $T_{2n+1}(0) = 0$.

28. Verify the special values (for $n = 0, 1, 2, \ldots$)
   (a) $U_n(1) = n + 1$,
   (b) $U_{2n}(0) = (-1)^n$,
   (c) $U_{2n+1}(0) = 0$.

29. Show that
$$\int_{-1}^{1} (1-x^2)^{-1/2} [T_n(x)]^2 \, dx = \begin{cases} \pi, & n = 0 \\ \dfrac{\pi}{2}, & n \geq 1 \end{cases}$$

30. Show that
$$\int_{-1}^{1} (1-x^2)^{1/2} [U_n(x)]^2 \, dx = \frac{\pi}{2}$$

In problems 31–38, verify the given relation for the Jacobi polynomials.

31. $P_n^{(a,b)}(-x) = (-1)^n P_n^{(b,a)}(x)$.

32. $P_n^{(a,b)}(1) = \binom{a+n+1}{n}$.

33. $P_n^{(a,b)}(-1) = (-1)^n \binom{b+n+1}{n}$.

34. $P_n^{(a,b)}(x) = \dfrac{(-1)^n}{2^n n!} (1-x)^{-a}(1+x)^{-b} \dfrac{d^n}{dx^n}[(1-x)^{a+n}(1+x)^{b+n}]$.

35. $\dfrac{d^k}{dx^k} P_n^{(a,b)}(x) = \dfrac{\Gamma(k+n+a+b+1)}{2^k \Gamma(n+a+b+1)} P_{n-k}^{(a+k,b+k)}(x)$.

36. $P_n^{(a,b-1)}(x) - P_n^{(a-1,b)}(x) = P_{n-1}^{(a,b)}(x)$.

37. $L_n^{(a)}(x) = \lim_{b \to \infty} P_n^{(a,b)}(1 - 2x/b)$.

38. $T_n(x) = \dfrac{2^{2n}(n!)^2}{(2n)!} P_n^{(-\frac{1}{2}, -\frac{1}{2})}(x)$.

6

Bessel Functions

6.1 Introduction

The German astronomer F.W. Bessel (1784–1846) first achieved fame by computing the orbit of Halley's comet. In addition to many other accomplishments in connection with his studies of planetary motion, he is credited with deriving the differential equation bearing his name.* It is known, however, that Bessel's equation was first investigated in 1703 by J. Bernoulli, who was studying the oscillatory behavior of a hanging chain. In fact, Bernoulli solved Bessel's equation by an infinite series that now defines the *Bessel function of the first kind*. Bessel functions were also met with by Euler and others who were concerned with various problems in mechanics. Nonetheless, it was Bessel in 1824 who carried out the first systematic study of the properties of these functions, and thus they are named in his honor.

Bessel functions are closely associated with problems possessing circular or cylindrical symmetry. For example, they arise in the study of free vibrations of a circular membrane and in finding the temperature distribution in a circular cylinder. They also occur in electromagnetic theory and numerous other areas of physics and engineering. In fact, Bessel functions occur so frequently in practice that they are undoubtedly the most important functions beyond the elementary ones.

---

*A short historical account of Bessel's problem of planetary motion is given in N.W. McLachlan, *Bessel Functions for Engineers*, London: Oxford, 1961, Chapter 1.

196 • Special Functions for Engineers and Applied Mathematicians

Because of their close association with cylindrical-shaped domains, all solutions of Bessel's equation are collectively called *cylinder functions*. The Bessel functions, of which there are several varieties, are certain special cases of cylinder functions. In addition to Bessel functions of the first kind, there are Bessel functions of the second and third kinds, modified Bessel functions of the first and second kinds, spherical Bessel functions, and so on.

6.2 Bessel Functions of the First Kind

Although Bessel functions arise in practice most frequently as solutions of certain DEs, it is both instructive and convenient to develop them from the same point of view that we adopted in introducing the orthogonal polynomials in Chapters 4 and 5, viz., by a generating function.

6.2.1 The Generating Function

By expanding the function

$$w(x,t) = \exp\left[\tfrac{1}{2}x\left(t - \frac{1}{t}\right)\right], \qquad t \neq 0 \tag{6.1}$$

in a series involving both positive and negative powers of t, we wish to establish the relation

$$w(x,t) = \sum_{n=-\infty}^{\infty} J_n(x) t^n \tag{6.2}$$

where $J_n(x)$ denotes the Bessel function we want to define.

To begin, we write $w(x,t)$ as the product of two exponential functions and expand each in a Maclaurin series to get

$$w(x,t) = e^{xt/2} \cdot e^{-x/2t}$$

$$= \sum_{j=0}^{\infty} \frac{(xt/2)^j}{j!} \cdot \sum_{k=0}^{\infty} \frac{(-x/2t)^k}{k!}$$

$$= \sum_{j=0}^{\infty} \sum_{k=0}^{\infty} \frac{(-1)^k (x/2)^{j+k}}{j! k!} t^{j-k}$$

We now make the change of index $n = j - k$. Because of the range of values on j and k, it follows that $-\infty < n < \infty$, and thus

$$w(x,t) = \sum_{n=-\infty}^{\infty} \sum_{k=0}^{\infty} \frac{(-1)^k (x/2)^{2k+n}}{k!(k+n)!} t^n \tag{6.3}$$

By defining the *Bessel function of the first kind* of order n by the series

$$J_n(x) = \sum_{k=0}^{\infty} \frac{(-1)^k (x/2)^{2k+n}}{k!(k+n)!}, \qquad -\infty < x < \infty \qquad (6.4)$$

we see that (6.3) leads to the desired generating-function relation

$$\exp\left[\tfrac{1}{2}x\left(t - \frac{1}{t}\right)\right] = \sum_{n=-\infty}^{\infty} J_n(x) t^n, \qquad t \neq 0 \qquad (6.5)$$

Since (6.5) involves both positive and negative values of n, we may wish to investigate the definition of $J_n(x)$ specifically when $n < 0$. The formal replacement of n with $-n$ in (6.4) yields

$$J_{-n}(x) = \sum_{k=0}^{\infty} \frac{(-1)^k (x/2)^{2k-n}}{k!(k-n)!}$$

$$= \sum_{k=n}^{\infty} \frac{(-1)^k (x/2)^{2k-n}}{k!(k-n)!}$$

where we have used the fact that $1/(k-n)! = 0$ ($k = 0, 1, \ldots, n-1$) by virtue of Theorem 2.1. Finally, the change of index $k = m + n$ gives us

$$J_{-n}(x) = \sum_{m=0}^{\infty} \frac{(-1)^{m+n} (x/2)^{2m+n}}{m!(m+n)!} \qquad (6.6)$$

from which it follows that

$$J_{-n}(x) = (-1)^n J_n(x), \qquad n = 0, 1, 2, \ldots \qquad (6.7)$$

Graphs of $J_n(x)$ for certain values of n are provided in Fig. 6.1. Observe that only $J_0(x)$ is nonzero when $x = 0$. To prove this, we simply set $x = 0$ in the generating-function relation (6.5) to get

$$1 = \sum_{n=-\infty}^{\infty} J_n(0) t^n$$

and by comparing like terms we deduce the results

$$J_0(0) = 1, \qquad J_n(0) = 0, \quad n \neq 0 \qquad (6.8)$$

6.2.2 Bessel Functions of Nonintegral Order

Thus far we have only discussed Bessel functions of integral order. We can generalize the series definition [Equation (6.4)] of the Bessel function $J_n(x)$

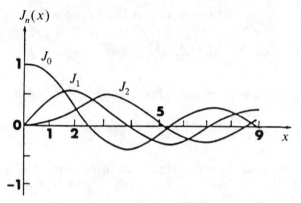

Figure 6.1 Graph of $J_n(x)$, $n = 0, 1, 2$

to include nonintegral values of n by replacing $(k + n)!$ with its gamma function equivalent. Hence, if p is any real number for which $p \geq 0$, we then define

$$J_p(x) = \sum_{k=0}^{\infty} \frac{(-1)^k (x/2)^{2k+p}}{k!\Gamma(k+p+1)} \qquad (6.9)$$

as the Bessel function of the first kind of order p.

The formal replacement of p with $-p$ in (6.9) yields

$$J_{-p}(x) = \sum_{k=0}^{\infty} \frac{(-1)^k (x/2)^{2k-p}}{k!\Gamma(k-p+1)} \qquad (6.10)$$

which for $p \neq 0, 1, 2, \ldots$ is not a multiple of $J_p(x)$. That is, since $J_{-p}(x)$ becomes infinite at $x = 0$ while $J_p(x)$ remains finite, the two functions are not proportional, and hence are linearly independent for nonintegral values of p. The ramifications of this observation will become clear in Section 6.5.

Although $J_p(x)$ and $J_{-p}(x)$ do not satisfy any generating-function relation, they are completely defined by their series representations and share most of the properties of $J_n(x)$ and $J_{-n}(x)$.

6.2.3 Recurrence Relations

There are many recurrence relations connecting the Bessel functions, analogous to those for the orthogonal polynomials. For example, suppose we multiply the series for $J_p(x)$ by x^p and then differentiate with respect to x.

This gives us

$$\frac{d}{dx}\left[x^p J_p(x)\right] = \frac{d}{dx} \sum_{k=0}^{\infty} \frac{(-1)^k x^{2k+2p}}{2^{2k+p} k! \Gamma(k+p+1)}$$

$$= \sum_{k=0}^{\infty} \frac{(-1)^k 2(k+p) x^{2k+2p-1}}{2^{2k+p} k! \Gamma(k+p+1)}$$

$$= x^p \sum_{k=0}^{\infty} \frac{(-1)^k (x/2)^{2k+(p-1)}}{k! \Gamma(k+p)} \quad (6.11)$$

or

$$\frac{d}{dx}\left[x^p J_p(x)\right] = x^p J_{p-1}(x) \quad (6.12)$$

Similarly, if we multiply $J_p(x)$ by x^{-p}, we find that (problem 14)

$$\frac{d}{dx}\left[x^{-p} J_p(x)\right] = -x^{-p} J_{p+1}(x) \quad (6.13)$$

If we carry out the differentiation in (6.12) and (6.13), and divide the results by the factors x^p and x^{-p}, respectively, we deduce that

$$J_p'(x) + \frac{p}{x} J_p(x) = J_{p-1}(x) \quad (6.14)$$

and

$$J_p'(x) - \frac{p}{x} J_p(x) = -J_{p+1}(x) \quad (6.15)$$

The substitution of $p = 0$ in (6.15) leads to the special result

$$J_0'(x) = -J_1(x) \quad (6.16)$$

Finally, the sum of (6.14) and (6.15) yields the relation

$$2J_p'(x) = J_{p-1}(x) - J_{p+1}(x) \quad (6.17)$$

whereas the difference of (6.14) and (6.15) gives us

$$\frac{2p}{x} J_p(x) = J_{p-1}(x) + J_{p+1}(x) \quad (6.18)$$

This last relation is the *three-term recurrence formula* for the Bessel functions.

Repeated application of the above recurrence relations can lead to the additional results*

$$\left(\frac{d}{x\,dx}\right)^m [x^p J_p(x)] = x^{p-m} J_{p-m}(x) \qquad (6.19)$$

and

$$\left(\frac{d}{x\,dx}\right)^m [x^{-p} J_p(x)] = (-1)^m x^{-p-m} J_{p+m}(x) \qquad (6.20)$$

where $m = 1, 2, 3, \ldots$.

6.2.4 Bessel's Differential Equation

By using the above recurrence formulas, we can derive a derivative relation involving only the Bessel function $J_p(x)$. To start, we rewrite Equation (6.14) in the form

$$x J_p'(x) - x J_{p-1}(x) + p J_p(x) = 0 \qquad (6.21)$$

and differentiate to find

$$x J_p''(x) + (p+1) J_p'(x) - x J_{p-1}'(x) - J_{p-1}(x) = 0 \qquad (6.22)$$

Multiplying (6.22) by x and subtracting (6.21) multiplied by p yields

$$x^2 J_p''(x) + x J_p'(x) - p^2 J_p(x) + (p-1) x J_{p-1}(x) - x^2 J_{p-1}'(x) = 0 \qquad (6.23)$$

Now if we rewrite Equation (6.15) in the form

$$x J_{p-1}'(x) = (p-1) J_{p-1}(x) - x J_p(x)$$

and use it to eliminate $J_{p-1}'(x)$ and $J_{p-1}(x)$ from (6.23), we obtain

$$x^2 J_p''(x) + x J_p'(x) + (x^2 - p^2) J_p(x) = 0 \qquad (6.24)$$

Hence, we deduce that the Bessel function $J_p(x)$ is a solution of the second-order linear DE†

$$x^2 y'' + x y' + (x^2 - p^2) y = 0 \qquad (6.25)$$

Equation (6.25) is called *Bessel's equation*. Among other areas of application, it arises in the solution of various partial differential equations

---

*We interpret $\left(\dfrac{d}{x\,dx}\right)^2 y = \dfrac{1}{x}\dfrac{d}{dx}\left(\dfrac{1}{x}\dfrac{dy}{dx}\right)$, and so on.

†Since only p^2 appears in (6.25), it is customary to make the assumption that $p \geq 0$.

of mathematical physics, particularly those problems displaying either circular or cylindrical symmetry (see Section 7.3).

EXERCISES 6.2

1. Show that the generating-function relation (6.5) can also be written in the form ($t \neq 0$)

$$\exp\left[\tfrac{1}{2}x\left(t - \frac{1}{t}\right)\right] = J_0(x) + \sum_{n=1}^{\infty} J_n(x)[t^n + (-1)^n t^{-n}]$$

2. Show that $J_n(x)$ is an even function for even n and an odd function for odd n, i.e.,

$$J_n(-x) = (-1)^n J_n(x), \qquad n = 0, \pm 1, \pm 2, \ldots$$

   (a) by using the generating function (6.5),
   (b) by using the series representation (6.4).

3. By using the series representation (6.4), show that

   (a) $J_1'(0) = \tfrac{1}{2}$, (b) $J_n'(0) = 0$ for $n > 1$.

4. For $w(x, t) = \exp[\tfrac{1}{2}x(t - 1/t)]$,

   (a) show that $w(x + y, t) = w(x, t)w(y, t)$.
   (b) From (a), deduce the *addition theorem*

$$J_n(x + y) = \sum_{k=-\infty}^{\infty} J_k(x) J_{n-k}(y)$$

   (c) From (b), derive the result

$$J_0(2x) = [J_0(x)]^2 + 2\sum_{k=1}^{\infty} (-1)^k [J_k(x)]^2$$

5. Given the generating function $w(x, t) = \exp[\tfrac{1}{2}x(t - 1/t)]$,

   (a) show that it satisfies the identity

$$\frac{\partial w}{\partial t} - \tfrac{1}{2}x\left(1 + \frac{1}{t^2}\right)w = 0$$

   (b) Using (a), derive the recurrence relation

$$\frac{2n}{x} J_n(x) = J_{n-1}(x) + J_{n+1}(x), \qquad n = 1, 2, 3, \ldots$$

6. Given the generating function $w(x,t) = \exp[\frac{1}{2}x(t - 1/t)]$,
   (a) show that it satisfies the identity
   $$\frac{\partial w}{\partial x} - \frac{1}{2}\left(t - \frac{1}{t}\right)w = 0$$
   (b) Using (a), derive the relation
   $$2J_n'(x) = J_{n-1}(x) - J_{n+1}(x), \qquad n = 1, 2, 3, \ldots$$

7. Show that ($k \neq 0$, $t \neq 0$)
   $$\exp\left[-\frac{x}{2t}\left(k - \frac{1}{k}\right)\right] \sum_{n=-\infty}^{\infty} J_n(x)k^n t^n = \sum_{n=-\infty}^{\infty} J_n(kx)t^n$$

8. From the product of the generating functions $w(x,t)w(-x,t)$,
   (a) show that
   $$1 = [J_0(x)]^2 + 2\sum_{n=1}^{\infty} [J_n(x)]^2$$
   (b) From (a), deduce that (for all x)
   $$|J_0(x)| \leq 1 \quad \text{and} \quad |J_n(x)| \leq \frac{1}{\sqrt{2}}, \quad n = 1, 2, 3, \ldots$$

9. Use the generating function (6.5) to derive the *Jacobi-Anger expansion*
   $$\exp(ix\sin\theta) = \sum_{n=-\infty}^{\infty} J_n(x)e^{in\theta}$$

10. Use the result of problem 9 to deduce that
    (a) $\cos(x\sin\theta) = J_0(x) + 2\sum_{n=1}^{\infty} J_{2n}(x)\cos(2n\theta)$,
    (b) $\sin(x\sin\theta) = 2\sum_{n=1}^{\infty} J_{2n-1}(x)\sin[(2n-1)\theta]$,
    (c) $\cos x = J_0(x) + 2\sum_{n=1}^{\infty} (-1)^n J_{2n}(x)$,
    (d) $\sin x = 2\sum_{n=1}^{\infty} (-1)^n J_{2n-1}(x)$.

11. Use the results of problem 10 to deduce that
    (a) $x = 2\sum_{n=1}^{\infty} (2n-1)J_{2n-1}(x)$.
    Hint: Differentiate problem 10(b).
    (b) $x\sin x = 2\sum_{n=1}^{\infty} (2n)^2 J_{2n}(x)$.

12. Set $t = e^\theta$ in the generating function (6.5) and deduce that

   (a) $\cosh(x\sinh\theta) = J_0(x) + 2\sum_{n=1}^{\infty} J_{2n}(x)\cosh(2n\theta),$

   (b) $\sinh(x\sinh\theta) = 2\sum_{n=1}^{\infty} J_{2n-1}(x)\sinh[(2n-1)\theta].$

13. Derive *Lommel's formula*

$$J_p(x)J_{1-p}(x) + J_{-p}(x)J_{p-1}(x) = \frac{2\sin p\pi}{\pi x}$$

14. Show that

$$\frac{d}{dx}[x^{-p}J_p(x)] = -x^{-p}J_{p+1}(x)$$

15. Show that

$$\frac{d}{dx}[J_p(kx)] = -kJ_{p+1}(kx) + \frac{p}{x}J_p(kx)$$

16. Show that

   (a) $\dfrac{d}{dx}[xJ_p(x)J_{p+1}(x)] = x\{[J_p(x)]^2 - [J_{p+1}(x)]^2\},$

   (b) $\dfrac{d}{dx}[x^2 J_{p-1}(x)J_{p+1}(x)] = 2x^2 J_p(x)J_p'(x).$

17. Show that

   (a) $\dfrac{1}{x}\dfrac{d}{dx}\left(\tfrac{1}{2}x^2\{[J_p(x)]^2 - J_{p-1}(x)J_{p+1}(x)\}\right) = [J_p(x)]^2.$

   (b) $\dfrac{d}{dx}\{[J_p(x)]^2 + [J_{p+1}(x)]^2\} = 2\left\{\dfrac{p}{x}[J_p(x)]^2 - \dfrac{p+1}{x}[J_{p+1}(x)]^2\right\}.$

18. Show directly that $y = J_{-p}(x)$ is a solution of

$$x^2 y'' + xy' + (x^2 - p^2)y = 0$$

19. Show directly that $y = J_p(kx)$ is a solution of

$$x^2 y'' + xy' + (k^2 x^2 - p^2)y = 0$$

20. Establish the following identities:

   (a) $J_{1/2}(x) = \sqrt{\dfrac{2}{\pi x}}\sin x,$ (b) $J_{-1/2}(x) = \sqrt{\dfrac{2}{\pi x}}\cos x,$

   (c) $J_{1/2}(x)J_{-1/2}(x) = \dfrac{\sin 2x}{\pi x},$

   (d) $[J_{1/2}(x)]^2 + [J_{-1/2}(x)]^2 = \dfrac{2}{\pi x}.$

21. By using the Cauchy product, show that
$$[J_0(x)]^2 = \sum_{n=0}^{\infty} \frac{(-1)^n (2n)!}{(n!)^4} \left(\frac{x}{2}\right)^{2n}$$

Hint: $\sum_{k=0}^{n} \binom{n}{k}^2 = \binom{2n}{n}$.

In problems 22–24, derive the given identity.

22. $J_0\left(\sqrt{x^2 - 2xt}\right) = \sum_{n=0}^{\infty} J_n(x) \frac{t^n}{n!}$.

23. $\left(\frac{x - 2t}{x}\right)^{-p/2} J_p\left(\sqrt{x^2 - 2xt}\right) = \sum_{n=0}^{\infty} J_{p+n}(x) \frac{t^n}{n!}$.

24. $e^{t\cos\phi} J_0(t \sin\phi) = \sum_{n=0}^{\infty} P_n(\cos\phi) \frac{t^n}{n!}$, where $P_n(x)$ is the nth Legendre polynomial.

25. A waveform with phase modulation distortion may be represented by*
$$s(t) = \cos[\omega_0 t + \varepsilon(t)]$$
where $\varepsilon(t)$ represents the "distortion term." In much of the analysis of such waveforms it suffices to approximate the distortion term by the first term of its Fourier series, i.e.,
$$\varepsilon(t) \simeq a \sin \omega_m t$$
where a denotes the peak phase error and ω_m is the fundamental frequency of the phase error. Thus, the original waveform becomes
$$s(t) \simeq \cos(\omega_0 t + a \sin \omega_m t)$$

(a) Show that this last form for $s(t)$ can be decomposed into its harmonic components with Bessel functions representing the corresponding amplitudes, i.e., show that
$$s(t) \simeq J_0(a)\cos\omega_0 t + \sum_{n=1}^{\infty} J_n(a)\big[\cos(\omega_0 t + n\omega_m t)$$
$$+ (-1)^n \cos(\omega_0 t - n\omega_m t)\big]$$

Hint: Use problem 10.

(b) Whenever the peak phase error satisfies $a \leq 0.4$ radians, we can use the approximations
$$J_0(a) \simeq 1, \quad J_1(a) \simeq \frac{a}{2}, \quad J_n(a) \simeq 0 \quad (n = 2, 3, 4, \ldots)$$

---

*For a further discussion of this kind of problem, see C.E. Cook and M. Bernfeld, *Radar Signals*, New York: Academic, 1967.

Show that under these conditions the phase modulation error term produces only the effect of "paired sidebands" with a frequency displacement of $\pm \omega_m$ with respect to ω_0, and a relative amplitude of $a/2$.

6.3 Integral Representations and Integrals of Bessel Functions

There are several integral representations of $J_p(x)$ that are especially useful in practice. Foremost among these is one involving the Bessel function of integral order. To derive it, we start with the generating-function relation

$$e^{\frac{1}{2}x(t-1/t)} = \sum_{k=-\infty}^{\infty} J_k(x) t^k$$

and set $t = e^{-i\phi}$ to get

$$e^{-ix\sin\phi} = \sum_{k=-\infty}^{\infty} J_k(x) e^{-ik\phi} \qquad (6.26)$$

where we have made the observation

$$t - \frac{1}{t} = e^{-i\phi} - e^{i\phi} = -2i\sin\phi$$

Next, we multiply both sides of (6.26) by $e^{in\phi}$ and integrate the result from 0 to π, obtaining

$$\int_0^\pi e^{i(n\phi - x\sin\phi)} d\phi = \sum_{k=-\infty}^{\infty} J_k(x) \int_0^\pi e^{i(n-k)\phi} d\phi \qquad (6.27)$$

assuming that termwise integration is permitted. Now using Euler's formula, we can express (6.27) in terms of sines and cosines, i.e.,

$$\int_0^\pi \cos(n\phi - x\sin\phi) d\phi + i \int_0^\pi \sin(n\phi - x\sin\phi) d\phi$$

$$= \sum_{k=-\infty}^{\infty} J_k(x) \int_0^\pi \cos(n-k)\phi \, d\phi + i \sum_{k=-\infty}^{\infty} J_k(x) \int_0^\pi \sin(n-k)\phi \, d\phi$$

$$(6.28)$$

Equating the real parts of (6.28), and using the result

$$\int_0^\pi \cos(n-k)\phi \, d\phi = \begin{cases} 0, & k \neq n \\ \pi, & k = n \end{cases} \qquad (6.29)$$

we find all terms of the infinite sum vanish except for the term corresponding to $k = n$, and thus we are left with the integral representation (for $n = 0, 1, 2, \ldots$)

$$J_n(x) = \frac{1}{\pi} \int_0^\pi \cos(n\phi - x\sin\phi)\, d\phi \qquad (6.30)$$

When $n = 0$, we get the special case

$$J_0(x) = \frac{1}{\pi} \int_0^\pi \cos(x\sin\phi)\, d\phi \qquad (6.31)$$

The representation (6.30) is restricted to Bessel functions of integral order. A less restrictive representation, due to S.D. Poisson (1781–1840), is given by

$$J_p(x) = \frac{(x/2)^p}{\sqrt{\pi}\,\Gamma(p + \tfrac{1}{2})} \int_{-1}^1 (1 - t^2)^{p - \tfrac{1}{2}} e^{ixt}\, dt, \qquad p > -\tfrac{1}{2}, \quad x > 0$$

$$(6.32)$$

where p is not restricted to integral values. To derive (6.32), we start with the relation

$$\int_{-1}^1 (1 - t^2)^{p - \tfrac{1}{2}} e^{ixt}\, dt = 2\int_0^1 (1 - t^2)^{p - \tfrac{1}{2}} \cos xt\, dt$$

$$= 2\sum_{k=0}^\infty \frac{(-1)^k x^{2k}}{(2k)!} \int_0^1 (1 - t^2)^{p - \tfrac{1}{2}} t^{2k}\, dt \qquad (6.33)$$

where we are using properties of even and odd functions and have expressed $\cos xt$ in a power series. The residual integral in (6.33) can be evaluated in terms of the beta function by making the change of variable $u = t^2$, from which we get (for $p > -\tfrac{1}{2}$)

$$\int_0^1 (1 - t^2)^{p - \tfrac{1}{2}} t^{2k}\, dt = \tfrac{1}{2} \int_0^1 (1 - u)^{p - \tfrac{1}{2}} u^{k - \tfrac{1}{2}}\, du$$

$$= \tfrac{1}{2} B(k + \tfrac{1}{2}, p + \tfrac{1}{2})$$

$$= \frac{\Gamma(k + \tfrac{1}{2})\Gamma(p + \tfrac{1}{2})}{2\Gamma(k + p + 1)} \qquad (6.34)$$

From the Legendre duplication formula, we have

$$\Gamma(k + \tfrac{1}{2}) = \frac{\sqrt{\pi}\,(2k)!}{2^{2k} k!}, \qquad k = 0, 1, 2, \ldots \qquad (6.35)$$

and by substituting the results of (6.34) and (6.35) into (6.33), we obtain

$$\int_{-1}^{1} (1-t^2)^{p-\frac{1}{2}} e^{ixt} dt = \sqrt{\pi}\,\Gamma(p+\tfrac{1}{2}) \sum_{k=0}^{\infty} \frac{(-1)^k (x/2)^{2k}}{k!\,\Gamma(k+p+1)}$$

$$= \sqrt{\pi}\,\Gamma(p+\tfrac{1}{2}) \left(\frac{x}{2}\right)^{-p} J_p(x) \qquad (6.36)$$

from which we deduce (6.32).

A variant of (6.32) results if we make the change of variable $t = \cos\theta$:

$$J_p(x) = \frac{(x/2)^p}{\sqrt{\pi}\,\Gamma(p+\tfrac{1}{2})} \int_0^{\pi} \cos(x\cos\theta) \sin^{2p}\theta\, d\theta, \qquad p > -\tfrac{1}{2},\quad x > 0 \qquad (6.37)$$

the verification of which is left to the reader (problem 2).

6.3.1 Indefinite Integrals Involving Bessel Functions

Many of the indefinite integrals that arise in practice are simple products of some Bessel function and x raised to a power. In particular, we find as a general rule that any integral of the form

$$I = \int x^m J_n(x)\, dx \qquad (6.38)$$

where m and n are integers such that $m + n > 0$, can be integrated in closed form when $m + n$ is odd, but will ultimately depend upon the residual integral $\int J_0(x)\, dx$ when $m + n$ is even.*

By starting with the identities [see (6.12) and (6.13)]

$$\frac{d}{dx}\left[x^p J_p(x) \right] = x^p J_{p-1}(x) \qquad (6.39)$$

and

$$\frac{d}{dx}\left[x^{-p} J_p(x) \right] = -x^{-p} J_{p+1}(x) \qquad (6.40)$$

we can derive two useful integration formulas for handling integrals of the form (6.38). Direct integration of (6.39) and (6.40) leads to

$$\int x^p J_{p-1}(x)\, dx = x^p J_p(x) + C \qquad (6.41)$$

---

*The integral $\int_0^x J_0(t)\, dt$ has been tabulated. See, for example, M. Abramowitz and I. Stegun, (Eds.), *Handbook of Mathematical Functions*, New York: Dover, 1965.

and

$$\int x^{-p} J_{p+1}(x)\, dx = -x^{-p} J_p(x) + C \qquad (6.42)$$

where C denotes a constant of integration.

Example 1: Reduce $\int x^2 J_2(x)\, dx$ to an integral involving only $J_0(x)$.

Solution: To use (6.42), we first write

$$\int x^2 J_2(x)\, dx = \int x^3 \left[x^{-1} J_2(x) \right] dx$$

and use integration by parts with

$$u = x^3, \qquad dv = x^{-1} J_2(x)\, dx$$
$$du = 3x^2\, dx, \qquad v = -x^{-1} J_1(x)$$

Thus we have

$$\int x^2 J_2(x)\, dx = -x^2 J_1(x) + 3 \int x J_1(x)\, dx$$

and a second integration by parts finally gives

$$\int x^2 J_2(x)\, dx = -x^2 J_1(x) - 3x J_0(x) + 3 \int J_0(x)\, dx$$

The last integral involving $J_0(x)$ cannot be evaluated in closed form, and so our integration is complete.

6.3.2 Definite Integrals Involving Bessel Functions

In practice we are often faced with the necessity of evaluating definite integrals involving Bessel functions in combinations with various elementary functions or, in some instances, special functions of other kinds. The usual procedure in such integrals is to replace the Bessel function by its series representation (or an integral representation) and then interchange the order in which the operations are carried out.

To illustrate the technique, let us consider the Laplace transform integral

$$I = \int_0^\infty e^{-ax} x^p J_p(bx)\, dx, \qquad p > -\tfrac{1}{2}, \quad a > 0, \quad b > 0 \qquad (6.43)$$

Here we replace $J_p(bx)$ by its series representation (6.9) and integrate the resulting series termwise to get

$$I = \sum_{k=0}^{\infty} \frac{(-1)^k (b/2)^{2k+p}}{k!\Gamma(k+p+1)} \int_0^{\infty} e^{-ax} x^{2k+2p} \, dx$$

$$= b^p \sum_{k=0}^{\infty} \frac{(-1)^k \Gamma(2k+2p+1)}{2^{2k+p} k! \Gamma(k+p+1)} (a^2)^{-(p+\frac{1}{2})-k} (b^2)^k \quad (6.44)$$

where the integral has been evaluated through properties of the gamma function. We wish to show that the series (6.44) is a binomial series, and hence it can be summed. Recalling the Legendre duplication formula and Equation (1.27) in Section 1.2.4, it follows that ($p > -\frac{1}{2}$)

$$\frac{(-1)^k \Gamma(2k+2p+1)}{2^{2k+p} k! \Gamma(k+p+1)} = \frac{(-1)^k 2^p}{\sqrt{\pi}} \frac{\Gamma(p+k+\frac{1}{2})}{k!}$$

$$= \frac{(-1)^k}{\sqrt{\pi}} 2^p \Gamma(p+\tfrac{1}{2}) \binom{p+k-\frac{1}{2}}{k}$$

$$= \frac{2^p \Gamma(p+\tfrac{1}{2})}{\sqrt{\pi}} \binom{-(p+\frac{1}{2})}{k} \quad (6.45)$$

Thus, (6.44) becomes

$$I = \frac{(2b)^p \Gamma(p+\tfrac{1}{2})}{\sqrt{\pi}} \sum_{k=0}^{\infty} \binom{-(p+\frac{1}{2})}{k} (a^2)^{-(p+\frac{1}{2})-k} (b^2)^k \quad (6.46)$$

and by summing this binomial series, we are led to*

$$\int_0^{\infty} e^{-ax} x^p J_p(bx) \, dx = \frac{(2b)^p \Gamma(p+\tfrac{1}{2})}{\sqrt{\pi} (a^2+b^2)^{p+\frac{1}{2}}}, \quad p > -\tfrac{1}{2}, \quad a > 0, \quad b > 0$$

$$(6.47)$$

Setting $p = 0$ in (6.47) yields the special result

$$\int_0^{\infty} e^{-ax} J_0(bx) \, dx = (a^2+b^2)^{-1/2}, \quad a > 0, \quad b > 0 \quad (6.48)$$

Strictly speaking, the validity of (6.48) rests upon the condition that $a > 0$ (or at least the real part of a positive if a is complex). Yet it is possible to justify a limiting procedure whereby the real part of a approaches zero. Thus, if we formally replace a in (6.48) with the pure

---

*Summing the series (6.46) requires that $a \neq b$, although the result (6.47) is valid even when $a = b$.

210 • Special Functions for Engineers and Applied Mathematicians

imaginary number ia, we get

$$\int_0^\infty e^{-iax} J_0(bx)\, dx = (b^2 - a^2)^{-1/2}$$

The separation of this expression into real and imaginary parts leads to

$$\int_0^\infty \cos(ax) J_0(bx)\, dx - i\int_0^\infty \sin(ax) J_0(bx)\, dx$$

$$= \begin{cases} (b^2 - a^2)^{-1/2}, & b > a \\ -i(a^2 - b^2)^{-1/2}, & b < a \end{cases} \qquad (6.49)$$

and by equating the real and imaginary parts of (6.49), we deduce the pair of integral formulas

$$\int_0^\infty \cos(ax) J_0(bx)\, dx = \begin{cases} (b^2 - a^2)^{-1/2}, & b > a \\ 0, & b < a \end{cases} \qquad (6.50)$$

and

$$\int_0^\infty \sin(ax) J_0(bx)\, dx = \begin{cases} 0, & b > a \\ (a^2 - b^2)^{-1/2}, & b < a \end{cases} \qquad (6.51)$$

These last two formulas are important in the theory of Fourier integrals. Both (6.50) and (6.51) diverge when $b = a$.

Example 2: Derive Weber's integral formula

$$\int_0^\infty x^{2m-p-1} J_p(x)\, dx = \frac{2^{2m-p-1}\Gamma(m)}{\Gamma(p - m + 1)}, \qquad 0 < m < \tfrac{1}{2}, \quad p > -\tfrac{1}{2}$$

Solution: Replacing $J_p(x)$ by its integral representation (6.37), we have

$$\int_0^\infty x^{2m-p-1} J_p(x)\, dx = \frac{2^{-p}}{\sqrt{\pi}\, \Gamma(p + \tfrac{1}{2})}$$

$$\times \int_0^\infty x^{2m-1} \int_0^\pi \cos(x\cos\theta)\sin^{2p}\theta\, d\theta\, dx$$

$$= \frac{2^{-p}}{\sqrt{\pi}\, \Gamma(p + \tfrac{1}{2})}$$

$$\times \int_0^\pi \sin^{2p}\theta \int_0^\infty x^{2m-1}\cos(x\cos\theta)\, dx\, d\theta$$

where we have reversed the order of integration. By making the substitution $t = x\cos\theta$ in the inner integral and using the result of problem 37 in Exercises 2.2, we obtain

$$\int_0^\infty x^{2m-1}\cos(x\cos\theta)\,dx = \cos^{-2m}\theta \int_0^\infty t^{2m-1}\cos t\,dt$$

$$= \cos^{-2m}\theta\,\Gamma(2m)\cos m\pi$$

$$= \pi^{-1/2}2^{2m-1}\Gamma(m)\Gamma(m+\tfrac{1}{2})\cos m\pi \cos^{-2m}\theta$$

The last step follows from the Legendre duplication formula. The remaining integral above now leads to

$$\int_0^\pi \sin^{2p}\theta \cos^{-2m}\theta\,d\theta = 2\int_0^{\pi/2}\sin^{2p}\theta\cos^{-2m}\theta\,d\theta$$

$$= \frac{\Gamma(p+\tfrac{1}{2})\Gamma(\tfrac{1}{2}-m)}{\Gamma(p-m+1)}$$

and hence, we deduce that

$$\int_0^\infty x^{2m-p-1}J_p(x)\,dx = \frac{2^{2m-p-1}\Gamma(m)\Gamma(m+\tfrac{1}{2})\Gamma(\tfrac{1}{2}-m)\cos m\pi}{\pi\Gamma(p-m+1)}$$

$$= \frac{2^{2m-p-1}\Gamma(m)}{\Gamma(p-m+1)}$$

where we are recalling the identity [problem 42(b) in Exercises 2.2]

$$\Gamma(\tfrac{1}{2}+m)\Gamma(\tfrac{1}{2}-m) = \pi\sec m\pi$$

(Although we won't show it, Weber's integral is valid for a much wider range of values on m and p than indicated above.)

EXERCISES 6.3

1. Using Equation (6.30), deduce the following results:
   (a) $[1+(-1)^n]J_n(x) = \dfrac{2}{\pi}\int_0^\pi \cos n\theta \cos(x\sin\theta)\,d\theta$ $(n=0,1,2,\dots)$.
   (b) $[1-(-1)^n]J_n(x) = \dfrac{2}{\pi}\int_0^\pi \sin n\theta \sin(x\sin\theta)\,d\theta$ $(n=0,1,2,\dots)$.
   (c) $J_{2k}(x) = \dfrac{1}{\pi}\int_0^\pi \cos 2k\theta \cos(x\sin\theta)\,d\theta$ $(k=0,1,2,\dots)$.
   (d) $J_{2k+1}(x) = \dfrac{1}{\pi}\int_0^\pi \sin[(2k+1)\theta]\sin(x\sin\theta)\,d\theta$ $(k=0,1,2,\dots)$.
   (e) $\int_0^\pi \cos[(2k+1)\theta]\cos(x\sin\theta)\,d\theta = 0$ $(k=0,1,2,\dots)$.
   (f) $\int_0^\pi \sin 2k\theta \sin(x\sin\theta)\,d\theta = 0$ $(k=0,1,2,\dots)$.

2. By setting $t = \cos\theta$ in (6.32), show that

$$J_p(x) = \frac{(x/2)^p}{\sqrt{\pi}\,\Gamma(p + \frac{1}{2})} \int_0^\pi \cos(x\cos\theta)\sin^{2p}\theta\,d\theta, \qquad p > -\tfrac{1}{2}, \quad x > 0$$

3. By writing $\cos xt$ in an infinite series and using termwise integration, deduce that

$$J_0(x) = \frac{2}{\pi} \int_0^1 \frac{\cos xt}{\sqrt{1 - t^2}}\,dt$$

4. Replacing $J_m(xt)$ by its series representation and using termwise integration, deduce the integral relation

$$J_p(x) = \frac{2(x/2)^{p-m}}{\Gamma(p-m)} \int_0^1 (1 - t^2)^{p-m-1} t^{m+1} J_m(xt)\,dt,$$
$$p > m > -1, \quad x > 0$$

In problems 5–16, use recurrence relations, integration by parts, etc., to verify the given result.

5. $\int x J_0(x)\,dx = x J_1(x) + C$.

6. $\int x^2 J_0(x)\,dx = x^2 J_1(x) + x J_0(x) - \int J_0(x)\,dx + C$.

7. $\int x^3 J_0(x)\,dx = (x^3 - 4x) J_1(x) + 2x^2 J_0(x) + C$.

8. $\int J_1(x)\,dx = -J_0(x) + C$.

9. $\int x J_1(x)\,dx = -x J_0(x) + \int J_0(x)\,dx + C$.

10. $\int x^2 J_1(x)\,dx = 2x J_1(x) - x^2 J_0(x) + C$.

11. $\int x^3 J_1(x)\,dx = 3x^2 J_1(x) - (x^3 - 3x) J_0(x) - 3\int J_0(x)\,dx + C$.

12. $\int J_3(x)\,dx = -J_2(x) - 2x^{-1} J_1(x) + C$.

13. $\int x^{-1} J_1(x)\,dx = -J_1(x) + \int J_0(x)\,dx + C$.

14. $\int x^{-2} J_2(x)\,dx = -\frac{2}{3x^2} J_1(x) - \tfrac{1}{3} J_1(x)$
    $+ \frac{1}{3x} J_0(x) + \frac{1}{3}\int J_0(x)\,dx + C$.

15. $\int J_0(x)\cos x\, dx = xJ_0(x)\cos x + xJ_1(x)\sin x + C$

16. $\int J_0(x)\sin x\, dx = xJ_0(x)\sin x - xJ_1(x)\cos x + C$

17. Show that
$$\int x\left\{[J_p(x)]^2 - [J_{p+1}(x)]^2\right\} dx = xJ_p(x)J_{p+1}(x) + C$$

Hint: Use the result of problem 16(a) in Exercises 6.2.

18. Show that
$$\int x[J_p(x)]^2\, dx = \tfrac{1}{2}x^2\left\{[J_p(x)]^2 - J_{p-1}(x)J_{p+1}(x)\right\} + C$$

Hint: Use the result of problem 17(a) in Exercises 6.2.

19. Show that (using repeated integration by parts)
$$\int J_0(x)\, dx = J_1(x) + \frac{J_2(x)}{x} + \frac{1\times 3}{x^2}J_3(x) + \cdots$$
$$+ \frac{(2n-2)!J_n(x)}{2^{n-1}(n-1)!x^{n-1}} + \frac{(2n)!}{2^n n!}\int \frac{J_n(x)}{x^n}\, dx$$

In problems 20–35, derive the given integral formula.

20. $\int_0^\infty J_0(bx)\, dx = \dfrac{1}{b},\ b>0$.

Hint: Let $a \to 0^+$ in Equation (6.48).

21. $\int_0^\infty J_{n+1}(x)\, dx = \int_0^\infty J_{n-1}(x)\, dx,\ n = 1, 2, 3, \ldots$.

Hint: Use Equation (6.17).

22. $\int_0^\infty J_n(x)\, dx = 1,\ n = 0, 1, 2, \ldots$.

Hint: Use problem 21.

23. $\int_0^\infty \dfrac{J_n(x)}{x}\, dx = \dfrac{1}{n},\ n = 1, 2, 3, \ldots$.

Hint: Use problems 21 and 22.

24. $\int_0^\infty e^{-ax}x^{p+1}J_p(bx)\, dx = \dfrac{2^{p+1}\Gamma(p+\tfrac{3}{2})}{\sqrt{\pi}}\dfrac{ab^p}{(a^2+b^2)^{p+\tfrac{3}{2}}},\ p > -1$,
$a > 0,\ b > 0$.

Hint: Differentiate both sides of Equation (6.47) with respect to a.

25. $\int_0^\infty x^2 e^{-ax} J_0(bx)\,dx = \dfrac{(2a^2 - b^2)}{(a^2 + b^2)^{5/2}}$, $a > 0$, $b > 0$.

Hint: Differentiate both sides of Equation (6.48) with respect to a.

26. $\int_0^\infty e^{-ax^2} x^{p+1} J_p(bx)\,dx = \dfrac{b^p}{(2a)^{p+1}} e^{-b^2/4a}$, $p > -1$, $a > 0$, $b > 0$.

27. $\int_0^\infty e^{-ax^2} x^{p+3} J_p(bx)\,dx = \dfrac{b^p}{2^{p+1} a^{p+2}}\left(p + 1 - \dfrac{b^2}{4a}\right) e^{-b^2/4a}$, $p > -1$, $a > 0$, $b > 0$.

Hint: Differentiate both sides of problem 26 with respect to a.

28. $\int_0^\infty x^{-1} \sin x\, J_0(bx)\,dx = \arcsin\left(\dfrac{1}{b}\right)$, $b > 1$.

Hint: Integrate Equation (6.50) with respect to a.

29. $\int_0^{\pi/2} J_0(x\cos\phi)\cos\phi\,d\phi = \dfrac{\sin x}{x}$.

30. $\int_0^{\pi/2} J_1(x\cos\phi)\,d\phi = \dfrac{1 - \cos x}{x}$.

31. $\int_0^a x(a^2 - x^2)^{-1/2} J_0(kx\sin\phi)\,d\phi = \dfrac{\sin(ka\sin\phi)}{k\sin\phi}$.

32. $\int_0^\pi e^{t\cos\phi} J_0(t\sin\phi)\sin\phi\,d\phi = 2$.

Hint: Use problem 24 in Exercises 6.2.

33. $\int_0^\infty e^{-t\cos\phi} J_0(t\sin\phi) t^n\,dt = n! P_n(\cos\phi)$, $0 \le \phi < \pi$, where $P_n(x)$ is the nth Legendre polynomial.

Hint: Use problem 24 in Exercises 6.2.

34. $\int_0^\infty x(x^2 + a^2)^{-1/2} J_0(bx)\,dx = \dfrac{1}{b} e^{-ab}$, $a \ge 0$, $b > 0$.

Hint: Use the integral representation

$$(x^2 + a^2)^{-1/2} = \dfrac{1}{\sqrt{\pi}} \int_0^\infty e^{-(x^2 + a^2)t} t^{-1/2}\,dt$$

and then interchange the order of integration.

35. $\int_0^\infty \dfrac{J_p(x)}{x^m}\,dx = \dfrac{\Gamma\left(\dfrac{p + 1 - m}{2}\right)}{2^m \Gamma\left(\dfrac{p + 1 + m}{2}\right)}$, $m > \tfrac{1}{2}$, $p - m > -1$.

36. The amplitude of a diffracted wave through a circular aperture is given by

$$U = k \int_0^a \int_0^{2\pi} e^{ibr\sin\theta} r\, d\theta\, dr$$

where k is a physical constant, a is the radius of the aperture, θ is the azimuthal angle in the plane of the aperture, and b is a constant inversely proportional to the wavelength of the incident wave. Show that the intensity of light in the diffraction pattern is given by

$$I = |U|^2 = \frac{4\pi^2 k^2 a^2}{b^2}[J_1(ab)]^2$$

6.4 Bessel Series

A *Bessel series*, which is a member of the class of generalized Fourier series,* has the form

$$f(x) = \sum_{n=1}^{\infty} c_n J_p(k_n x), \qquad 0 < x < b, \quad p > -\tfrac{1}{2} \qquad (6.52)$$

where the c's are constants to be determined and the k_n ($n = 1, 2, 3, \ldots$) are solutions of the equation†

$$J_p(k_n b) = 0, \qquad n = 1, 2, 3, \ldots \qquad (6.53)$$

The theory of Bessel series closely parallels that of Legendre series. For example, the Bessel functions satisfy an orthogonality relation, and the constants (Fourier coefficients) c_n are defined by a formula similar to that for Legendre series. The conditions under which the series (6.52) converges will be stated (see Theorem 6.1 below), although we will not present the formal proof.

6.4.1 Orthogonality

The theory of all generalized Fourier series rests heavily upon the *orthogonality property* of the particular special functions. In the case of Bessel functions, we have ($p > -1$)

$$\int_0^b x J_p(k_m x) J_p(k_n x)\, dx = 0, \qquad m \neq n \qquad (6.54)$$

where k_m and k_n are distinct roots satisfying (6.53).

---

*See the discussion in Section 4.4 on generalized Fourier series.

†The Bessel function $J_p(x)$ has an infinite number of zeros for $x > 0$. See Theorem 5.6 in L.C. Andrews, *Ordinary Differential Equations with Applications*, Glenview, Ill.: Scott, Foresman, 1982.

In order to prove (6.54), we first note that since $y = J_p(x)$ is a solution of Bessel's equation

$$x^2 y'' + xy' + (x^2 - p^2) y = 0 \qquad (6.55)$$

it follows that $y = J_p(kx)$ satisfies the more general equation (see problem 19 in Exercises 6.2)

$$x^2 y'' + xy' + (k^2 x^2 - p^2) y = 0 \qquad (6.56)$$

For our purposes we wish to rewrite (6.56) in the more useful form

$$x \frac{d}{dx}(xy') + (k^2 x^2 - p^2) y = 0 \qquad (6.57)$$

and hence, $J_p(k_m x)$ and $J_p(k_n x)$ satisfy respectively the DEs

$$x \frac{d}{dx}\left[x \frac{d}{dx} J_p(k_m x)\right] + (k_m^2 x^2 - p^2) J_p(k_m x) = 0 \qquad (6.58)$$

$$x \frac{d}{dx}\left[x \frac{d}{dx} J_p(k_n x)\right] + (k_n^2 x^2 - p^2) J_p(k_n x) = 0 \qquad (6.59)$$

If we multiply (6.58) by $x^{-1} J_p(k_n x)$ and (6.59) by $x^{-1} J_p(k_m x)$, subtract the resulting equations, and integrate from 0 to b, we find upon rearranging the terms

$$(k_m^2 - k_n^2) \int_0^b x J_p(k_m x) J_p(k_n x)\, dx = \int_0^b J_p(k_m x) \frac{d}{dx}\left[x \frac{d}{dx} J_p(k_n x)\right] dx$$

$$- \int_0^b J_p(k_n x) \frac{d}{dx}\left[x \frac{d}{dx} J_p(k_m x)\right] dx$$

Carrying out the integrations (by parts) on the right-hand side and dividing by the factor $k_m^2 - k_n^2$ leads to

$$\int_0^b x J_p(k_m x) J_p(k_n x)\, dx$$

$$= \frac{x \left[J_p(k_m x) \frac{d}{dx} J_p(k_n x) - J_p(k_n x) \frac{d}{dx} J_p(k_m x) \right] \Big|_{x=0}^{x=b}}{k_m^2 - k_n^2}$$

(6.60)

By hypothesis, $k_m \neq k_n$ and $J_p(k_m b) = J_p(k_n b) = 0$, and thus the right-hand side of (6.60) vanishes, which proves the orthogonality property (6.54).

When $k_m = k_n$, the resulting integral

$$I = \int_0^b x \left[J_p(k_n x) \right]^2 dx$$

is also of interest to us. To deduce its value we take the limit of (6.60) as

$k_m \to k_n$. Because the right-hand side of (6.60) approaches the indeterminate form $0/0$ in the limit, we need to employ L'Hôpital's rule, which leads to (treating k_m as the variable and all other parameters constant)

$$I = \frac{x}{2k_n}\left[\frac{d}{dx}J_p(k_n x)\frac{d}{dk_n}J_p(k_n x) - J_p(k_n x)\frac{d}{dk_n}\frac{d}{dx}J_p(k_n x)\right]\bigg|_{x=0}^{x=b} \tag{6.61}$$

Now, using the recurrence relations (see problem 15 in Exercises 6.2)

$$\frac{d}{dx}J_p(kx) = \frac{p}{x}J_p(kx) - kJ_{p+1}(kx) \tag{6.62a}$$

$$\frac{d}{dk}J_p(kx) = \frac{p}{k}J_p(kx) - xJ_{p+1}(kx) \tag{6.62b}$$

we find that (6.61) reduces to

$$I = \left\{\frac{1}{2}\frac{p^2}{k_n^2}[J_p(k_n x)]^2 + \frac{1}{2}x^2[J_{p+1}(k_n x)]^2 \right.$$

$$\left. - \frac{p}{k_n}J_p(k_n x)J_{p+1}(k_n x)\right\}\bigg|_{x=0}^{x=b}$$

or finally

$$\int_0^b x[J_p(k_n x)]^2\,dx = \tfrac{1}{2}b^2[J_{p+1}(k_n b)]^2 \tag{6.63}$$

6.4.2 A Convergence Theorem

Returning now to the series

$$f(x) = \sum_{n=1}^{\infty} c_n J_p(k_n x), \quad 0 < x < b, \quad p > -\tfrac{1}{2} \tag{6.64}$$

where $J_p(k_n b) = 0$ $(n = 1, 2, 3, \ldots)$, let us assume the validity of this representation and attempt to formally find the Fourier coefficients. To begin, we multiply both sides of (6.64) by $xJ_p(k_m x)$ and integrate from 0 to b. Under the assumption that termwise integration is permitted, we obtain

$$\int_0^b xf(x)J_p(k_m x)\,dx = \sum_{n=1}^{\infty} c_n \overbrace{\int_0^b xJ_p(k_m x)J_p(k_n x)\,dx}^{0\,(n \neq m)}$$

$$= c_m \int_0^b x[J_p(k_m x)]^2\,dx \tag{6.65}$$

218 • Special Functions for Engineers and Applied Mathematicians

and hence deduce that (changing the index back to n)

$$c_n = \frac{2}{b^2[J_{p+1}(k_n b)]^2} \int_0^b x f(x) J_p(k_n x)\, dx, \qquad n = 1, 2, 3, \ldots \quad (6.66)$$

Theorem 6.1. If f is a piecewise smooth function in the interval $0 \le x \le b$, then the Bessel series (6.64) with constants defined by (6.66) converges pointwise to $f(x)$ at points of continuity of f, and to $\frac{1}{2}[f(x^+) + f(x^-)]$ at points of discontinuity of f.*

Example 3: Find the Bessel series for

$$f(x) = \begin{cases} x, & 0 < x < 1 \\ 0, & 1 < x < 2 \end{cases}$$

corresponding to the set of functions $\{J_1(k_n x)\}$, where k_n satisfies $J_1(2k_n) = 0$ ($n = 1, 2, 3, \ldots$).

Solution: The series we seek is

$$f(x) = \sum_{n=1}^{\infty} c_n J_1(k_n x), \qquad 0 < x < 2$$

where

$$c_n = \frac{1}{2[J_2(2k_n)]^2} \int_0^2 x f(x) J_1(k_n x)\, dx$$

$$= \frac{1}{2[J_2(2k_n)]^2} \int_0^1 x^2 J_1(k_n x)\, dx \qquad (\text{let } t = k_n x)$$

$$= \frac{1}{2[J_2(2k_n)]^2} \frac{1}{k_n^3} \int_0^{k_n} t^2 J_1(t)\, dt$$

Recalling the formula $t^2 J_1(t) = (d/dt)[t^2 J_2(t)]$, we find that

$$\int_0^{k_n} t^2 J_1(t)\, dt = \int_0^{k_n} \frac{d}{dt}[t^2 J_2(t)]\, dt = k_n^2 J_2(k_n)$$

and thus

$$c_n = \frac{J_2(k_n)}{2k_n[J_2(2k_n)]^2}, \qquad n = 1, 2, 3, \ldots$$

---
*The series always converges to zero for $x = b$, and converges to zero at $x = 0$ if $p > 0$.

The desired Bessel series is therefore given by

$$f(x) = \frac{1}{2} \sum_{n=1}^{\infty} \frac{J_2(k_n)}{k_n [J_2(2k_n)]^2} J_1(k_n x)$$

Generalizations of the Bessel series can be developed where the k_n ($n = 1, 2, 3, \ldots$) satisfy the more general condition

$$hJ_p(k_n b) + k_n J_p'(k_n b) = 0 \quad (h \text{ constant}) \qquad (6.67)$$

The theory in such cases requires only a slight modification of that presented here and is taken up in the exercises.

EXERCISES 6.4

In problems 1 and 2, verify the series relation given that $J_0(k_n) = 0$ ($n = 1, 2, 3, \ldots$).

1. $\frac{1}{8}(1 - x^2) = \sum_{n=1}^{\infty} \frac{J_0(k_n x)}{k_n^3 J_1(k_n)}$, $0 \leq x \leq 1$.

2. $\log x = -2 \sum_{n=1}^{\infty} \frac{J_0(k_n x)}{[k_n J_1(k_n)]^2}$, $0 < x \leq 1$.

In problems 3–5, find the Bessel series for $f(x)$ in terms of $\{J_0(k_n x)\}$, given that $J_0(k_n) = 0$ ($n = 1, 2, 3, \ldots$).

3. $f(x) = 0.1 J_0(k_3 x)$, $0 < x < 1$.
4. $f(x) = 1$, $0 < x < 1$.
5. $f(x) = x^4$, $0 < x < 1$.
6. If $p \geq -\frac{1}{2}$ and $J_p(k_n) = 0$ ($n = 1, 2, 3, \ldots$), show that

$$x^p = 2 \sum_{n=1}^{\infty} \frac{J_p(k_n x)}{k_n J_{p+1}(k_n)}, \qquad 0 < x < 1$$

7. If $p > -\frac{1}{2}$ and $J_p(k_n) = 0$ ($n = 1, 2, 3, \ldots$), show that

(a) $x^{p+1} = 2^2(p + 1) \sum_{n=1}^{\infty} \frac{J_{p+1}(k_n x)}{k_n^2 J_{p+1}(k_n)}$, $0 < x < 1$,

(b) $x^{p+2} = 2^3(p + 1)(p + 2) \sum_{k=1}^{\infty} \frac{J_{p+2}(k_n x)}{k_n^3 J_{p+1}(k_n)}$, $0 < x < 1$.

8. Expand $f(x) = x^{-p}$, $0 < x < 1$, in the series

$$x^{-p} = \sum_{n=1}^{\infty} c_n J_p(k_n x), \quad 0 < x < 1$$

where $J_p(k_n) = 0$ ($n = 1, 2, 3, \ldots$ and $p \geq 0$).

9. Given that ($p > -\frac{1}{2}$)

$$J_p'(k_n b) = 0, \quad n = 1, 2, 3, \ldots$$

show that

(a) $\int_0^b x J_p(k_m x) J_p(k_n x) \, dx = 0$, $m \neq n$,

(b) $\int_0^b x [J_p(k_n x)]^2 \, dx = \dfrac{k_n^2 b^2 - p^2}{2k_n^2} [J_p(k_n b)]^2$.

10. Given that ($p > -\frac{1}{2}$)

$$h J_p(k_n b) + k_n J_p'(k_n b) = 0, \quad n = 1, 2, 3, \ldots \quad (h \text{ constant})$$

show that

(a) $\int_0^b x J_p(k_m x) J_p(k_n x) \, dx = 0$, $m \neq n$,

(b) $\int_0^b x [J_p(k_n x)]^2 \, dx = \dfrac{(k_n^2 + h^2) b^2 - p^2}{2k_n^2} [J_p(k_n b)]^2$.

11. Under the assumption that ($p > 0$)

$$J_p'(k_n) = 0, \quad n = 1, 2, 3, \ldots$$

use the result of problem 9 to derive the Bessel series

$$x^p = 2 \sum_{n=1}^{\infty} \frac{k_n J_{p+1}(k_n)}{(k_n^2 - p^2)[J_p(k_n)]^2} J_p(k_n x), \quad 0 < x < 1$$

12. Does the expansion in problem 11 hold when $p = 0$? Explain.

6.5 Bessel Functions of the Second and Third Kinds

We have previously shown that

$$y_1 = J_p(x) = \sum_{k=0}^{\infty} \frac{(-1)^k (x/2)^{2k+p}}{k! \Gamma(k + p + 1)}, \quad p \geq 0 \quad (6.68)$$

is a solution of Bessel's equation

$$x^2 y'' + xy' + (x^2 - p^2)y = 0 \qquad (6.69)$$

Because $J_{-p}(x)$ satisfies the same recurrence relations as $J_p(x)$, it follows that $J_{-p}(x)$ is also a solution of (6.69). Moreover, for p not an integer, we have already established that $J_{-p}(x)$ is linearly independent of $J_p(x)$, and hence, under these conditions a general solution of (6.69) is given by

$$y = C_1 J_p(x) + C_2 J_{-p}(x), \qquad p \neq n \quad (n = 0, 1, 2, \ldots) \qquad (6.70)$$

where C_1 and C_2 are arbitrary constants.

For $p = n$ ($n = 0, 1, 2, \ldots$), the solutions $J_n(x)$ and $J_{-n}(x)$ are related by [see Section 6.2.1]

$$J_{-n}(x) = (-1)^n J_n(x), \qquad n = 0, 1, 2, \ldots \qquad (6.71)$$

and thus are *not* linearly independent. Therefore, (6.70) cannot represent a general solution of (6.69) in this case.

For purposes of constructing a general solution of (6.69), it is preferable to find a second solution whose independence of $J_p(x)$ is not restricted to certain values of p. Hence, we introduce the function

$$Y_p(x) = \frac{(\cos p\pi) J_p(x) - J_{-p}(x)}{\sin p\pi} \qquad (6.72)$$

called the *Bessel function of the second kind* of order p. Because $Y_p(x)$ is a linear combination of $J_p(x)$ and $J_{-p}(x)$, it is clearly a solution of (6.69). Furthermore, it is linearly independent of $J_p(x)$ when p is not an integer. (Why?) When $p = n$ ($n = 0, 1, 2, \ldots$), however, it requires further investigation. That is, when $p = n$ we find that (6.72) assumes the indeterminate form $0/0$. Nonetheless, the limit as $p \to n$ does exist and we define (see Section 6.5.1)

$$Y_n(x) = \lim_{p \to n} Y_p(x) \qquad (6.73)$$

The function $Y_n(x)$ is linearly independent of $J_n(x)$,* and we conclude therefore that for arbitrary values of p, the general solution of (6.69) is

$$y = C_1 J_p(x) + C_2 Y_p(x) \qquad (6.74)$$

6.5.1 Series Expansion for $Y_n(x)$

We wish to derive an expression for the Bessel function of the second kind when p takes on integer values. Because the limit (6.73) leads to the

---

*The Wronskian of J_p and Y_p is $2/\pi x$ (see problem 8), and thus the functions are linearly independent for all p.

indeterminate form 0/0, we must apply L'Hôpital's rule, from which we deduce

$$Y_n(x) = \lim_{p \to n} Y_p(x)$$

$$= \lim_{p \to n} \frac{(-\pi \sin p\pi) J_p(x) + (\cos p\pi) \frac{\partial}{\partial p} J_p(x) - \frac{\partial}{\partial p} J_{-p}(x)}{\pi \cos p\pi}$$

$$= \lim_{p \to n} \frac{1}{\pi} \left[\frac{\partial}{\partial p} J_p(x) - (-1)^n \frac{\partial}{\partial p} J_{-p}(x) \right] \qquad (6.75)$$

The derivative of the Bessel function with respect to its order leads to ($x > 0$)

$$\frac{\partial}{\partial p} J_p(x)$$

$$= \sum_{k=0}^{\infty} \frac{(-1)^k}{k!} \left\{ \frac{(x/2)^{2k+p} \log(x/2)}{\Gamma(k+p+1)} - \frac{(x/2)^{2k+p} \Gamma'(k+p+1)}{[\Gamma(k+p+1)]^2} \right\}$$

$$= \sum_{k=0}^{\infty} \frac{(-1)^k (x/2)^{2k+p}}{k! \Gamma(k+p+1)} [\log(x/2) - \psi(k+p+1)]$$

where $\psi(x)$ is the digamma function (see Section 2.5). We can further write this last expression as

$$\frac{\partial}{\partial p} J_p(x) = J_p(x) \log(x/2) - \sum_{k=0}^{\infty} \frac{(-1)^k (x/2)^{2k+p}}{k! \Gamma(k+p+1)} \psi(k+p+1) \qquad (6.76)$$

By a similar analysis, it follows that

$$\frac{\partial}{\partial p} J_{-p}(x) = -J_{-p}(x) \log(x/2) + \sum_{k=0}^{\infty} \frac{(-1)^k (x/2)^{2k-p}}{k! \Gamma(k-p+1)} \psi(k-p+1) \qquad (6.77)$$

At this point we wish to first consider the special case when $p \to 0$. Here we see that (6.75) reduces to

$$Y_0(x) = \frac{2}{\pi} \lim_{p \to 0} \frac{\partial}{\partial p} J_p(x),$$

or by using (6.76), we obtain ($x > 0$)

$$Y_0(x) = \frac{2}{\pi} J_0(x) \log(x/2) - \frac{2}{\pi} \sum_{k=0}^{\infty} \frac{(-1)^k (x/2)^{2k}}{(k!)^2} \psi(k+1) \qquad (6.78)$$

Another form of (6.78) can be obtained by making the observation*

$$\sum_{k=0}^{\infty} \frac{(-1)^k (x/2)^{2k}}{(k!)^2} \psi(k+1)$$

$$= -\gamma + \sum_{k=1}^{\infty} \frac{(-1)^k (x/2)^{2k}}{(k!)^2} \left(-\gamma + 1 + \frac{1}{2} + \cdots + \frac{1}{k}\right)$$

$$= -\gamma \sum_{k=0}^{\infty} \frac{(-1)^k (x/2)^{2k}}{(k!)^2}$$

$$+ \sum_{k=1}^{\infty} \frac{(-1)^k (x/2)^{2k}}{(k!)^2} \left(1 + \frac{1}{2} + \cdots + \frac{1}{k}\right) \quad (6.79)$$

from which we deduce ($x > 0$)

$$Y_0(x) = \frac{2}{\pi} J_0(x) [\log(x/2) + \gamma]$$

$$- \frac{2}{\pi} \sum_{k=1}^{\infty} \frac{(-1)^k (x/2)^{2k}}{(k!)^2} \left(1 + \frac{1}{2} + \cdots + \frac{1}{k}\right) \quad (6.80)$$

The derivation of the series for $Y_n(x)$, $n = 1, 2, 3, \ldots$, is a little more difficult to obtain. Proceeding as before and taking the limit in (6.75) by using (6.76) and (6.77), we find

$$Y_n(x) = \frac{1}{\pi} [J_n(x) + (-1)^n J_{-n}(x)] \log(x/2)$$

$$- \frac{1}{\pi} \left[\sum_{k=0}^{\infty} \frac{(-1)^k (x/2)^{2k+n}}{k! \Gamma(k+n+1)} \psi(k+n+1) \right.$$

$$\left. + (-1)^n \sum_{k=0}^{\infty} \frac{(-1)^k (x/2)^{2k-n}}{k! \Gamma(k-n+1)} \psi(k-n+1) \right] \quad (6.81)$$

Recalling that

$$|\Gamma(k-n+1)| \to \infty, \quad k = 0, 1, \ldots, n-1$$

and

$$|\psi(k-n+1)| \to \infty, \quad k = 0, 1, \ldots, n-1$$

we see that the first n terms in the last series in (6.81) become inde-

---

*γ is Euler's constant.

224 • Special Functions for Engineers and Applied Mathematicians

terminate. However, it can be shown that (see problem 9)

$$\lim_{p \to n} \frac{\psi(k-p+1)}{\Gamma(k-p+1)} = (-1)^{n-k}(n-k-1)!, \quad k = 0, 1, \ldots, n-1 \tag{6.82}$$

and therefore

$$(-1)^n \sum_{k=0}^{\infty} \frac{(-1)^k (x/2)^{2k-n}}{k!\Gamma(k-n+1)} \psi(k-n+1)$$

$$= \sum_{k=0}^{n-1} \frac{(n-k-1)!}{k!} (x/2)^{2k-n}$$

$$+ (-1)^n \sum_{k=n}^{\infty} \frac{(-1)^k (x/2)^{2k-n}}{k!\Gamma(k-n+1)} \psi(k-n+1) \tag{6.83}$$

Finally, making the change of index $m = k - n$ in the last sum in (6.83), we obtain the desired result (for $n = 1, 2, 3, \ldots$ and $x > 0$)

$$Y_n(x) = \frac{2}{\pi} J_n(x) \log(x/2) - \frac{1}{\pi} \sum_{k=0}^{n-1} \frac{(n-k-1)!}{k!} (x/2)^{2k-n}$$

$$- \frac{1}{\pi} \sum_{m=0}^{\infty} \frac{(-1)^m (x/2)^{2m+n}}{m!(m+n)!} [\psi(m+n+1) + \psi(m+1)] \tag{6.84}$$

Graphs of $Y_n(x)$ for various values of n are shown in Fig. 6.2. Observe the logarithmic behavior as $x \to 0^+$. Also note that these functions have oscillatory characteristics similar to those of $J_n(x)$.

6.5.2 Hankel Functions

Another class of Bessel functions is the class of *Bessel functions of the third kind*, or *Hankel functions*, defined by

$$H_p^{(1)}(x) = J_p(x) + iY_p(x) \tag{6.85}$$

and

$$H_p^{(2)}(x) = J_p(x) - iY_p(x) \tag{6.86}$$

The primary motivation for introducing the Hankel functions is that these linear combinations of $J_p(x)$ and $Y_p(x)$ lend themselves more readily to the

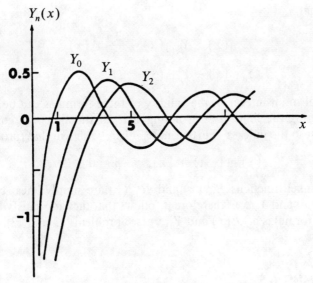

Figure 6.2 Graph of $Y_n(x)$, $n = 0, 1, 2$

development of asymptotic formulas for large x, from which we can deduce the asymptotic formulas for $J_p(x)$ and $Y_p(x)$ (see Section 6.9.2). Also, the Hankel functions are occasionally encountered directly in applications.

It follows from their definition that the Hankel functions are solutions of Bessel's equation (6.69). Moreover, they are linear independent solutions of this DE. Thus we can choose to write the general solution of Bessel's equation in the alternate form

$$y = C_1 H_p^{(1)}(x) + C_2 H_p^{(2)}(x) \qquad (6.87)$$

where C_1 and C_2 are arbitrary constants.

6.5.3 Recurrence Relations

Because $Y_p(x)$ is a linear combination of $J_p(x)$ and $J_{-p}(x)$ for nonintegral p, it follows that $Y_p(x)$ satisfies the same recurrence formulas as $J_p(x)$ and $J_{-p}(x)$. For example, it is easily established that

$$\frac{d}{dx}\left[x^p Y_p(x)\right] = x^p Y_{p-1}(x) \qquad (6.88)$$

$$\frac{d}{dx}\left[x^{-p} Y_p(x)\right] = -x^{-p} Y_{p+1}(x) \qquad (6.89)$$

and also that

$$Y_{p-1}(x) + Y_{p+1}(x) = \frac{2p}{x} Y_p(x) \tag{6.90}$$

$$Y_{p-1}(x) - Y_{p+1}(x) = 2Y_p'(x) \tag{6.91}$$

For p equal to an integer n, the validity of these formulas can be deduced by considering the limit $p \to n$, noting that all functions are continuous with respect to the index p. Furthermore, it can be shown that (problem 14)

$$Y_{-n}(x) = (-1)^n Y_n(x), \quad n = 0, 1, 2, \ldots \tag{6.92}$$

The Hankel functions $H_p^{(1)}(x)$ and $H_p^{(2)}(x)$ are simply linear combinations of $J_p(x)$ and $Y_p(x)$. Therefore it follows that they too satisfy the same recurrence formulas as $J_p(x)$ and $Y_p(x)$ (see problems 12 and 13).

EXERCISES 6.5

In problems 1–4, write the general solution of the DE in terms of Bessel functions.

1. $x^2 y'' + xy' + (x^2 - \frac{1}{4}) y = 0$.
2. $xy'' + y' + xy = 0$.
3. $16 x^2 y'' + 16 xy' + (16 x^2 - 1) y = 0$.
4. $x^2 y'' + xy' + (4x^2 - 1) y = 0$.

    Hint: Let $t = 2x$.

5. Show that the change of variable $y = u(x)/\sqrt{x}$ reduces Bessel's equation (6.69) to

$$u'' + \left[1 + \frac{1 - 4p^2}{4x^2}\right] u = 0$$

6. Use the result of problem 5 to find a general solution of Bessel's equation (6.69) when $p = \frac{1}{2}$ that does not involve Bessel functions.
7. The Wronskian of the solutions of the second-order DE

$$y'' + a(x) y' + b(x) y = 0$$

    is given by (*Abel's formula*)

$$W(y_1, y_2)(x) = C \exp\left(-\int a(x)\, dx\right)$$

    for some constant C. Use this result to deduce that the Wronskian of the solutions of Bessel's equation is

$$W(y_1, y_2)(x) = \frac{C}{x}$$

8. From the result of problem 7, show that

   (a) $W(J_p, J_{-p})(x) = -\dfrac{2 \sin p\pi}{\pi x}$, $p \neq$ integer.

   Hint: Use the relation $C = \lim_{x \to 0^+} x W(J_p, J_{-p})(x)$.

   (b) From (a), deduce that $W(J_p, Y_p)(x) = 2/\pi x$.

9. Using the identities $\Gamma(x)\Gamma(1 - x) = \pi \csc \pi x$ and $\psi(1 - x) - \psi(x) = \pi \cot \pi x$, show that

   $$\lim_{p \to n} \frac{\psi(k - p + 1)}{\Gamma(k - p + 1)} = (-1)^{n-k}(n - k - 1)!,$$

   $$k = 0, 1, \ldots, n - 1$$

10. Show that

    (a) $\dfrac{d}{dx}[x^p Y_p(x)] = x^p Y_{p-1}(x)$,

    (b) $\dfrac{d}{dx}[x^{-p} Y_p(x)] = -x^{-p} Y_{p+1}(x)$.

11. From the results of problem 10, deduce that

    (a) $Y_{p-1}(x) + Y_{p+1}(x) = \dfrac{2p}{x} Y_p(x)$,

    (b) $Y_{p-1}(x) - Y_{p+1}(x) = 2 Y_p'(x)$.

12. Show that the identities in problem 10 for $Y_p(x)$ are also true for $H_p^{(1)}(x)$ and $H_p^{(2)}(x)$.

13. Show that the identities in problem 11 for $Y_p(x)$ are also true for $H_p^{(1)}(x)$ and $H_p^{(2)}(x)$.

14. Verify that

    $$Y_{-n}(x) = (-1)^n Y_n(x), \qquad n = 0, 1, 2, \ldots$$

15. By making the change of variable $t = bx$, show that ($b > 0$)

    $$y = C_1 J_p(bx) + C_2 Y_p(bx)$$

    is the general solution of

    $$x^2 y'' + xy' + (b^2 x^2 - p^2) y = 0, \qquad p \geq 0$$

228 • **Special Functions for Engineers and Applied Mathematicians**

16. Show that the boundary value problem ($p \geq 0$)
$$x^2 y'' + xy' + (k^2 x^2 - p^2) y = 0, \quad 0 < x < 1$$
$$y(x) \text{ finite as } x \to 0^+, \quad y(1) = 0$$
has only the set of solutions $y_n(x) = J_p(k_n x)$, $n = 1, 2, 3, \ldots$, where the k's are chosen to satisfy the relation
$$J_p(k) = 0, \quad k > 0$$
Hint: See problem 15.

17. Solve Bessel's equation
$$x^2 y'' + xy' + (x^2 - p^2) y = 0, \quad p \geq 0$$
by assuming a power-series solution of the form (*Frobenius method*)*
$$y = x^s \sum_{n=0}^{\infty} c_n x^n$$
and

(a) show that one solution corresponding to $s = p$ is
$$y_1(x) = J_p(x)$$

(b) For $p = 0$, show that the method of Frobenius leads to the general solution
$$y = (A + B \log x) \sum_{k=0}^{\infty} \frac{(-1)^k (x/2)^{2k}}{(k!)^2}$$
$$+ B \sum_{k=1}^{\infty} \frac{(-1)^{k-1} (x/2)^{2k}}{(k!)^2} \left(1 + \frac{1}{2} + \cdots + \frac{1}{k} \right)$$
where A and B are arbitrary constants.

6.6 Differential Equations Related to Bessel's Equation

Elementary problems are regarded as solved when their solutions can be expressed in terms of tabulated functions, such as trigonometric and exponential functions. The same can be said of many problems of a more complicated nature when their solutions can be expressed in terms of Bessel functions, since extensive tables of Bessel functions have been compiled for various values of x and p.†

---

*For an introductory discussion of the Frobenius method, see L.C. Andrews, *Ordinary Differential Equations with Applications*, Glenview, Ill.: Scott, Foresman, 1982, Chapter 9.

†For example, see M. Abramowitz and I. Stegun (eds.), *Handbook of Mathematical Functions*, Dover Pub. Co., New York (1965), Chapters 9 and 10.

A fairly large number of DEs occurring in physics and engineering are specializations of the form

$$x^2 y'' + (1 - 2a)xy' + [b^2 c^2 x^{2c} + (a^2 - c^2 p^2)] y = 0,$$
$$p \geq 0, \quad b > 0 \quad (6.93)$$

the general solution of which, expressed in terms of Bessel functions, is

$$y = x^a [C_1 J_p(bx^c) + C_2 Y_p(bx^c)] \quad (6.94)$$

where C_1 and C_2 are arbitrary constants.

To derive the solution formula (6.94) requires two transformations of variables. First, let us set

$$y = x^a z \quad (6.95)$$

from which we obtain

$$xy' = x^{a+1} z' + a x^a z$$
$$x^2 y'' = x^{a+2} z'' + 2a x^{a+1} z' + a(a-1) x^a z$$

Then substituting these expressions into (6.93) and simplifying, we get

$$x^2 z'' + xz' + (b^2 c^2 x^{2c} - c^2 p^2) z = 0 \quad (6.96)$$

Next, we make the change of independent variable

$$t = x^c \quad (6.97)$$

from which it follows, through application of the chain rule, that

$$xz' = cx^c \frac{dz}{dt}$$

$$x^2 z'' = c(c-1) x^c \frac{dz}{dt} + c^2 x^{2c} \frac{d^2 z}{dt^2}$$

Hence, Equation (6.96) becomes

$$t^2 \frac{d^2 z}{dt^2} + t \frac{dz}{dt} + (b^2 t^2 - p^2) z = 0 \quad (6.98)$$

whose general solution is (see problem 15 in Exercises 6.5)

$$z(t) = C_1 J_p(bt) + C_2 Y_p(bt) \quad (6.99)$$

Transforming back to the original variables x and y leads us to the desired result (6.94).

Remark: For those cases when p is not an integer, we can express the general solution (6.94) in the alternate form

$$y = x^a \left[C_1 J_p(bx^c) + C_2 J_{-p}(bx^c) \right]$$

Example 4: Find the general solution of *Airy's equation**

$$y'' + xy = 0$$

Solution: In order to compare this equation with (6.93), we must multiply through by x^2, putting it in the form

$$x^2 y'' + x^3 y = 0$$

Thus, we see that

$$1 - 2a = 0, \quad b^2 c^2 = 1, \quad 2c = 3, \quad a^2 - c^2 p^2 = 0$$

from which we calculate $a = \frac{1}{2}$, $b = \frac{2}{3}$, $c = \frac{3}{2}$, and $p = \frac{1}{3}$. The general solution therefore has the form

$$y = x^{1/2} \left[C_1 J_{1/3}\left(\tfrac{2}{3} x^{3/2}\right) + C_2 Y_{1/3}\left(\tfrac{2}{3} x^{3/2}\right) \right]$$

or, since p is not an integer, we also can represent the general solution in the form

$$y = x^{1/2} \left[C_1 J_{1/3}\left(\tfrac{2}{3} x^{3/2}\right) + C_2 J_{-1/3}\left(\tfrac{2}{3} x^{3/2}\right) \right]$$

EXERCISES 6.6

In problems 1–12, express the general solution in terms of Bessel functions.

1. $xy'' + y' + \frac{1}{4} y = 0$.
2. $4x^2 y'' + 4xy' + (x^2 - n^2) y = 0$.
3. $x^2 y'' + xy' + 4(x^4 - k^2) y = 0$.
4. $xy'' - y' + xy = 0$.
5. $xy'' + (1 + 2n) y' + xy = 0$.
6. $x^2 y'' + (x^2 + \frac{1}{4}) y = 0$.
7. $x^2 y'' - 7xy' + (36 x^6 - \frac{175}{16}) y = 0$.

---

*The solutions of this DE, called *Airy functions*, are important in the theory of diffraction of radio waves around the earth's surface.

8. $y'' + y = 0$.
9. $y'' + k^2 x^2 y = 0$.
10. $y'' + k^2 x^4 y = 0$.
11. $4x^2 y'' + (1 + 4x) y = 0$.
12. $x^2 y'' + 5xy' + (9x^2 - 12) y = 0$.
13. Given the DE
$$y'' + ae^{mx} y = 0, \quad m > 0$$
(a) show that the substitution $t = e^{mx}$ transforms it into
$$t \frac{d^2 y}{dt^2} + \frac{dy}{dt} + \frac{a}{m^2} y = 0$$
(b) Solve the DE in (a) in terms of Bessel functions.
(c) Write the general solution of the original DE in terms of Bessel functions.

14. Given the DE
$$x^2 y'' + x(1 - 2x \tan x) y' - (x \tan x + n^2) y = 0$$
(a) show that the transformation $y = u(x) \sec x$ leads to an equation in u solvable in terms of Bessel functions.
(b) Write the general solution for y in terms of Bessel functions.

15. A particle of variable mass $m = (a + bt)^{-1}$, where a and b are positive constants, starting from rest at a distance r_0 from the origin O, is attracted to O by a force always directed toward O and whose magnitude is $k^2 mr$ ($k > 0$). The equation of motion is given by
$$\frac{d}{dt}\left(m \frac{dr}{dt}\right) = -k^2 mr$$
Solve this equation for r subject to the prescribed initial conditions.

Hint: Make the change of variable $bx = a + bt$, transforming the equation of motion to $x^2 r'' - xr' + k^2 x^2 r = 0$.

16. In a problem on the stability of a tapered strut, the displacement y satisfies the boundary-value problem
$$y'' + \left(\frac{K^2}{4x}\right) y = 0, \quad y'(a) = 0, \ y'(b) = 0 \quad (0 < a < b)$$
For solutions to exist, show that the constant K must satisfy
$$J_0(K\sqrt{a}) Y_0(K\sqrt{b}) = J_0(K\sqrt{b}) Y_0(K\sqrt{a}), \quad K > 0$$

232 • Special Functions for Engineers and Applied Mathematicians

17. The small deflections of a uniform column of length b bending under its own weight are governed by

$$\theta'' + K^2 x \theta = 0, \qquad \theta'(0) = 0, \quad \theta(b) = 0$$

where θ is the angle of deflection from the vertical and K is a positive constant.

(a) Show that the solution of the DE satisfying the first boundary condition at $x = 0$ is

$$\theta(x) = Cx^{1/2} J_{-1/3}\left(\tfrac{2}{3} K x^{3/2}\right)$$

where C is an arbitrary constant.

(b) Show that the shortest column length for which buckling may occur (denoted by b_0) is $b_0 \simeq 1.99 K^{-2/3}$.

Hint: The first zero of $J_{-1/3}(u)$ is $u \simeq 1.87$.

18. An axial load P is applied to a column whose circular cross section is tapered so that the moment of inertia is $I(x) = (x/a)^4$. If the column is simply supported at the ends $x = 1$ and $x = a$ ($a > 1$), the deflections are governed by

$$x^4 y'' + k^2 y = 0, \qquad y(1) = 0, \quad y(a) = 0$$

where $k^2 = Pa^4/E$ (constant).

(a) Express the general solution (not satisfying the boundary conditions) in terms of Bessel functions.

(b) By making the substitution $y = xu(x)$ followed by $x = 1/t$, show that the general solution of the DE can also be expressed in terms of sines and cosines.

(c) Apply the prescribed boundary conditions to the solution in (b) and show that the first buckling mode is described by

$$y(x) = x \sin\left[\frac{a\pi}{a-1}\left(1 - \frac{1}{x}\right)\right]$$

Remark: For additional applications like problems 16–18, consult N.W. McLachlan, *Bessel Functions for Engineers*, 2nd ed., London: Oxford U.P., 1961.

6.7 Modified Bessel Functions

In the previous section we found that the general solution of

$$x^2 y'' + xy' + (b^2 x^2 - p^2) y = 0, \qquad p \geq 0, \quad b > 0 \qquad (6.100)$$

is given by
$$y = C_1 J_p(bx) + C_2 Y_p(bx) \qquad (6.101)$$
where C_1 and C_2 are any constants. The DE
$$x^2 y'' + xy' - (x^2 + p^2) y = 0, \qquad p \geq 0 \qquad (6.102)$$
which bears great resemblance to Bessel's equation, is *Bessel's modified equation*. It is of the form (6.100) with $b^2 = -1$, and so we can formally write the solution of (6.102) as
$$y = C_1 J_p(ix) + C_2 Y_p(ix) \qquad (6.103)$$

The disadvantage of the general solution (6.103) is that it is expressed in terms of complex functions, and in most situations we prefer real functions. The problem is similar to stating that
$$y = C_1 e^{ix} + C_2 e^{-ix}$$
is the general solution of $y'' + y = 0$. In order to avoid the imaginary arguments in (6.103), we first define the *modified Bessel function of the first kind* of order p,
$$I_p(x) = i^{-p} J_p(ix) = \sum_{m=0}^{\infty} \frac{(x/2)^{2m+p}}{m! \Gamma(m + p + 1)} \qquad (6.104)$$

When p is not an integer, the function $I_{-p}(x)$ [obtained by replacing p with $-p$ in (6.104)] is another solution of (6.102) which is linearly independent of $I_p(x)$, since $J_p(ix)$ and $J_{-p}(ix)$ are linearly independent. However, when $p = n$ ($n = 0, 1, 2, \ldots$), we find that
$$I_{-n}(x) = i^n J_{-n}(ix)$$
$$= i^n (-1)^n J_n(ix)$$
$$= (-1)^{2n} I_n(x)$$
or
$$I_{-n}(x) = I_n(x), \qquad n = 0, 1, 2, \ldots \qquad (6.105)$$

Rather than use $Y_p(ix)$ to define a second linearly independent solution of (6.102) which is not restricted to certain values of p, in most applications it is preferable to introduce *Macdonald's function*, or the *modified Bessel function of the second kind* of order p,
$$K_p(x) = \frac{\pi}{2} \frac{I_{-p}(x) - I_p(x)}{\sin p\pi} \qquad (6.106)$$

When p takes on integer values, we define
$$K_n(x) = \lim_{p \to n} K_p(x), \qquad n = 0, 1, 2, \ldots \qquad (6.107)$$

Following a procedure analogous to that in Section 6.5.1 (see problems 15 and 16), it can be shown that $(x > 0)$

$$K_0(x) = -I_0(x)[\gamma + \log(x/2)] + \sum_{k=1}^{\infty} \frac{(x/2)^{2k}}{(k!)^2}\left(1 + \frac{1}{2} + \cdots + \frac{1}{k}\right)$$

(6.108)

and for $n = 1, 2, 3, \ldots$ $(x > 0)$

$$K_n(x) = (-1)^{n-1}I_n(x)\log(x/2) + \frac{1}{2}\sum_{k=0}^{n-1}\frac{(-1)^k(n-k-1)!}{k!}(x/2)^{2k-n}$$

$$+ \frac{(-1)^n}{2}\sum_{m=0}^{\infty}\frac{(x/2)^{2m+n}}{m!(m+n)!}[\psi(m+n+1) + \psi(m+1)]$$

(6.109)

Thus, for all values of p $(p \geq 0)$ we write the general solution of Bessel's modified equation (6.102) as

$$y = C_1 I_p(x) + C_2 K_p(x) \tag{6.110}$$

The graphs of $I_n(x)$ and $K_n(x)$ for $n = 0, 1,$ and 2 are shown in Figs. 6.3 and 6.4. Observe the negative exponential decay of $K_n(x)$ as $x \to +\infty$, while the graphs of $I_n(x)$ appear to grow exponentially for large values of x. None of the modified Bessel functions displays the oscillatory characteristics that are associated with $J_n(x)$ and $Y_n(x)$. Because $J_n(x)$ and $Y_n(x)$ have characteristics more closely associated with the sine and cosine, and $I_n(x)$ and $K_n(x)$ have characteristics similar to the hyperbolic sine and hyperbolic cosine, the former are sometimes referred to as *circular Bessel functions* and the latter as *hyperbolic Bessel functions*.

An important relation between the modified Bessel function of the second kind and the Hankel functions can be derived through relations between $I_p(x)$ and $K_p(x)$ and the Bessel functions of the first and second kinds. We start with the relation

$$H_p^{(1)}(ix) = J_p(ix) + iY_p(ix)$$

$$= J_p(ix) + \frac{i}{\sin p\pi}[\cos p\pi J_p(ix) - J_{-p}(ix)]$$

$$= \frac{J_{-p}(ix) - e^{-ip\pi}J_p(ix)}{i\sin p\pi}$$

$$= \frac{e^{-ip\pi/2}I_{-p}(x) - e^{-ip\pi/2}I_p(x)}{i\sin p\pi}$$

$$= \frac{2}{i\pi}e^{-ip\pi/2}K_p(x)$$

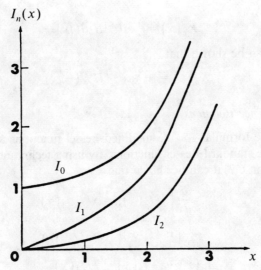

Figure 6.3 Graph of $I_n(x)$, $n = 0, 1, 2$

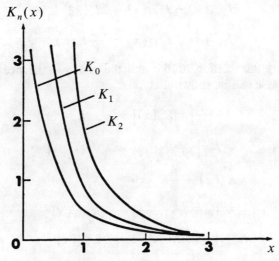

Figure 6.4 Graph of $K_n(x)$, $n = 0, 1, 2$

or equivalently,
$$K_p(x) = \tfrac{1}{2}\pi i^{p+1}H_p^{(1)}(ix) \qquad (6.111a)$$

Similarly, it can be shown that
$$K_p(x) = -\tfrac{1}{2}\pi i^{1-p}H_p^{(2)}(-ix) \qquad (6.111b)$$

6.7.1 Recurrence Relations

The recurrence formulas for the modified Bessel functions are very similar to those of the standard Bessel functions. By using techniques analogous to those in Section 6.2, it can be shown that

$$\frac{d}{dx}\left[x^p I_p(x)\right] = x^p I_{p-1}(x) \qquad (6.112)$$

$$\frac{d}{dx}\left[x^{-p} I_p(x)\right] = x^{-p} I_{p+1}(x) \qquad (6.113)$$

$$I_p'(x) + \frac{p}{x}I_p(x) = I_{p-1}(x) \qquad (6.114)$$

$$I_p'(x) - \frac{p}{x}I_p(x) = I_{p+1}(x) \qquad (6.115)$$

$$I_{p-1}(x) + I_{p+1}(x) = 2I_p'(x) \qquad (6.116)$$

$$I_{p-1}(x) - I_{p+1}(x) = \frac{2p}{x}I_p(x) \qquad (6.117)$$

Also, by using the relation (6.106) and the above recurrence formulas for $I_p(x)$, it likewise can be shown that

$$\frac{d}{dx}\left[x^p K_p(x)\right] = -x^p K_{p-1}(x) \qquad (6.118)$$

$$\frac{d}{dx}\left[x^{-p} K_p(x)\right] = -x^{-p} K_{p+1}(x) \qquad (6.119)$$

$$K_p'(x) + \frac{p}{x}K_p(x) = -K_{p-1}(x) \qquad (6.120)$$

$$K_p'(x) - \frac{p}{x}K_p(x) = -K_{p+1}(x) \qquad (6.121)$$

$$K_{p-1}(x) + K_{p+1}(x) = -2K_p'(x) \qquad (6.122)$$

$$K_{p-1}(x) - K_{p+1}(x) = -\frac{2p}{x}K_p(x) \qquad (6.123)$$

6.7.2 Integral Representations

By starting with the integral representation
$$J_p(x) = \frac{(x/2)^p}{\sqrt{\pi}\,\Gamma(p+\tfrac{1}{2})}\int_{-1}^{1}(1-t^2)^{p-\tfrac{1}{2}}e^{ixt}\,dt, \qquad p > -\tfrac{1}{2}, \quad x > 0$$

replacing x with ix, and multiplying both sides by i^{-p}, we can derive the integral representation

$$I_p(x) = \frac{(x/2)^p}{\sqrt{\pi}\,\Gamma(p + \tfrac{1}{2})} \int_{-1}^{1} (1 - t^2)^{p-\frac{1}{2}} e^{-xt}\, dt, \qquad p > -\tfrac{1}{2}, \quad x > 0$$

(6.124)

Other integral representations for the modified Bessel function $I_p(x)$ can be derived in a similar manner.

A result for $K_p(x)$ which is similar to (6.124), but more complicated to derive, is the integral representation

$$K_p(x) = \frac{\sqrt{\pi}\,(x/2)^p}{\Gamma(p + \tfrac{1}{2})} \int_{1}^{\infty} (t^2 - 1)^{p-\frac{1}{2}} e^{-xt}\, dt, \qquad p > -\tfrac{1}{2}, \quad x > 0$$

(6.125)

Another representation for $K_p(x)$, which is valid for arbitrary p, is given by

$$K_p(x) = \frac{1}{2}\left(\frac{x}{2}\right)^p \int_0^\infty e^{-t-(x^2/4t)} t^{-p-1}\, dt, \qquad x > 0 \qquad (6.126)$$

The derivations of (6.125) and (6.126) without the use of complex variable theory are quite involved and therefore will be omitted.*

EXERCISES 6.7

1. Verify directly that $y_1 = I_p(x)$ and $y_2 = I_{-p}(x)$ are solutions of Bessel's modified equation (6.102).

   In problems 2–5, write the general solution of the DE in terms of modified Bessel functions.

2. $x^2 y'' + xy' - (x^2 + 1)y = 0$.   3. $xy'' + y' - xy = 0$.

4. $4x^2 y'' + 4xy' - (4x^2 + 1)y = 0$.   5. $x^2 y'' + xy' - (4x^2 + 1)y = 0$.

6. If y_1 and y_2 are any two solutions of Bessel's modified equation (6.102), show that for some constant C, the Wronskian is

$$W(y_1, y_2)(x) = \frac{C}{x}$$

   Hint: See problem 7 in Exercises 6.5

---

*The derivations of (6.125) and (6.126) are discussed in N.N. Lebedev, *Special Functions and Their Applications*, New York: Dover, 1972, pp. 116–120.

7. Using the result of problem 6, deduce that

   (a) $W(I_p, I_{-p})(x) = -\dfrac{2 \sin p\pi}{\pi x}$,

   (b) $W(I_p, K_p)(x) = -\dfrac{1}{x}$.

8. Show that
$$I_p(x)K_{p+1}(x) + I_{p+1}(x)K_p(x) = \dfrac{1}{x}$$

9. Show that $K_{-p}(x) = K_p(x)$.

10. Show that

    (a) $I_{1/2}(x) = \sqrt{\dfrac{2}{\pi x}}\sinh x$,   (b) $I_{-1/2}(x) = \sqrt{\dfrac{2}{\pi x}}\cosh x$,

    (c) $K_{1/2}(x) = \sqrt{\dfrac{2}{\pi x}}\,e^{-x}$,   (d) $[I_{-1/2}(x)]^2 - [I_{1/2}(x)]^2 = \dfrac{2}{\pi x}$.

11. Show that

    (a) $\dfrac{d}{dx}[x^p I_p(x)] = x^p I_{p-1}(x)$,

    (b) $\dfrac{d}{dx}[x^{-p} I_p(x)] = x^{-p} I_{p+1}(x)$.

12. Using the results of problem 11, show that

    (a) $I_p'(x) = I_{p-1}(x) - \dfrac{p}{x}I_p(x)$,

    (b) $I_p'(x) = \dfrac{p}{x}I_p(x) + I_{p+1}(x)$,

    (c) $I_p'(x) = \tfrac{1}{2}[I_{p-1}(x) + I_{p+1}(x)]$,

    (d) $I_{p-1}(x) - I_{p+1}(x) = \dfrac{2p}{x}I_p(x)$.

13. Verify that

    (a) $\dfrac{d}{dx}[x^p K_p(x)] = -x^p K_{p-1}(x)$,

    (b) $\dfrac{d}{dx}[x^{-p} K_p(x)] = -x^{-p} K_{p+1}(x)$.

14. Using the results of problem 13, show that

    (a) $K_p'(x) = -K_{p-1}(x) - \dfrac{p}{x}K_p(x)$,

    (b) $K_p'(x) = \dfrac{p}{x}K_p(x) - K_{p+1}(x)$,

    (c) $K_p'(x) = -\tfrac{1}{2}[K_{p-1}(x) + K_{p+1}(x)]$,

    (d) $K_{p-1}(x) - K_{p+1}(x) = -\dfrac{2p}{x}K_p(x)$.

15. Show that

$$K_0(x) = -\lim_{p\to 0} \frac{\partial}{\partial p} I_p(x)$$

and use this result to deduce that $(x > 0)$

$$K_0(x) = -I_0(x)[\log(x/2) + \gamma] + \sum_{k=1}^{\infty} \frac{(x/2)^{2k}}{(k!)^2}\left(1 + \frac{1}{2} + \cdots + \frac{1}{k}\right)$$

16. Show that (for $n = 1, 2, 3, \ldots$)

$$K_n(x) = \frac{(-1)^n}{2} \lim_{p\to n}\left[\frac{\partial}{\partial p} I_{-p}(x) - \frac{\partial}{\partial p} I_p(x)\right]$$

and deduce that $(x > 0)$

$$K_n(x) = (-1)^{n-1} I_n(x)\log(x/2) + \frac{1}{2}\sum_{k=0}^{n-1} \frac{(-1)^k(n-k-1)!}{k!}(x/2)^{2k-n}$$
$$+ \frac{(-1)^n}{2}\sum_{m=0}^{\infty} \frac{(x/2)^{2m+n}}{m!(m+n)!}[\psi(m+n+1) + \psi(m+1)]$$

17. Given

$$F(x) = x^p \int_1^{\infty} (t^2 - 1)^{p-1/2} e^{-xt}\, dt, \qquad p > -\tfrac{1}{2}, \quad x > 0$$

verify that $F(x)$ is a solution of Bessel's modified equation (6.102), and hence conclude that

$$F(x) = AI_p(x) + BK_p(x)$$

for some constants A and B.

18. Derive the generating-function relation

$$\exp\left[\frac{1}{2}x\left(t + \frac{1}{t}\right)\right] = \sum_{n=-\infty}^{\infty} I_n(x) t^n, \qquad t \neq 0$$

19. Use the result of problem 7 in Exercises 6.2 to deduce that

$$I_n(x) = \sum_{m=0}^{\infty} \frac{x^m}{m!} J_{n+m}(x), \qquad n = 0, 1, 2, \ldots$$

20. Show that

$$xI_1(x) = 4\sum_{n=1}^{\infty} nI_{2n}(x)$$

Hint: Use problem 12(d).

240 • Special Functions for Engineers and Applied Mathematicians

21. Use problem 18 to show that

(a) $\exp(x\cos\theta) = \sum\limits_{n=-\infty}^{\infty} I_n(x)\cos n\theta$,

(b) $e^x = I_0(x) + 2\sum\limits_{n=1}^{\infty} I_n(x)$,

(c) $e^{-x} = I_0(x) + 2\sum\limits_{n=1}^{\infty} (-1)^n I_n(x)$.

22. Use problem 21 to verify the identities

(a) $1 = I_0(x) + 2\sum\limits_{n=1}^{\infty} (-1)^n I_{2n}(x)$,

(b) $\cosh x = I_0(x) + 2\sum\limits_{n=1}^{\infty} I_{2n}(x)$,

(c) $\sinh x = 2\sum\limits_{n=1}^{\infty} I_{2n-1}(x)$.

23. If b in the DE

$$x^2 y'' + (1-2a)xy' + \left[b^2 c^2 x^{2c} + (a^2 - c^2 p^2)\right] y = 0, \qquad p \geq 0$$

is allowed to be pure imaginary, say $b = i\beta$, show that the general solution can be expressed as ($\beta > 0$)

$$y = C_1 x^a I_p(\beta x^c) + C_2 x^a K_p(\beta x^c)$$

In problems 24–28, use the result of problem 23 to express the general solution of each DE in terms of modified Bessel functions.

24. $y'' - y = 0$.
25. $y'' - xy = 0$.
26. $x^2 y'' + xy' - (4 + 36x^4)y = 0$.
27. $xy'' - 3y' - 9x^5 y = 0$.
28. $y'' - k^2 x^4 y = 0$.

29. Evaluate:

(a) $\int x I_0(x)\,dx$,

(b) $\int x^2 I_0(x)\,dx$,

(c) $\int x I_1(x)\,dx$,

(d) $\int x^2 I_1(x)\,dx$.

In problems 30–35, verify the integral relation.

30. $I_p(x) = \dfrac{(x/2)^p}{\sqrt{\pi}\,\Gamma(p+\frac{1}{2})} \int_0^\pi e^{\pm x\cos\theta} \sin^{2p}\theta\,d\theta, \quad x > 0$.

31. $I_p(x) = \dfrac{(x/2)^p}{\sqrt{\pi}\,\Gamma(p+\frac{1}{2})} \int_0^\pi \cosh(x\cos\theta)\sin^{2p}\theta\,d\theta, \quad x > 0$.

32. $\int_0^\infty \dfrac{x^{p+1}J_p(bx)}{(x^2+a^2)^{m+1}}\,dx = \dfrac{a^{p-m}b^m}{2^m\Gamma(m+1)}K_{p-m}(ab)$, $a>0, b>0$,
$-1 < p < 2m + \tfrac{3}{2}$.

33. $\int_0^\infty \dfrac{xJ_0(bx)}{\sqrt{a^2+b^2}}\,dx = \dfrac{1}{b}e^{-ab}$, $a \geq 0, b > 0$.

    Hint: Use problem 32.

34. $\int_0^\infty \dfrac{K_m\!\left(a\sqrt{x^2+y^2}\right)}{(x^2+y^2)^{m/2}}J_p(bx)x^{p+1}\,dx$

    $= \dfrac{b^p}{a^m}\left(\dfrac{\sqrt{a^2+b^2}}{y}\right)^{m-p-1}K_{m-p-1}\!\left(y\sqrt{a^2+b^2}\right)$, $a > 0, b > 0$,

    $y > 0, p > -1$.

35. $\int_0^\infty \dfrac{\exp\!\left(-a\sqrt{x^2+y^2}\right)}{\sqrt{x^2+y^2}}J_0(bx)x\,dx = \dfrac{\exp\!\left(-y\sqrt{a^2+b^2}\right)}{\sqrt{a^2+b^2}}$, $a > 0$,

    $y > 0$.

    Hint: Use problem 34.

6.8 Other Bessel Functions

In addition to the Bessel functions introduced thus far, there are a host of related functions also belonging to the same general family. This family includes spherical Bessel functions, Kelvin's functions, Struve functions, Lommel functions, and the Anger and Weber functions. Our treatment in this section, however, will only involve the spherical Bessel functions; some of the related functions mentioned above will be introduced in the exercises.

6.8.1 Spherical Bessel Functions

Spherical Bessel functions are commonly associated with solving the Helmholtz partial differential equation (PDE) in spherical coordinates.* The separation of variables solution technique applied to this PDE leads to an ordinary DE in the radial variable which has the form

$$x^2 y'' + 2xy' + \left[k^2 x^2 - n(n+1)\right] y = 0, \qquad n = 0, 1, 2, \ldots$$

(6.127)

where the constant k enters directly from the Helmholtz equation and the

---

*See Section 7.2.3.

integer n is a separation constant which often has the physical interpretation of angular momentum. We recognize (6.127) as a special case of (6.93) for which $a = -\frac{1}{2}$, $b = k$, $c = 1$, and $p = n + \frac{1}{2}$. Hence, the general solution of (6.127) can be expressed as

$$y = C_1 x^{-1/2} J_{n+\frac{1}{2}}(kx) + C_2 x^{-1/2} Y_{n+\frac{1}{2}}(kx) \qquad (6.128)$$

Because this combination of Bessel functions arises so often in practice, it is customary to define new functions

$$j_n(x) = \sqrt{\frac{\pi}{2x}} J_{n+\frac{1}{2}}(x) \qquad (6.129)$$

$$y_n(x) = \sqrt{\frac{\pi}{2x}} Y_{n+\frac{1}{2}}(x) \qquad (6.130)$$

called, respectively, *spherical Bessel functions of the first and second kind* of order n. *Spherical Hankel functions* can then be defined by

$$h_n^{(1)}(x) = j_n(x) + i y_n(x) \qquad (6.131)$$

$$h_n^{(2)}(x) = j_n(x) - i y_n(x) \qquad (6.132)$$

Recurrence relations analogous to those for Bessel functions can be derived directly from the Bessel relations for the spherical Bessel functions. If we let $f_n(x)$ denote any of $j_n(x)$, $y_n(x)$, $h_n^{(1)}(x)$, or $h_n^{(2)}(x)$, we readily find that (see the exercises)

$$\frac{d}{dx}\left[x^{n+1} f_n(x)\right] = x^{n+1} f_{n-1}(x) \qquad (6.133)$$

$$\frac{d}{dx}\left[x^{-n} f_n(x)\right] = -x^{-n} f_{n+1}(x) \qquad (6.134)$$

$$f_n'(x) = f_{n-1}(x) - \frac{n+1}{x} f_n(x) \qquad (6.135)$$

$$f_n'(x) = \frac{n}{x} f_n(x) - f_{n+1}(x) \qquad (6.136)$$

$$(2n+1) f_n'(x) = n f_{n-1}(x) - (n+1) f_{n+1}(x) \qquad (6.137)$$

$$f_{n-1}(x) + f_{n+1}(x) = \frac{2n+1}{x} f_n(x) \qquad (6.138)$$

By using the series representation

$$J_{n+\frac{1}{2}}(x) = \sum_{m=0}^{\infty} \frac{(-1)^m (x/2)^{2m+n+\frac{1}{2}}}{m!\,\Gamma(m+n+\frac{3}{2})} \qquad (6.139)$$

together with the Legendre duplication formula

$$\sqrt{\pi}\,\Gamma(2x) = 2^{2x-1}\Gamma(x)\Gamma(x + \tfrac{1}{2})$$

it follows that

$$j_n(x) = 2^n x^n \sum_{m=0}^{\infty} \frac{(-1)^m (m+n)!\, x^{2m}}{m!\,(2m + 2n + 1)!} \qquad (6.140)$$

Because of the factorials occurring in both numerator and denominator in (6.140), it becomes an awkward expression for numerical computations. However, it turns out that the spherical Bessel functions are closely related to the trigonometric functions. For instance, by setting $n = 0$ in (6.140), we find

$$j_0(x) = \sum_{m=0}^{\infty} \frac{(-1)^m x^{2m}}{(2m+1)!} \qquad (6.141)$$

and thus deduce that

$$j_0(x) = \frac{\sin x}{x} \qquad (6.142)$$

With the aid of the above recurrence formulas, it can easily be shown that (see problem 7)

$$\begin{aligned} j_1(x) &= \frac{\sin x}{x^2} - \frac{\cos x}{x} \\ j_2(x) &= \left(\frac{3}{x^3} - \frac{1}{x}\right)\sin x - \frac{3}{x^2}\cos x \\ &\vdots \end{aligned} \qquad (6.143)$$

Similarly, it follows that (see problem 8)

$$\begin{aligned} y_0(x) &= -\frac{\cos x}{x} \\ y_1(x) &= -\frac{\cos x}{x^2} - \frac{\sin x}{x} \\ y_2(x) &= -\left(\frac{3}{x^3} - \frac{1}{x}\right)\cos x - \frac{3}{x^2}\sin x \\ &\vdots \end{aligned} \qquad (6.144)$$

Other properties of these functions are taken up in the exercises. The graphs of some of the spherical Bessel functions are shown in Figures 6.5 and 6.6.

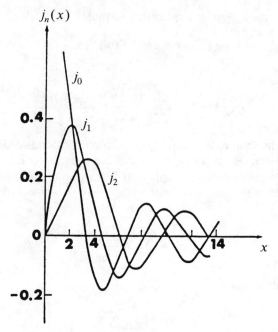

Figure 6.5 Graph of $j_n(x)$, $n = 0, 1, 2$

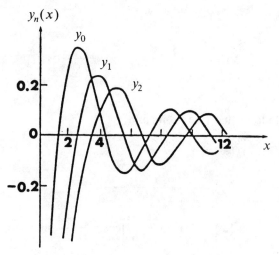

Figure 6.6 Graph of $y_n(x)$, $n = 0, 1, 2$

EXERCISES 6.8

In problems 1–6, verify the given recurrence relation.

1. $\dfrac{d}{dx}[x^{n+1}j_n(x)] = x^{n+1}j_{n-1}(x).$

2. $\dfrac{d}{dx}[x^{-n}j_n(x)] = -x^{-n}j_{n+1}(x).$

3. $j_n'(x) = j_{n-1}(x) - \dfrac{n+1}{x}j_n(x).$

4. $j_n'(x) = \dfrac{n}{x}j_n(x) - j_{n+1}(x).$

5. $(2n+1)j_n'(x) = nj_{n-1}(x) - (n+1)j_{n+1}(x).$

6. $j_{n-1}(x) + j_{n+1}(x) = \dfrac{2n+1}{x}j_n(x).$

7. Show that

    (a) $j_1(x) = \dfrac{\sin x}{x^2} - \dfrac{\cos x}{x},$

    (b) $j_2(x) = \left(\dfrac{3}{x^3} - \dfrac{1}{x}\right)\sin x - \dfrac{3}{x^2}\cos x.$

    Hint: Use any of the recurrence formulas in problems 1–6.

8. Show that

    (a) $y_0(x) = -\dfrac{\cos x}{x},$

    (b) $y_1(x) = -\dfrac{\cos x}{x^2} - \dfrac{\sin x}{x},$

    (c) $y_2(x) = -\left(\dfrac{3}{x^3} - \dfrac{1}{x}\right)\cos x - \dfrac{3}{x^2}\sin x.$

9. Show that (for $n = 0, 1, 2, \ldots$)
$$j_n(x) = \dfrac{(x/2)^n}{2n!}\int_0^\pi \cos(x\cos\theta)\sin^{2n+1}\theta\, d\theta, \qquad x > 0$$

10. Show that the Wronskian of the spherical Bessel functions is given by
$$W(j_n, y_n)(x) = \dfrac{1}{x^2}$$

11. The *spherical modified Bessel functions of the first and second kinds* are defined, respectively, by ($n = 0, 1, 2, \ldots$)
$$i_n(x) = \sqrt{\dfrac{\pi}{2x}}\, I_{n+\frac{1}{2}}(x) \quad \text{and} \quad k_n(x) = \sqrt{\dfrac{2}{\pi x}}\, K_{n+\frac{1}{2}}(x)$$

Show that

(a) $i_0(x) = \dfrac{\sinh x}{x}$. (b) $k_0(x) = \dfrac{e^{-x}}{x}$.

12. Show that the Wronskian of the spherical modified Bessel functions is given by (see problem 11)

$$W(i_n, k_n)(x) = -\dfrac{1}{x^2}$$

13. The *Struve function* of order p is defined by

$$H_p(x) = \dfrac{2(x/2)^p}{\sqrt{\pi}\,\Gamma(p+\tfrac{1}{2})} \int_0^{\pi/2} \sin(x\cos\theta)\sin^{2p}\theta\,d\theta,$$

$$p > -\tfrac{1}{2}, \quad x > 0$$

Show that it has the series representation*

$$H_p(x) = \sum_{k=0}^{\infty} \dfrac{(-1)^k (x/2)^{2k+p+1}}{\Gamma(k+\tfrac{3}{2})\Gamma(k+p+\tfrac{3}{2})}$$

14. Show that (see problem 13)

(a) $H_{-1/2}(x) = J_{1/2}(x)$,

(b) $\displaystyle\int_{-\pi/2}^{\pi/2} e^{-ix\cos\theta}\cos\theta\,d\theta = 2 - \pi[H_1(x) + iJ_1(x)]$.

15. Verify that $H_p(x)$ is a particular solution of (see problem 13)

$$x^2 y'' + xy' + (x^2 - p^2)y = \dfrac{(x/2)^{p+1}}{\sqrt{\pi}\,\Gamma(p+\tfrac{1}{2})}, \qquad p > -\tfrac{1}{2}$$

16. The *integral Bessel function* of order p is defined by

$$\mathrm{Ji}_p(x) = \int_{\infty}^{x} \dfrac{J_p(t)}{t}\,dt$$

Verify that

(a) $p\,\mathrm{Ji}_p(x) = p\displaystyle\int_0^x \dfrac{J_p(t)}{t}\,dt - 1$,

(b) $p\,\mathrm{Ji}_p(x) = \displaystyle\int_0^x J_{p-1}(t)\,dt - J_p(x) - 1$.

Hint: Recall problem 35 in Exercises 6.3.

---

*The series representation is actually valid for *all* p.

17. Referring to problem 16, show that

(a) $\text{Ji}_0(x) = \log(x/2) + \gamma + \sum_{k=1}^{\infty} \frac{(-1)^k (x/2)^{2k}}{2k(k!)^2}$,

(b) $\text{Ji}_n(x) = -\frac{1}{n} + \sum_{k=0}^{\infty} \frac{(-1)^k (x/2)^{2k+n}}{(2k+n)k!(n+k)!}$, $n = 1, 2, 3, \ldots$.

18. *Kelvin's functions* (named after Lord Kelvin) are defined by

$$\text{ber}(x) + i\,\text{bei}(x) = I_0(xe^{\frac{1}{4}\pi i})$$

$$\text{ker}(x) + i\,\text{kei}(x) = K_0(xe^{\frac{1}{4}\pi i})$$

Verify that $y = C_1[\text{ber}(x) + i\,\text{bei}(x)] + C_2[\text{ker}(x) + i\,\text{kei}(x)]$ is a general solution of

$$x^2 y'' + xy' - ix^2 y = 0$$

19. Show that (see problem 18)

(a) $\text{ber}(x) = \sum_{k=0}^{\infty} \frac{(-1)^k (x/2)^{4k}}{[(2k)!]^2}$,

(b) $\text{bei}(x) = \sum_{k=0}^{\infty} \frac{(-1)^k (x/2)^{4k+2}}{[(2k+1)!]^2}$.

20. Using the results of problem 19, show that

$$\text{ber}^2(x) + \text{bei}^2(x) = \sum_{k=0}^{\infty} \frac{(x/2)^{4k}}{(k!)^2 (2k+1)!}$$

21. The *Anger function* is defined by

$$\mathbf{J}_p(x) = \frac{1}{\pi} \int_0^{\pi} \cos(p\theta - x\sin\theta)\, d\theta, \qquad x \geq 0$$

Show that

(a) $2\mathbf{J}_p(x) = \mathbf{J}_{p-1}(x) - \mathbf{J}_{p+1}(x)$,

(b) $\mathbf{J}_{p-1}(x) + \mathbf{J}_{p+1}(x) = \frac{2p}{x}\mathbf{J}_p(x) - \frac{2}{\pi x}\sin p\pi$.

22. The *Weber function* is defined by

$$\mathbf{E}_p(x) = \frac{1}{\pi} \int_0^{\pi} \sin(p\theta - x\sin\theta)\, d\theta, \qquad x \geq 0$$

Show that

(a) $2\mathbf{E}_p'(x) = \mathbf{E}_{p-1}(x) - \mathbf{E}_{p+1}(x)$.

(b) $\mathbf{E}_{p-1}(x) + \mathbf{E}_{p+1}(x) = \dfrac{2p}{x}\mathbf{E}_p(x) - \dfrac{2}{\pi x}(1 - \cos p\pi)$.

6.9 Asymptotic Formulas

For numerical computations it is usually convenient to use simplified asymptotic formulas when the argument of the Bessel function is either very small or very large. In fact, it can be shown that almost all computations involving Bessel functions can be performed with the use of these asymptotic formulas.*

6.9.1 Small Arguments

Let us first examine the cases where x is a positive small number, i.e., $x \ll 1$. Here we simply utilize the first term or so of the series representation. For example, retaining only the first term of the series (6.9) for $J_p(x)$, we obtain the asymptotic formula

$$J_p(x) \sim \frac{(x/2)^p}{\Gamma(p+1)}, \qquad x \to 0^+, \qquad p \neq -1, -2, -3, \ldots \quad (6.145)$$

Similarly, for the modified Bessel function $I_p(x)$ we obtain the same asymptotic formula

$$I_p(x) \sim \frac{(x/2)^p}{\Gamma(p+1)}, \qquad x \to 0^+, \qquad p \neq -1, -2, -3, \ldots \quad (6.146)$$

For the Bessel function of the second kind, we start with the series representation (6.80) for $Y_0(x)$, and write

$$Y_0(x) = \frac{2}{\pi} J_0(x) \left[\log\left(\frac{x}{2}\right) + \gamma \right] + \cdots$$

However, since $J_0(x) \simeq 1$ and $|\log x| \gg \gamma - \log 2$ for $x \ll 1$, we deduce the result

$$Y_0(x) \sim \frac{2}{\pi} \log x, \qquad x \to 0^+ \quad (6.147)$$

In the general case where $p > 0$, we start with

$$Y_p(x) = \frac{\cos(p\pi) J_p(x) - J_{-p}(x)}{\sin(p\pi)}$$

---

*For example, see Chapter 11 in G. Arfken, *Mathematical Methods for Physicists*, New York: Academic, 1970.

Here we make the observation that for small x, $J_p(x) \simeq 0$ and thus

$$Y_p(x) \sim -\frac{J_{-p}(x)}{\sin(p\pi)}$$

$$\sim -\frac{(x/2)^{-p}}{\sin(p\pi)\Gamma(1-p)}, \qquad x \to 0^+$$

Recalling the identity

$$\Gamma(x)\Gamma(1-x) = \frac{\pi}{\sin(\pi x)}$$

we finally arrive at

$$Y_p(x) \sim -\frac{\Gamma(p)}{\pi}\left(\frac{2}{x}\right)^p, \qquad p > 0, \quad x \to 0^+ \qquad (6.148)$$

Without providing the details, it can also readily be shown that (see the exercises)

$$K_0(x) \sim -\log x, \qquad x \to 0^+ \qquad (6.149)$$

$$K_p(x) \sim \frac{\Gamma(p)}{2}\left(\frac{2}{x}\right)^p, \qquad p > 0, \quad x \to 0^+ \qquad (6.150)$$

$$j_n(x) \sim \frac{x^n}{(2n+1)!!}, \qquad x \to 0^+ \qquad (6.151)$$

$$y_n(x) \sim -\frac{(2n-1)!!}{x^{n+1}}, \qquad x \to 0^+ \qquad (6.152)$$

6.9.2 Large Arguments

In order to derive asymptotic formulas for large arguments, we start with the integral representation

$$K_p(x) = \frac{\sqrt{\pi}\,(x/2)^p}{\Gamma(p+\tfrac{1}{2})}\int_1^\infty e^{-xt}(t^2-1)^{p-\tfrac{1}{2}}\,dt, \qquad p > -\tfrac{1}{2}, \quad x > 0 \qquad (6.153)$$

The substitution $t = 1 + u/x$ leads to

$$K_p(x) = \frac{\sqrt{\pi}\,(x/2)^p}{\Gamma(p+\tfrac{1}{2})} e^{-x}\int_0^\infty e^{-u}\left(\frac{u^2}{x^2} + \frac{2u}{x}\right)^{p-\tfrac{1}{2}} \frac{du}{x}$$

$$= \frac{\sqrt{\pi}\,(x/2)^p}{\Gamma(p+\tfrac{1}{2})}\left(\frac{2}{x}\right)^{p-\tfrac{1}{2}} \frac{e^{-x}}{x}\int_0^\infty e^{-u}\left(1 + \frac{u}{2x}\right)^{p-\tfrac{1}{2}} u^{p-\tfrac{1}{2}}\,du \qquad (6.154)$$

Now, for $x \gg u$, we use the approximation $(1 + u/2x)^{p-\frac{1}{2}} \simeq 1$, and thus

$$K_p(x) \sim \sqrt{\frac{\pi}{2x}} \frac{e^{-x}}{\Gamma(p+\frac{1}{2})} \int_0^\infty e^{-u} u^{p-\frac{1}{2}} du,$$

from which we deduce

$$K_p(x) \sim \sqrt{\frac{\pi}{2x}} e^{-x}, \qquad p \geq 0, \quad x \to \infty \qquad (6.155)$$

Based upon the asymptotic formula (6.155), we can derive asymptotic formulas for the other Bessel functions. That is, starting with the relation [from Equation (6.111a)]

$$K_p(x) = \tfrac{1}{2}\pi i^{p+1} H_p^{(1)}(ix)$$

it then follows that (replacing x by $-ix$)

$$H_p^{(1)}(x) = \frac{2}{\pi} i^{-(p+1)} K_p(-ix) \qquad (6.156)$$

Assuming the validity of (6.155) for complex arguments, we are led to

$$H_p^{(1)}(x) \sim \frac{2}{\pi} i^{-(p+1)} \left(\frac{\pi}{-2ix}\right)^{1/2} e^{ix}$$

$$\sim \left(\frac{2}{\pi x}\right)^{1/2} i^{-(p+\frac{1}{2})} e^{ix}, \qquad x \to \infty \qquad (6.157)$$

and by writing $i = e^{i\pi/2}$, we obtain

$$H_p^{(1)}(x) \sim \left(\frac{2}{\pi x}\right)^{1/2} \exp\left\{i\left[x - \frac{(p+\frac{1}{2})\pi}{2}\right]\right\}$$

$$\sim \left(\frac{2}{\pi x}\right)^{1/2} \left\{\cos\left[x - \frac{(p+\frac{1}{2})\pi}{2}\right] + i\sin\left[x - \frac{(p+\frac{1}{2})\pi}{2}\right]\right\},$$

$$x \to \infty \quad (6.158)$$

Finally, recalling the relation

$$H_p^{(1)}(x) = J_p(x) + iY_p(x) \qquad (6.159)$$

we see that equating real and imaginary parts of (6.158) and (6.159) leads to the set of asymptotic formulas ($p \geq 0$)

$$J_p(x) \sim \sqrt{\frac{2}{\pi x}} \cos\left[x - \frac{(p+\frac{1}{2})\pi}{2}\right], \qquad x \to \infty \qquad (6.160)$$

and

$$Y_p(x) \sim \sqrt{\frac{2}{\pi x}} \sin\left[x - \frac{(p+\frac{1}{2})\pi}{2}\right], \qquad x \to \infty \qquad (6.161)$$

Also, from the relation
$$I_p(x) = i^{-p} J_p(ix)$$
we find that
$$I_p(x) \sim \frac{e^x}{\sqrt{2\pi x}}, \qquad p \geq 0, \quad x \to \infty \tag{6.162}$$

Lastly, again using (6.160) and (6.161), it can be shown that
$$j_n(x) \sim \frac{1}{x} \sin\left(x - \frac{n\pi}{2}\right), \qquad x \to \infty \tag{6.163}$$
and
$$y_n(x) \sim -\frac{1}{x} \cos\left(x - \frac{n\pi}{2}\right), \qquad x \to \infty \tag{6.164}$$

EXERCISES 6.9

In problems 1–4, derive the asymptotic formula for small arguments.

1. $K_0(x) \sim -\log x$, $x \to 0^+$.

2. $K_p(x) \sim \dfrac{\Gamma(p)}{2}\left(\dfrac{2}{x}\right)^p$, $p > 0$, $x \to 0^+$.

3. $j_n(x) \sim \dfrac{x^n}{(2n+1)!!}$, $x \to 0^+$.

    Hint: See problem 15 in Exercises 2.2 for the definition of the !! notation.

4. $y_n(x) \sim -\dfrac{(2n-1)!!}{x^{n+1}}$, $x \to 0^+$.

5. By expressing the factor $(1 + u/2x)^{p-\frac{1}{2}}$ in a binomial series in Equation (6.154), show that ($p > -\frac{1}{2}$)
$$K_p(x) \sim \sqrt{\frac{\pi}{2x}} \frac{e^{-x}}{\Gamma(p+\frac{1}{2})} \sum_{n=0}^{\infty} \binom{p-\frac{1}{2}}{n} \frac{\Gamma(p+n+\frac{1}{2})}{(2x)^n}, \qquad x \to \infty$$

6. Show that

    (a) $j_n(x) \sim \dfrac{1}{x} \sin\left(x - \dfrac{n\pi}{2}\right)$, $x \to \infty$,

    (b) $y_n(x) \sim -\dfrac{1}{x} \cos\left(x - \dfrac{n\pi}{2}\right)$, $x \to \infty$.

7. Show that

    (a) $\operatorname{ber}(x) \sim \dfrac{1}{\sqrt{2\pi x}} e^{x/\sqrt{2}} \cos\left(\dfrac{x}{\sqrt{2}} - \dfrac{\pi}{8}\right)$, $x \to \infty$,

    (b) $\operatorname{bei}(x) \sim \dfrac{1}{\sqrt{2\pi x}} e^{x/\sqrt{2}} \sin\left(\dfrac{x}{\sqrt{2}} - \dfrac{\pi}{8}\right)$, $x \to \infty$.

7

Boundary-Value Problems

7.1 Introduction

It was during the nineteenth century that problems of heat conduction and electromagnetic theory were first formulated in terms of partial differential equations (PDEs), the solutions of which often led to one or more special function. Since that time, the study of special functions has been closely linked with the study of differential equations, particularly those arising in mathematical physics. It turns out that the geometry of the problem, rather than the PDE itself, has the most influence on which special function will arise in the solution of such PDEs. For example, we find Legendre functions associated with problems displaying spherical symmetry, Bessel functions for circular or cylindrical domains, and Hermite polynomials for parabolic cylinders.

Finding the solution of a differential equation subject to certain boundary conditions is what we call a *boundary-value problem*. The study of this class of problems reduces primarily to the study of the *heat equation*, the *wave equation*, and *Laplace's equation* (or the *potential equation*). In the space of a single chapter we cannot hope to do justice to this vast subject. Therefore our approach will be largely heuristic, even though many of the mathematical concepts that we use have deep mathematical significance. Nonetheless, it is believed that the few examples presented here will illustrate to some degree how this important field of application can provide a unifying framework in which to introduce many of the special functions.

7.2 Spherical Domains: Legendre Functions

The flow of heat in thermally conducting regions of space is described by solutions of the *heat equation*

$$\nabla^2 u = a^{-2}\frac{\partial u}{\partial t} \tag{7.1}$$

where u denotes the temperature at all spatial points and time within the region of interest and a^2 is a physical constant. The quantity $\nabla^2 u$ is called the *Laplacian*, and in rectangular coordinates takes the form

$$\nabla^2 u = \frac{\partial^2 u}{\partial x^2} + \frac{\partial^2 u}{\partial y^2} + \frac{\partial^2 u}{\partial z^2}$$

The fundamental problem of heat conduction in solids is the solution of (7.1) when the distribution of temperature throughout the solid is known at time $t = 0$ and at the surfaces of the solid certain boundary conditions are prescribed.

When *steady-state* conditions prevail (no heat flow), we have $\partial u/\partial t = 0$, and in this case (7.1) reduces to *Laplace's equation*

$$\nabla^2 u = 0 \tag{7.2}$$

Laplace's equation also arises in the study of the gravitational potential in free space, the electrostatic potential in a uniform dielectric, and the electric potential in the theory of steady flow of currents in solid conductors, among other areas.

In this section we wish to discuss solutions of Laplace's equation and related equations in spherical domains. For solution purposes it is convenient to formulate such problems in spherical coordinates (r, θ, ϕ) as shown in Fig. 7.1.

7.2.1 Electric Potential Due to a Sphere

Suppose that on the surface of a hollow sphere of unit radius, a fixed distribution of electric potential is maintained in such a way that it is

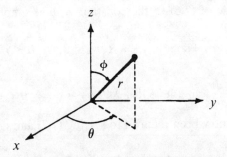

Figure 7.1 Spherical Coordinates

independent of the polar azimuthal angle θ shown in Fig. 7.1. In the absence of any further charges within the sphere, we wish to find the potential distribution $u(r,\phi)$ within the sphere. Laplace's equation (7.2) is the governing equation for this problem; in spherical coordinates (independent of θ) it becomes (see problems 11 and 12)

$$\frac{\partial}{\partial r}\left(r^2 \frac{\partial u}{\partial r}\right) + \frac{1}{\sin\phi}\frac{\partial}{\partial \phi}\left(\sin\phi \frac{\partial u}{\partial \phi}\right) = 0 \qquad (7.3)$$

If the electric potential on the spherical shell is described by $f(\phi)$, then we impose the boundary condition

$$u(1,\phi) = f(\phi), \qquad 0 < \phi < \pi \qquad (7.4)$$

The solution of (7.3) subject to the boundary condition (7.4) is known as a *Dirichlet problem*, or *boundary-value problem of the first kind*.

To solve (7.3), we start with the assumption that the solution can be expressed in the product form

$$u(r,\phi) = R(r)\Phi(\phi) \qquad (7.5)$$

(this is called the *method of separation of variables*). The direct substitution of (7.5) into Laplace's equation (7.3) leads to*

$$\frac{d}{dr}\left[r^2 R'(r)\Phi(\phi)\right] + \frac{1}{\sin\phi}\frac{d}{d\phi}\left[\sin\phi R(r)\Phi'(\phi)\right] = 0$$

which, by rearranging and dividing by the product $R(r)\Phi(\phi)$, becomes

$$\frac{\frac{d}{dr}\left[r^2 R'(r)\right]}{R(r)} = -\frac{\frac{1}{\sin\phi}\frac{d}{d\phi}\left[\sin\phi \Phi'(\phi)\right]}{\Phi(\phi)} \qquad (7.6)$$

We now make the observation that the left-hand side of (7.6) involves only functions of r and the right-hand side only functions of ϕ. Thus we have "separated the variables." Because r and ϕ are independent variables, it follows that the only way (7.6) can be valid is if both sides are constant. Equating each side of (7.6) to the constant λ, and simplifying, we obtain two ordinary DEs:

$$r^2 R''(r) + 2rR'(r) - \lambda R(r) = 0, \qquad 0 < r < 1 \qquad (7.7)$$

and

$$\frac{1}{\sin\phi}\frac{d}{d\phi}\left[\sin\phi \Phi'(\phi)\right] + \lambda\Phi(\phi) = 0, \qquad 0 < \phi < \pi \qquad (7.8)$$

Our problem has now been reduced to solving (7.7) and (7.8).

---

*All partial derivatives become ordinary derivatives under the assumption (7.5), and thus we can resort to the prime notation for derivatives when convenient.

By setting $x = \cos\phi$ in (7.8), we get the more recognizable form (see problem 14 in Exercises 4.2)

$$\frac{d}{dx}\left[(1-x^2)\frac{d\Phi}{dx}\right] + \lambda\Phi = 0, \quad -1 < x < 1 \tag{7.9}$$

Physical considerations demand that the potential u everywhere on and inside the sphere remain bounded. The only bounded solutions of (7.9) occur when λ assumes one of the values (called *eigenvalues*)

$$\lambda_n = n(n+1), \quad n = 0, 1, 2, \ldots \tag{7.10}$$

and in this case we recognize (7.9) as Legendre's equation. Hence, the bounded solutions are given by the Legendre polynomials*

$$\Phi_n(\phi) \equiv P_n(x) = P_n(\cos\phi), \quad n = 0, 1, 2, \ldots \tag{7.11}$$

With the separation constant λ defined by (7.10), Equation (7.7) becomes

$$r^2 R''(r) + 2r R'(r) - n(n+1) R(r) = 0 \tag{7.12}$$

This DE is a *Cauchy-Euler equation* with general solution (see problems 4 and 5)

$$R_n(r) = a_n r^n + b_n r^{-(n+1)}, \quad n = 0, 1, 2, \ldots \tag{7.13}$$

where a_n and b_n denote arbitrary constants. To avoid infinite values of $R_n(r)$ at $r = 0$, we must select $b_n = 0$ for all n. Therefore,

$$R_n(r) = a_n r^n, \quad n = 0, 1, 2, \ldots \tag{7.14}$$

and by forming the product of (7.11) and (7.14) we generate the family of solutions

$$u_n(r, \phi) = a_n r^n P_n(\cos\phi), \quad n = 0, 1, 2, \ldots$$

Finally, summing over all possible values of n (*superposition principle*[†]), we get

$$u(r, \phi) = \sum_{n=0}^{\infty} a_n r^n P_n(\cos\phi) \tag{7.15}$$

Equation (7.15) represents a bounded solution of Laplace's equation (7.3) for any choice of the constants a_n. To satisfy the boundary condition

---

*Recall that the Legendre functions $Q_n(x)$ are not bounded at $x = \pm 1$, i.e., for $\phi = 0$ or $\phi = \pi$ (see Section 4.6).

[†] The *superposition principle* states that if $u_1, u_2, \ldots, u_n, \ldots$, are all solutions of a homogeneous linear PDE, then $u = \Sigma u_n$ is also a solution.

(7.4), however, we must select constants a_n such that

$$u(1,\phi) = f(\phi) = \sum_{n=0}^{\infty} a_n P_n(\cos\phi), \qquad 0 < \phi < \pi \qquad (7.16)$$

This last expression is recognized as a *Legendre series* for $f(\phi)$, and therefore the Fourier coefficients of (7.16) are given by (see problem 42 in Exercises 4.4)

$$a_n = (n + \tfrac{1}{2}) \int_0^\pi f(\phi) P_n(\cos\phi) \sin\phi \, d\phi, \qquad n = 0,1,2,\ldots \qquad (7.17)$$

Example 1: Find the electric potential inside a unit sphere when the boundary potential is prescribed by

$$u(1,\phi) = f(\phi) = \begin{cases} U_0, & 0 \le \phi < \dfrac{\pi}{2} \\ -U_0, & \dfrac{\pi}{2} < \phi \le \pi \end{cases}$$

Solution: The solution is given by Equation (7.15), where the constants are determined from (7.17), which yields

$$a_n = (n+\tfrac{1}{2})U_0\left[\int_0^{\pi/2} P_n(\cos\phi)\sin\phi \, d\phi - \int_{\pi/2}^{\pi} P_n(\cos\phi)\sin\phi \, d\phi\right]$$

$$= (n+\tfrac{1}{2})U_0\left[\int_0^1 P_n(x)\,dx - \int_{-1}^0 P_n(x)\,dx\right]$$

The last step follows from the change of variables $x = \cos\phi$. Owing to the even-odd character of the Legendre polynomials, we see that the replacement of x by $-x$ in the last integral leads to the conclusion

$$a_n = 0, \qquad n = 0,2,4,\ldots$$

and

$$a_n = (2n+1)U_0\int_0^1 P_n(x)\,dx, \qquad n = 1,3,5,\ldots$$

Recalling Example 2 in Section 4.4.3, the evaluation of this integral yields (setting $n = 2k+1$)

$$a_{2k+1} = U_0 \frac{(-1)^k (2k)!(4k+3)}{2^{2k+1}k!(k+1)!}, \qquad k = 0,1,2,\ldots$$

and hence, the solution we seek becomes

$$u(r,\phi) = U_0 \sum_{k=0}^{\infty} \frac{(-1)^k(2k)!(4k+3)}{2^{2k+1}k!(k+1)!} r^{2k+1} P_{2k+1}(\cos\phi)$$

For problems involving electric potentials it is also natural to inquire about the potential outside the sphere ($r > 1$). To determine the potential in this region we must again solve Laplace's equation (7.3) subject to the boundary condition (7.4). In this case, however, our boundedness condition is not prescribed at $r = 0$ (which is outside the region of interest), but for $r \to \infty$. Hence, this time we set $a_n = 0$ ($n = 0, 1, 2, \ldots$) in Equation (7.13) and obtain

$$R_n(r) = \frac{b_n}{r^{n+1}}, \qquad n = 0, 1, 2, \ldots \tag{7.18}$$

Combining (7.11) and (7.18) by the superposition principle leads to

$$u(r, \phi) = \sum_{n=0}^{\infty} \frac{b_n}{r^{n+1}} P_n(\cos \phi) \tag{7.19}$$

The determination of the constants b_n from the boundary condition (7.4) leads to the same integral as before [see (7.17)].

7.2.2 Steady-State Temperatures in a Sphere

Let us now consider the case where the temperature distribution on the surface of a homogeneous solid sphere of unit radius is maintained at a fixed distribution independent of time. Assuming the sphere is void of any heat sources, we wish to determine the (steady-state) temperature distribution everywhere within the sphere. The general form of Laplace's equation in spherical coordinates is given by (see problem 12)

$$\frac{\partial}{\partial r}\left(r^2 \frac{\partial u}{\partial r}\right) + \frac{1}{\sin \phi} \frac{\partial}{\partial \phi}\left(\sin \phi \frac{\partial u}{\partial \phi}\right) + \frac{1}{\sin^2 \phi} \frac{\partial^2 u}{\partial \theta^2} = 0 \tag{7.20}$$

and the temperatures on the surface of the sphere are prescribed by the boundary condition

$$u(1, \theta, \phi) = f(\theta, \phi), \qquad -\pi < \theta < \pi, \quad 0 < \phi < \pi \tag{7.21}$$

To solve (7.20) by the separation-of-variables method, we initially assume the product form

$$u(r, \theta, \phi) = P(r, \phi)\Theta(\theta) \tag{7.22}$$

The substitution of (7.22) into (7.20) and subsequent division by the product $P(r, \phi)\Theta(\theta)$ leads to*

$$\frac{\sin^2 \phi \frac{\partial}{\partial r}\left(r^2 \frac{\partial P}{\partial r}\right) + \sin \phi \frac{\partial}{\partial \phi}\left(\sin \phi \frac{\partial P}{\partial \phi}\right)}{P} = -\frac{\Theta''}{\Theta} \tag{7.23}$$

---

*For notational convenience, we will no longer display the arguments of the functions involved in the separation of variables.

In (7.23) we have separated the variables r, ϕ from θ, and thus by equating both sides to the constant λ, we obtain

$$\Theta'' + \lambda\Theta = 0, \qquad -\pi < \theta < \pi \qquad (7.24)$$

and

$$\frac{\partial}{\partial r}\left(r^2 \frac{\partial P}{\partial r}\right) + \frac{1}{\sin\phi}\frac{\partial}{\partial \phi}\left(\sin\phi \frac{\partial P}{\partial \phi}\right) - \frac{\lambda}{\sin^2\phi} P = 0 \qquad (7.25)$$

To preserve the single-valuedness of the temperature distribution $u(r, \theta, \phi)$, we must require that $\Theta(\theta)$ be a *periodic function* with period 2π. Hence, we impose the *periodic boundary conditions*

$$\Theta(-\pi) = \Theta(\pi), \qquad \Theta'(-\pi) = \Theta'(\pi) \qquad (7.26)$$

The solution of (7.24) satisfying the periodic conditions (7.26) demands that λ be restricted to the values

$$\lambda_m = m^2, \qquad m = 0, 1, 2, \ldots \qquad (7.27)$$

and thus we obtain the solutions

$$\Theta_m(\theta) = \begin{cases} a_0, & m = 0 \\ a_m \cos m\theta + b_m \sin m\theta, & m = 1, 2, 3, \ldots \end{cases} \qquad (7.28)$$

Equation (7.25) is still a PDE, and so we apply the separation-of-variables method once more in the hopes of reducing (7.25) to a system of ordinary DEs. Writing $\lambda = m^2$ and setting

$$P(r, \phi) = R(r)\Phi(\phi) \qquad (7.29)$$

we find that

$$\frac{\frac{d}{dr}(r^2 R')}{R} = -\frac{\frac{1}{\sin\phi}\frac{d}{d\phi}(\sin\phi \, \Phi')}{\Phi} + \frac{m^2}{\sin^2\phi} = \mu$$

and consequently,

$$r^2 R'' + 2rR' - \mu R = 0, \qquad 0 < r < 1 \qquad (7.30)$$

and

$$\frac{1}{\sin\phi}\frac{d}{d\phi}(\sin\phi \, \Phi') + \left[\mu - \frac{m^2}{\sin^2\phi}\right]\Phi = 0, \qquad 0 < \phi < \pi \qquad (7.31)$$

where μ is the new separation constant.

The change of variable $x = \cos\phi$ in (7.31) puts it in the form

$$\frac{d}{dx}\left[(1-x^2)\frac{d\Phi}{dx}\right] + \left[\mu - \frac{m^2}{1-x^2}\right]\Phi = 0, \quad -1 < x < 1 \quad (7.32)$$

The temperature distribution throughout the sphere must remain bounded, and this condition requires that μ be restricted to the set of values

$$\mu_n = n(n+1), \quad n = 0, 1, 2, \ldots \quad (7.33)$$

However, for these values of μ we see that (7.32) is the *associated Legendre equation* (Section 4.7), and its bounded solutions are the *associated Legendre functions* defined by

$$\Phi_{mn}(\phi) \equiv P_n^m(x) = P_n^m(\cos\phi), \quad m, n = 0, 1, 2, \ldots \quad (7.34)$$

For $\mu = n(n+1)$, the bounded solutions of (7.30) are of the form

$$R_n(r) = c_n r^n$$

and thus we see that $u_{mn}(r, \theta, \phi) = R_n(r)\Theta_m(\theta)\Phi_{mn}(\phi)$ gives us the family of solutions

$$u_{mn}(r, \theta, \phi)$$
$$= \begin{cases} A_{0n} r^n P_n(\cos\phi), & m = 0 \\ (A_{mn}\cos m\theta + B_{mn}\sin m\theta) r^n P_n^m(\cos\phi), & m = 1, 2, 3, \ldots \end{cases}$$
(7.35)

where $A_{0n} = a_0 c_n$, $A_{mn} = a_m c_n$, and $B_{mn} = b_m c_n$. Finally, summing over all such solutions by invoking the superposition principle, we arrive at

$$u(r, \theta, \phi) = \sum_{n=0}^{\infty} A_{0n} r^n P_n(\cos\phi)$$
$$+ \sum_{m=1}^{\infty} \sum_{n=0}^{\infty} (A_{mn}\cos m\theta + B_{mn}\sin m\theta) r^n P_n^m(\cos\phi)$$
(7.36)

The constants A_{0m}, A_{mn}, and B_{mn} have to be selected in such a way that the boundary condition (7.21) is satisfied, which leads to

$$f(\theta, \phi) = \sum_{n=0}^{\infty} A_{0n} P_n(\cos\phi)$$
$$+ \sum_{m=1}^{\infty} \sum_{n=0}^{\infty} (A_{mn}\cos m\theta + B_{mn}\sin m\theta) P_n^m(\cos\phi) \quad (7.37)$$

This last relation is what is known as a *generalized Fourier series in two variables*. Although the theory associated with such series follows in a natural way from the theory of one variable, it goes beyond the intended scope of this text. The interested reader might consult one of the standard texts in PDEs, such as A.G. Webster, *Partial Differential Equations of Mathematical Physics*, New York: Dover, 1955. As a final observation here, we note that for the special case where the prescribed temperatures are independent of the angle θ, the temperatures inside the sphere will also be independent of θ. This condition necessitates that we allow only $m = 0$ in the solution (7.36), and in this case our solution (7.36) reduces to the result (7.15).

7.2.3 Solutions of the Helmholtz Equation

The Legendre functions are prominent not only in problems featuring Laplace's equation, but also with other equations. For example, solutions of the *Helmholtz equation**

$$\nabla^2 \psi + k^2 \psi = 0 \tag{7.38}$$

which is of particular importance in mathematical physics, also lead to Legendre functions.

By assuming that (7.38) has solutions of the form

$$\psi(r, \theta, \phi) = R(r)\Theta(\theta)\Phi(\phi) \tag{7.39}$$

the separation-of-variables technique leads to the system of ordinary DEs (see problem 16)

$$\Theta'' + \lambda \Theta = 0 \tag{7.40}$$

$$\frac{1}{\sin \phi} \frac{d}{d\phi}(\sin \phi \, \Phi') + \left[\mu - \frac{\lambda}{\sin^2 \phi}\right]\Phi = 0 \tag{7.41}$$

and

$$\frac{d}{dr}(r^2 R') + (k^2 r^2 - \mu)R = 0 \tag{7.42}$$

where λ and μ denote separation constants. If we require that the solutions be bounded and periodic with period 2π, we once again arrive at the conclusion that the separation constants are restricted to

$$\lambda_m = m^2, \quad m = 0, 1, 2, \ldots \tag{7.43}$$

---

*The Helmholtz equation arises in the separation-of-variables technique applied to both the heat and the wave equation. See problems 13 and 14.

and
$$\mu_n = n(n+1), \quad n = 0, 1, 2, \ldots \tag{7.44}$$

With these restrictions, we see that (7.40) has the solutions given by (7.28) and that (7.41) is the associated Legendre equation with bounded solutions $\Phi_{mn}(\phi) = P_n^m(\cos\phi)$.

Also, for $\mu = n(n+1)$, Equation (7.42) is recognized as a special case of Bessel's equation whose bounded solutions are (see Section 6.8.1)

$$R_n(r) = j_n(kr) \tag{7.45}$$

where $j_n(kr)$ is the *spherical Bessel function of the first kind*. We conclude, therefore, that all bounded periodic solutions of the Helmholtz equation (7.38) in spherical coordinates are various linear combinations of the family of solutions

$$\psi_{mn}(r,\theta,\phi)$$
$$= \begin{cases} A_{0n} j_n(kr) P_n(\cos\phi), & m = 0 \\ (A_{mn}\cos m\theta + B_{mn}\sin m\theta) j_n(kr) P_n^m(\cos\phi), & m = 1, 2, 3, \ldots \end{cases} \tag{7.46}$$

where $n = 0, 1, 2, \ldots$.

EXERCISES 7.2

1. Find the electric potential in the *interior* of the unit sphere assuming the potential on the surface is

   (a) $f(\phi) = 1$,   (b) $f(\phi) = \cos\phi$,
   (c) $f(\phi) = \cos^2\phi$,   (d) $f(\phi) = \cos 2\phi$.

2. Find the electric potential in the *exterior* of the unit sphere assuming the potential on the surface is prescribed as given in problem 1.

3. The base $\phi = \tfrac{1}{2}\pi$, $r < 1$, of a solid hemisphere $r \le 1$, $0 \le \phi \le \tfrac{1}{2}\pi$, is kept at temperature $u = 0$, while $u = T_0$ on the hemispherical surface $r = 1$, $0 < \phi < \tfrac{1}{2}\pi$. Show that the steady-state temperature distribution is given by

$$u(r,\phi) = T_0 \sum_{n=0}^{\infty} (-1)^n \left(\frac{4n+3}{2n+2}\right) \frac{(2n)!}{2^{2n}(n!)^2} r^{2n+1} P_{2n+1}(\cos\phi)$$

4. The DE
$$ax^2y'' + bxy' + cy = 0$$
where a, b, and c are constants, is called a *Cauchy-Euler equation*.
   (a) Show that the change of variable $x = e^t$ leads to
$$xy' = Dy, \quad x^2y'' = D(D-1)y, \quad D = \frac{d}{dt}$$
   (b) Using the result of part (a), show that the Cauchy-Euler equation can be transformed into the constant-coefficient DE
$$[aD^2 + (b-a)D + c]y = 0$$

5. Use the result of problem 4 to verify that (7.13) is the general solution of the Cauchy-Euler equation
$$r^2R'' + 2rR' - n(n+1)R = 0$$

6. Solve the electric-potential problem in Section 7.2.1 for a sphere of radius c.

7. If the potential on the surface of a sphere of unit radius is kept at constant potential U_0, show that at points far from the spherical surface the potential is (approximately) given by
$$u(r,\phi) \simeq \frac{3U_0}{2r^2}, \quad r \gg 1$$

8. For a long period of time, the temperature $u(r,\phi)$ on the surface of a sphere of radius c has been maintained at $u(c,\phi) = T_0(1 - \cos^2\phi)$, where T_0 is a constant and ϕ is the cone angle in spherical coordinates. Find the temperature inside the sphere.

9. A spherical shell has an inner radius of 1 unit and an outer radius of 2 units. The prescribed temperatures on the inner and outer surfaces are given respectively by
$$u(1,\phi) = 30 + 10\cos\phi, \quad u(2,\phi) = 50 - 20\cos\phi$$
Determine the steady-state temperature everywhere within the spherical shell.

10. Show that the Laplacian $\nabla^2 u$ in cylindrical coordinates defined by
$$x = r\cos\theta, \quad y = r\sin\theta, \quad z = z$$
is given by
$$\nabla^2 u = \frac{\partial^2 u}{\partial r^2} + \frac{1}{r}\frac{\partial u}{\partial r} + \frac{1}{r^2}\frac{\partial^2 u}{\partial \theta^2} + \frac{\partial^2 u}{\partial z^2}$$
Hint: Start with $\nabla^2 u = \dfrac{\partial^2 u}{\partial x^2} + \dfrac{\partial^2 u}{\partial y^2} + \dfrac{\partial^2 u}{\partial z^2}$ and use the chain rule.

11. Show that the Laplacian $\nabla^2 u$ in spherical coordinates defined by
$$x = r\cos\theta \sin\phi, \qquad y = r\sin\theta \sin\phi, \qquad z = r\cos\phi,$$
is given by
$$\nabla^2 u = \frac{\partial^2 u}{\partial r^2} + \frac{2}{r}\frac{\partial u}{\partial r} + \frac{1}{r^2}\frac{\partial^2 u}{\partial \phi^2} + \frac{\cot\phi}{r^2}\frac{\partial u}{\partial \phi} + \frac{1}{r^2 \sin^2\phi}\frac{\partial^2 u}{\partial \theta^2}$$

12. Show that the Laplacian in problem 11 can also be expressed as
$$\nabla^2 u = \frac{1}{r^2}\left[\frac{\partial}{\partial r}\left(r^2 \frac{\partial u}{\partial r}\right) + \frac{1}{\sin\phi}\frac{\partial}{\partial \phi}\left(\sin\phi \frac{\partial u}{\partial \phi}\right) + \frac{1}{\sin^2\phi}\frac{\partial^2 u}{\partial \theta^2}\right]$$

13. Show that by assuming the product form $u(r, \theta, \phi, t) = \psi(r, \theta, \phi) W(t)$, the heat equation
$$\nabla^2 u = a^{-2} \frac{\partial u}{\partial t}$$
reduces to the two equations
$$W' + \lambda a^2 W = 0$$
$$\nabla^2 \psi + \lambda \psi = 0$$

14. Show that by assuming the product form $u(r, \theta, \phi, t) = \psi(r, \theta, \phi) W(t)$, the wave equation
$$\nabla^2 u = c^{-2} \frac{\partial^2 u}{\partial t^2}$$
reduces to the two equations
$$W'' + \lambda c^2 W = 0$$
$$\nabla^2 \psi + \lambda \psi = 0$$

15. Let $u(r, \phi, t)$ denote the temperature distribution in a sphere of unit radius (independent of θ) whose surface is maintained at 0°C, and whose initial temperature distribution is described by
$$u(r, \phi, 0) = f(r, \phi)$$

(a) Show that the temperature distribution throughout the sphere is of the general form (see problem 13)
$$u(r, \phi, t) = \sum_{m=1}^{\infty} \sum_{n=0}^{\infty} C_{mn} P_n(\cos\phi) j_n(k_{mn} r) e^{-a^2 k_{mn}^2 t},$$
where k_{mn} denotes the mth solution of $j_n(k) = 0$, $n = 0, 1, 2, \ldots$.

(b) If the initial temperature distribution is constant, i.e., $f(r,\phi) = T_0$, show that the solution in (a) is independent of ϕ and that it reduces to

$$u(r,t) = \frac{2T_0}{\pi} \sum_{m=1}^{\infty} (-1)^{m+1} \left(\frac{\sin m\pi r}{mr} \right) e^{-a^2 m^2 \pi^2 t}$$

16. Verify that the separation-of-variables technique applied to the Helmholtz equation (7.38) by the substitution (7.39) leads to the three equations (7.40)–(7.42).

7.3 Circular and Cylindrical Domains: Bessel Functions

Heat conduction in a circular plate and the vibrations of a circular membrane are mathematically similar problems. Among other similarities, both problems lead to solutions in terms of Bessel functions. Parallelisms of this nature exist for many geometries, and thus for purposes of illustrating the mathematical techniques, it generally doesn't matter whether we solve a heat-conduction problem or a vibration problem.

In this section we wish to discuss two problems, each of which leads to a different kind of Bessel function. One problem involves the vibrations of a circular membrane, and the other the steady-state temperature distribution in a cylinder.

7.3.1 Radial Symmetric Vibrating Membrane

We wish to determine the small displacements u of a thin circular membrane (such as a drumhead) of unit radius whose edge is rigidly fixed. The governing equation for this problem is the *wave equation*

$$\nabla^2 u = c^{-2} \frac{\partial^2 u}{\partial t^2} \qquad (7.47)$$

where c is a physical constant having the dimensions of velocity.

The shape of the region (Fig. 7.2) suggests the use of polar coordinates. Moreover, if the displacement u depends only upon the radial distance r from the center of the membrane and on time t, then (7.47) expressed in terms of polar coordinates becomes

$$\frac{\partial^2 u}{\partial r^2} + \frac{1}{r}\frac{\partial u}{\partial r} = c^{-2}\frac{\partial^2 u}{\partial t^2}, \qquad 0 < r < 1, \quad t > 0 \qquad (7.48)$$

Since we have assumed the membrane is rigidly fixed on the boundary, we

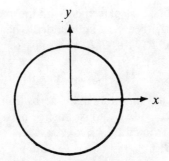

Figure 7.2 A Circular Membrane

impose the boundary condition
$$u(1,t) = 0, \quad t > 0 \tag{7.49}$$
If the membrane is also initially deflected to the form $f(r)$ with velocity $g(r)$, we prescribe the additional (initial) conditions
$$u(r,0) = f(r), \quad \frac{\partial u}{\partial t}(r,0) = g(r), \quad 0 < r < 1 \tag{7.50}$$

Expressing u in the product form $u(r,t) = R(r)W(t)$ and substituting into (7.48), the separation-of-variables technique leads to
$$\frac{R'' + \frac{1}{r}R'}{R} = \frac{W''}{c^2 W} = -\lambda \tag{7.51}$$
where $-\lambda$ is the separation constant. (The choice of the negative sign is conventional.) Hence, (7.51) is equivalent to the system of ordinary DEs
$$rR'' + R' + \lambda r R = 0 \tag{7.52}$$
$$W'' + \lambda c^2 W = 0 \tag{7.53}$$
The boundary condition (7.49) becomes
$$u(1,t) = R(1)W(t) = 0$$
from which we deduce
$$R(1) = 0 \tag{7.54}$$

Equation (7.52) is recognized as Bessel's equation of order zero, with general solution (upon setting $\lambda = k^2 > 0$*)
$$R(r) = C_1 J_0(kr) + C_2 Y_0(kr) \tag{7.55}$$

---

*Equation (7.52) subject to the condition (7.54) has no nontrivial bounded solutions for $\lambda \leq 0$. See problem 14.

266 • **Special Functions for Engineers and Applied Mathematicians**

In order to maintain finite displacements of the membrane at $r = 0$, we must set $C_2 = 0$, since Y_0 becomes unbounded when the argument is zero. The remaining solution, $R(r) = C_1 J_0(kr)$, must then satisfy the boundary condition (7.54), i.e.,

$$R(1) = C_1 J_0(k) = 0 \tag{7.56}$$

The Bessel function J_0 has infinitely many zeros (not evenly spaced) on the positive axis. Thus, for $C_1 \neq 0$, (7.56) is satisfied by selecting k as one of the zeros of J_0, which are denoted by $k_1, k_2, \ldots, k_n, \ldots$. With k so restricted, we set $\lambda = k_n^2$ ($n = 1, 2, 3, \ldots$) in (7.53) to obtain

$$W'' + k_n^2 c^2 W = 0 \tag{7.57}$$

which has the general solution

$$W_n(t) = a_n \cos k_n ct + b_n \sin k_n ct \tag{7.58}$$

Combining our results, we have the family of solutions

$$u_n(r, t) = (a_n \cos k_n ct + b_n \sin k_n ct) J_0(k_n r) \tag{7.59}$$

for $n = 1, 2, 3, \ldots$. These solutions are called *standing waves*, since each can be viewed as having fixed shape $J_0(k_n r)$ with varying amplitude $W_n(t)$. The zeros of a standing wave, i.e., curves where $J_0(k_n r) = 0$, are referred to as *nodal lines*. Clearly, the number of nodal lines depends upon the value of n. For example, when $n = 1$ there is no nodal line for $0 < r < 1$. When $n = 2$, there is one nodal line, when $n = 3$ there are two nodal lines, and so forth (see Fig. 7.3.).

By forming a linear combination of the solutions (7.59) through the superposition principle, we obtain

$$u(r, t) = \sum_{n=1}^{\infty} (a_n \cos k_n ct + b_n \sin k_n ct) J_0(k_n r) \tag{7.60}$$

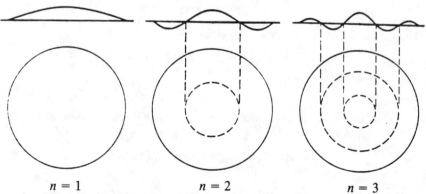

Figure 7.3 Nodal Lines for a Circular Membrane

The constants a_n and b_n are selected in such a way that the initial conditions (7.50) are satisfied. Hence,

$$u(r,0) = f(r) = \sum_{n=1}^{\infty} a_n J_0(k_n r), \qquad 0 < r < 1 \qquad (7.61)$$

and

$$\frac{\partial u}{\partial t}(r,0) = g(r) = \sum_{n=1}^{\infty} k_n c b_n J_0(k_n r), \qquad 0 < r < 1 \qquad (7.62)$$

which are recognized as Bessel series for $f(r)$ and $g(r)$, respectively, where

$$a_n = \frac{2}{[J_1(k_n)]^2} \int_0^1 r f(r) J_0(k_n r) \, dr, \qquad n = 1, 2, 3, \ldots \qquad (7.63)$$

and

$$k_n c b_n = \frac{2}{[J_1(k_n)]^2} \int_0^1 r g(r) J_0(k_n r) \, dr, \qquad n = 1, 2, 3, \ldots \qquad (7.64)$$

7.3.2 Radial Symmetric Problem in a Cylinder

Let us consider a solid homogeneous cylinder with unit radius and height π units (see Fig. 7.4). If the temperatures on the surfaces of the cylinder are prescribed in such a way that they are a function of only the radial distance r and height z, the temperatures inside the cylinder will also depend on only these variables. The problem described is one of steady-states, which is governed by Laplace's equation. In cylindrical coordinates, Laplace's equa-

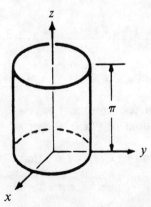

Figure 7.4 A Solid Cylinder

tion has the form (see problem 10, Exercises 7.2)

$$\frac{\partial^2 u}{\partial r^2} + \frac{1}{r}\frac{\partial u}{\partial r} + \frac{1}{r^2}\frac{\partial^2 u}{\partial \theta^2} + \frac{\partial^2 u}{\partial z^2} = 0 \qquad (7.65)$$

but under the conditions just stated, it follows that $\partial^2 u/\partial \theta^2 = 0$, and thus (7.65) reduces to

$$\frac{\partial^2 u}{\partial r^2} + \frac{1}{r}\frac{\partial u}{\partial r} + \frac{\partial^2 u}{\partial z^2} = 0, \qquad 0 < r < 1, \quad 0 < z < \pi \qquad (7.66)$$

We will assume the boundary temperatures are prescribed by

$$u(r,0) = 0, \qquad u(r,\pi) = 0, \qquad u(1,z) = f(z) \qquad (7.67)$$

If we set $u(r, z) = R(r)Z(z)$, then separation of variables leads to

$$rR'' + R' - \lambda rR = 0 \qquad (7.68)$$

$$Z'' + \lambda Z = 0, \qquad Z(0) = 0, \quad Z(\pi) = 0 \qquad (7.69)$$

where λ is once again the separation constant. By assuming $\lambda = k^2 > 0$, the general solution of (7.69) is

$$Z(z) = C_1 \cos kz + C_2 \sin kz \qquad (7.70)$$

Imposing the first boundary condition in (7.69), we see that

$$Z(0) = C_1 = 0 \qquad (7.71)$$

whereas the second condition leads to

$$Z(\pi) = C_2 \sin k\pi = 0 \qquad (7.72)$$

For $C_2 \neq 0$, we can satisfy this relation by choosing $k = n$ ($n = 1, 2, 3, \ldots$). Hence, we find that

$$\lambda = k^2 = n^2, \qquad n = 1, 2, 3, \ldots \qquad (7.73)$$

and

$$Z_n(z) = \sin nz, \qquad n = 1, 2, 3, \ldots \qquad (7.74)$$

where we set $C_1 = 1$ for convenience. For $\lambda = 0$ and $\lambda < 0$ there are no further (nontrivial) solutions of (7.69), the proof of which we leave to the exercises.

For values of λ given by (7.73), we see that (7.68) becomes

$$rR'' + R' - n^2 rR = 0 \qquad (7.75)$$

which we recognize as *Bessel's modified equation of order zero*, with general

solution
$$R_n(r) = c_n I_0(nr) + d_n K_0(nr) \tag{7.76}$$
However, because K_0 is unbounded at $r = 0$, we must select $d_n = 0$ for all n. Then, combining solutions (7.74) and (7.76) through use of the superposition principle, we obtain
$$u(r,z) = \sum_{n=1}^{\infty} c_n I_0(nr)\sin nz \tag{7.77}$$

The remaining task at this point is the determination of the constants c_n. By imposing the last boundary condition in (7.67), we have
$$u(1,z) = f(z) = \sum_{n=1}^{\infty} c_n I_0(n)\sin nz, \qquad 0 < z < \pi \tag{7.78}$$
which is a Fourier sine series for the function $f(z)$. Hence,
$$c_n I_0(n) = \frac{2}{\pi}\int_0^\pi f(z)\sin nz\, dz, \qquad n = 1,2,3,\ldots \tag{7.79}$$

EXERCISES 7.3

1. If the initial conditions (7.50) are given by $f(r) = A J_0(k_1 r)$ (A constant) and $g(r) = 0$, show that the subsequent displacements of the membrane are described by
$$u(r,t) = A J_0(k_1 r)\cos(k_1 ct)$$
where $J_0(k_1) = 0$.

2. If the initial conditions (7.50) are given by $f(r) = 0$, $g(r) = 1$, show that the subsequent displacements of the membrane are described by
$$u(r,t) = \frac{2}{c}\sum_{n=1}^{\infty} \frac{\sin(k_n ct)}{k_n^2 J_1(k_n)} J_0(k_n r)$$
where $J_0(k_n) = 0$ ($n = 1,2,3,\ldots$).

3. Solve the problem described in Section 7.3.1 for a circular membrane of radius ρ.

4. The temperature distribution in a circular plate, independent of the polar angle θ, is described by solutions of
$$\frac{\partial^2 u}{\partial r^2} + \frac{1}{r}\frac{\partial u}{\partial r} = a^{-2}\frac{\partial u}{\partial t}, \qquad 0 < r < 1, \quad t > 0$$
$$\frac{\partial u}{\partial r}(1,t) = 0, \qquad u(r,0) = f(r)$$

Show that solutions are of the form
$$u(r,t) = c_0 + \sum_{n=1}^{\infty} c_n J_0(k_n r) e^{-a^2 k_n^2 t}$$
where $J_0'(k_n) = 0$ $(n = 1, 2, 3, \ldots)$.

5. Solve explicitly for the c's in problem 4 when the initial temperature distribution is (see problem 9 in Exercise 6.4)
   (a) $f(r) = J_0(k_1 r)$, where $J_0'(k_1) = 0$,
   (b) $f(r) = T_0$ (constant),
   (c) $f(r) = 1 - r^2$.

6. Over a long, solid cylinder of unit radius at uniform temperature T_1 is fitted a long, hollow cylinder $1 \le r \le 2$ of the same material at uniform temperature T_2. Show that the temperature distribution throughout the two cylinders is given by
$$u(r,t) = T_2 + \frac{1}{2}(T_1 - T_2) \sum_{n=1}^{\infty} \frac{J_1(k_n)}{k_n [J_1(2k_n)]^2} J_0(k_n r) e^{-a^2 k_n^2 t}$$
where $J_0(2k_n) = 0$ $(n = 1, 2, 3, \ldots)$. What temperature is approached in the limit as $t \to \infty$?

7. The temperature distribution $u(r,t)$ in a thin circular plate with heat exchanges from its faces into the surrounding medium at $0°C$ satisfies the boundary-value problem
$$\frac{\partial^2 u}{\partial r^2} + \frac{1}{r}\frac{\partial u}{\partial r} - bu = \frac{\partial u}{\partial t}, \qquad 0 < r < 1, \quad t > 0$$
$$u(1,t) = 0, \qquad u(r,0) = 1$$
where b is a positive constant. Show that the solution is given by
$$u(r,t) = 2e^{-bt} \sum_{n=1}^{\infty} \frac{J_0(k_n r)}{k_n J_1(k_n)} e^{-k_n^2 t}$$
where $J_0(k_n) = 0$ $(n = 1, 2, 3, \ldots)$.

8. Solve the problem described in Section 7.3.2 when $f(z) = \sin 3z$.

9. A right circular cylinder is 1 m long and 2 m in diameter. One end and its lateral surface are maintained at a temperature of $0°C$ and its other end at $100°C$. Calculate the first three terms in the series solution giving the temperature distribution at an interior point.

10. Show that the solutions of
$$\frac{\partial^2 u}{\partial r^2} + \frac{1}{r}\frac{\partial u}{\partial r} + \frac{\partial^2 u}{\partial z^2} = 0, \qquad 0 < r < 1, \quad 0 < z < a$$
$$u(r,0) = 0, \quad u(r,a) = f(r), \qquad u(1,z) = 0$$

are given by
$$u(r,z) = \sum_{n=1}^{\infty} c_n J_0(k_n r) \frac{\sinh k_n z}{\sinh k_n a}$$
where $J_0(k_n) = 0$ $(n = 1, 2, 3, \ldots)$. Find an expression for c_n.

11. Find the solution forms for the boundary value problem
$$\frac{\partial^2 u}{\partial r^2} + \frac{1}{r}\frac{\partial u}{\partial r} + \frac{\partial^2 u}{\partial z^2} = 0, \quad 0 < r < 1, \quad 0 < z < a$$
$$u(r, 0) = f(r), \quad u(r, a) = 0, \quad u(1, z) = 0$$

12. Suppose a cylindrical column of unit radius is considered to be of infinite height extending along the z-axis. If the lateral surface is maintained at zero temperature and the initial temperature distribution inside the column is prescribed by $f(r, \theta)$, show that the subsequent temperature distribution of the column has the form
$$u(r, \theta, t) = \sum_{m=1}^{\infty} \sum_{n=0}^{\infty} (a_{mn}\cos n\theta + b_{mn}\sin n\theta) J_n(k_{mn}r) e^{-a^2 k_{mn}^2 t}$$
where $J_n(k_{mn}) = 0$ $(m = 1, 2, 3, \ldots, n = 0, 1, 2, \ldots)$.

13. A long hollow cylinder of inner radius a and outer radius b is initially heated to a temperature distribution $u(r, 0) = f(r)$. If both the inner and outer surfaces are maintained at temperature zero, show that the temperature distribution throughout the cylinder has solutions of the form
$$u(r, t) = \sum_{n=1}^{\infty} c_n [Y_0(k_n a) J_0(k_n r) - J_0(k_n a) Y_0(k_n r)] e^{-a^2 k_n^2 t}$$
where $Y_0(k_n a) J_0(k_n b) - J_0(k_n a) Y_0(k_n b) = 0$ $(n = 1, 2, 3, \ldots)$.

14. Given the boundary-value problem
$$rR'' + R' + \lambda r R = 0, \quad R(1) = 0$$
show that $R(r) = 0$ is the only bounded solution for

(a) $\lambda = 0$,
(b) $\lambda = -k^2 < 0$.

15. Given the boundary-value problem
$$Z'' + \lambda Z = 0, \quad Z(0) = 0, \quad Z(\pi) = 0$$
show that $Z(z) = 0$ is the only solution for

(a) $\lambda = 0$,
(b) $\lambda = -k^2 < 0$.

8

The Hypergeometric Function

8.1 Introduction

Because of the many relations connecting the special functions to each other, and to the elementary functions, it is natural to inquire whether more general functions can be developed so that the special functions and elementary functions are merely specializations of these general functions. General functions of this nature have in fact been developed and are collectively referred to as *functions of the hypergeometric type*. There are several varieties of these functions, but the most common are the standard *hypergeometric function* (which we discuss in this chapter) and the *confluent hypergeometric function* (Chapter 9). Still, other generalizations exist, such as *MacRobert's E-function* and *Meijer's G-function*, for which even *generalized hypergeometric functions* are certain specializations (Chapter 10).

The major development of the theory of the hypergeometric function was carried out by Gauss and published in his famous memoir of 1812, a memoir that is also noted as being the real beginning of rigor in mathematics.* Some important results concerning the hypergeometric function had been developed earlier by Euler and others, but it was Gauss who made the first systematic study of the series that defines this function.

---

*C.F. Gauss, Disquisitiones Generales circa Seriem Infinitam..., *Comment. Soc. Reg. Sci. Gottingensis Recent.*, **2** (1812).

8.2 The Pochhammer Symbol

In dealing with certain product forms, factorials, and gamma functions, it is useful to introduce the abbreviation

$$(a)_0 = 1, \quad (a)_n = a(a+1)\cdots(a+n-1), \quad n = 1, 2, 3, \ldots \tag{8.1}$$

called the *Pochhammer symbol*. Using properties of the gamma function, it follows that this symbol can also be defined by

$$(a)_n = \frac{\Gamma(a+n)}{\Gamma(a)}, \quad n = 0, 1, 2, \ldots \tag{8.2}$$

Remark: For typographical convenience the symbol $(a)_n$ is sometimes replaced by *Appel's symbol* (a, n).

The Pochhammer symbol $(a)_n$ is important in most of the following material in this text. Because of its close association with the gamma function, it clearly satisfies a large number of identities. Some of the special properties are listed in Theorem 8.1 below, while other relations are taken up in the exercises.

Theorem 8.1. The Pochhammer symbol $(a)_n$ satisfies the identities:

(1) $(1)_n = n!$,
(2) $(a+n)(a)_n = a(a+1)_n$,
(3) $\binom{-a}{n} = \frac{(-1)^n}{n!}(a)_n$,
(4) $(a)_{n+k} = (a)_k(a+k)_n = (a)_n(a+n)_k$ (addition formula),
(5) $(a)_{k-n} = (-1)^n(a)_k/(1-a-k)_n$,
(6) $(a)_{2n} = 2^{2n}(\tfrac{1}{2}a)_n(\tfrac{1}{2} + \tfrac{1}{2}a)_n$ (duplication formula).

(Partial) proof: We will prove only parts (1), (2), and (3). The remaining proofs are left to the exercises.
From the definition, it follows that

(1): $\qquad (1)_n = 1 \times 2 \times \cdots \times n = n!,$

(2): $\qquad (a+n)(a)_n = a(a+1)\cdots(a+n-1)(a+n)$
$\qquad\qquad\qquad\quad = a(a+1)_n,$

(3):
$$\binom{-a}{n} = \frac{-a(-a-1)\cdots(-a-n+1)}{n!}$$
$$= \frac{(-1)^n}{n!} a(a+1)\cdots(a+n-1)$$
$$= \frac{(-1)^n}{n!}(a)_n. \blacksquare$$

From the definition, we see that the parameter a can be either positive or negative, but generally we assume $a \neq 0$. An exception to this is the special value $(0)_0 = 1$. If a is a negative integer, we find that (see problem 17)

$$(-k)_n = \begin{cases} \dfrac{(-1)^n k!}{(k-n)!}, & 0 \leq n \leq k \\ 0, & n > k \end{cases} \qquad (8.3)$$

Part (5) of Theorem 8.1 can be used to give meaning to the Pochhammer symbol for negative index: by setting $k = 0$ we obtain

$$(a)_{-n} = \frac{(-1)^n}{(1-a)_n}, \qquad n = 1, 2, 3, \ldots \qquad (8.4)$$

Like the binomial coefficient, the Pochhammer symbol plays a very important role in combinatorial problems, probability theory, and algorithm development. In developing certain relations it is more convenient to use the Pochhammer symbol than it is to use the binomial coefficient. The use of this symbol (and the hypergeometric function) in the evaluation of certain series and combinatorial relations is illustrated in Section 8.5.

The Pochhammer symbol and binomial coefficient are related directly by the formula given in part (3) of Theorem 8.1. A more complex relation between these symbols is developed in the next example.

Example 1: Based on the properties of the Pochhammer symbol listed in Theorem 8.1, show that

$$\binom{a+k-1}{n} = \frac{(-1)^n (1-a)_n (a)_k}{n!(a-n)_k}, \qquad k = 1, 2, 3, \ldots$$

Solution: From (3) and (5) of Theorem 8.1, we first obtain

$$\binom{a+k-1}{n} = \frac{(-1)^n}{n!}(1-a-k)_n$$
$$= \frac{(a)_k}{n!(a)_{k-n}}$$

Replacing n by $-n$ in part (4) of Theorem 8.1, we find
$$(a)_{k-n} = (a)_{-n}(a-n)_k$$
$$= \frac{(-1)^n(a-n)_k}{(1-a)_n}$$
where the last step is a consequence of Equation (8.4). Combining the above results leads to the desired relation
$$\binom{a+k-1}{n} = \frac{(-1)^n(1-a)_n(a)_k}{n!(a-n)_k}$$

EXERCISES 8.2

In problems 1–16, verify the identity.

1. $(-n)_n = (-1)^n n!$.
2. $(a-n)_n = (-1)^n(1-a)_n$.
3. $(a)_{n+1} = a(a+1)_n$.
4. $(a)_{n+k} = (a)_k(a+k)_n$.
5. $(a+1)_n - n(a+1)_{n-1} = (a)_n$.
6. $(a-1)_n + n(a)_{n-1} = (a)_n$.
7. $(n+k)! = n!(n+1)_k$.
8. $\Gamma(a+1-n) = \dfrac{(-1)^n\Gamma(a+1)}{(-a)_n}$.
9. $(a+n)_{k-n}(a+k)_{n-k} = 1$.
10. $(a+k)_{n-k} = (-1)^{n-k}(1-a-n)_{n-k}$.
11. $(a)_{k-n} = \dfrac{(-1)^n(a)_k}{(1-a-k)_n}$.
12. $(a)_{2n} = 2^{2n}(\tfrac{1}{2}a)_n(\tfrac{1}{2}+\tfrac{1}{2}a)_n$.
13. $(2n)! = 2^{2n}(\tfrac{1}{2})_n n!$.
14. $(2n+1)! = 2^{2n}(\tfrac{3}{2})_n n!$.
15. $\dbinom{2a}{2n} = \dfrac{(-a)_n(\tfrac{1}{2}-a)_n}{(\tfrac{1}{2})_n n!}$.
16. $(a)_{3n} = 3^{3n}(\tfrac{1}{3}a)_n(\tfrac{1}{3}+\tfrac{1}{3}a)_n(\tfrac{2}{3}+\tfrac{1}{3}a)_n$.

17. Show that $(k = 1, 2, 3, \ldots)$
$$(-k)_n = \begin{cases} \dfrac{(-1)^n k!}{(k-n)!}, & 0 \le n \le k \\ 0, & n > k \end{cases}$$

18. Show that
$$(1+n)_n = 4^n(\tfrac{1}{2})_n$$

19. Show that
$$\binom{n+a-1}{n} = \frac{(a)_n}{n!}$$

8.3 The Function $F(a, b; c; x)$

The series defined by*

$$1 + \frac{ab}{c}x + \frac{a(a+1)b(b+1)}{c(c+1)}\frac{x^2}{2!} + \cdots = \sum_{n=0}^{\infty} \frac{(a)_n (b)_n}{(c)_n} \frac{x^n}{n!} \quad (8.5)$$

is called the *hypergeometric series*. It gets its name from the fact that for $a = 1$ and $c = b$ the series reduces to the elementary *geometric series*

$$1 + x + x^2 + \cdots = \sum_{n=0}^{\infty} x^n \quad (8.6)$$

Denoting the general term of (8.5) by $u_n(x)$ and applying the ratio test, we see that

$$\lim_{n\to\infty}\left|\frac{u_{n+1}(x)}{u_n(x)}\right| = \lim_{n\to\infty}\left|\frac{(a)_{n+1}(b)_{n+1}x^{n+1}}{(c)_{n+1}(n+1)!} \cdot \frac{(c)_n n!}{(a)_n(b)_n x^n}\right|$$

$$= |x| \lim_{n\to\infty}\left|\frac{(a+n)(b+n)}{(c+n)(n+1)}\right|$$

where we have made use of property (4) of Theorem 8.1. Completing the limit process reveals that

$$\lim_{n\to\infty}\left|\frac{u_{n+1}(x)}{u_n(x)}\right| = |x| \quad (8.7)$$

under the assumption that none of a, b, or c is zero or a negative integer. Therefore, we conclude that the series (8.5) converges under these circumstances for all $|x| < 1$ and diverges for all $|x| > 1$. For $|x| = 1$, it can be shown that a sufficient condition for convergence of the series is $c - a - b > 0$.†

---

*Throughout our discussion the parameters a, b, c are assumed to be real.
†See E.D. Rainville, *Special Functions*, New York: Chelsea, 1971, p. 46.

The function

$$F(a,b;c;x) = \sum_{n=0}^{\infty} \frac{(a)_n (b)_n}{(c)_n} \frac{x^n}{n!}, \qquad |x| < 1 \qquad (8.8)$$

defined by the hypergeometric series is called the *hypergeometric function*. It is also commonly denoted by the symbol

$$_2F_1(a,b;c;x) \equiv F(a,b;c;x) \qquad (8.9)$$

where the 2 and 1 refer to the number of numerator and denominator parameters, respectively, in its series representation. The semicolons separate the numerator parameters a and b (which are themselves separated by a comma), the denominator parameter c, and the argument x.

If c is zero or a negative integer, the series (8.8) generally does not exist, and hence the function $F(a,b;c;x)$ is not defined. However, if either a or b (or both) is zero or a negative integer, the series is finite and thus converges for *all* x. That is, if $a = -m$ ($m = 0, 1, 2, \ldots$) then $(-m)_n = 0$ when $n > m$, and in this case (8.8) reduces to the *hypergeometric polynomial* defined by

$$F(-m,b;c;x) = \sum_{n=0}^{m} \frac{(-m)_n (b)_n}{(c)_n} \frac{x^n}{n!}, \qquad -\infty < x < \infty \qquad (8.10)$$

8.3.1 Elementary Properties

There are several properties of the hypergeometric function that are immediate consequences of its definition (8.8). First, we note the *symmetry property* of the parameters a and b, i.e.,

$$F(a,b;c;x) = F(b,a;c;x). \qquad (8.11)$$

Second, by differentiating the series (8.8) termwise, we find that

$$\frac{d}{dx} F(a,b;c;x) = \underbrace{\sum_{n=1}^{\infty} \frac{(a)_n (b)_n}{(c)_n} \frac{x^{n-1}}{(n-1)!}}_{n \to n+1}$$

$$= \sum_{n=0}^{\infty} \frac{(a)_{n+1} (b)_{n+1}}{(c)_{n+1}} \frac{x^n}{n!}$$

$$= \frac{ab}{c} \sum_{n=0}^{\infty} \frac{(a+1)_n (b+1)_n}{(c+1)_n} \frac{x^n}{n!}$$

and hence,

$$\frac{d}{dx} F(a,b;c;x) = \frac{ab}{c} F(a+1, b+1; c+1; x) \qquad (8.12)$$

Repeated application of (8.12) leads to the general formula (see problem 1)

$$\frac{d^k}{dx^k}F(a,b;c;x) = \frac{(a)_k(b)_k}{(c)_k}F(a+k,b+k;c+k;x),$$

$$k = 1, 2, 3, \ldots \quad (8.13)$$

The parameters a, b, and c in the definition of the hypergeometric function play much the same role in the relationships of this function in that the parameters n or p did for the Legendre polynomials and Bessel functions. The usual nomenclature for the hypergeometric functions in which one parameter changes by $+1$ or -1 is "contiguous functions." There are six contiguous functions, defined by $F(a \pm 1, b; c; x)$, $F(a, b \pm 1; c; x)$, and $F(a, b; c \pm 1; x)$. Gauss was the first to show that between $F(a, b; c; x)$ and any two contiguous functions there exists a linear relation with coefficients at most linear in x. The six contiguous functions, taken two at a time, lead to a total of fifteen recurrence relations of this kind, i.e., $\binom{6}{2} = 15.$*

In order to derive one of the fifteen recurrence relations, we first observe that

$$x\frac{d}{dx}F(a,b;c;x) + aF(a,b;c;x)$$
$$= \sum_{n=0}^{\infty} \frac{(a)_n(b)_n}{(c)_n}\frac{nx^n}{n!} + \sum_{n=0}^{\infty} \frac{a(a)_n(b)_n}{(c)_n}\frac{x^n}{n!}$$
$$= \sum_{n=0}^{\infty} \frac{(a+n)(a)_n(b)_n}{(c)_n}\frac{x^n}{n!}$$
$$= a\sum_{n=0}^{\infty} \frac{(a+1)_n(b)_n}{(c)_n}\frac{x^n}{n!}$$

from which we deduce

$$x\frac{d}{dx}F(a,b;c;x) + aF(a,b;c;x) = aF(a+1,b;c;x) \quad (8.14)$$

Similarly, from the symmetry property (8.11),

$$x\frac{d}{dx}F(a,b;c;x) + bF(a,b;c;x) = bF(a,b+1;c;x) \quad (8.15)$$

and by subtracting (8.15) from (8.14), it follows at once that

$$(a-b)F(a,b;c;x) = aF(a+1,b;c;x) - bF(a,b+1;c;x)$$
$$(8.16)$$

---

*For a listing of all 15 relations, see A. Erdelyi et al., *Higher Transcendental Functions*, Vol. I, New York: McGraw-Hill, 1953, pp. 103–104.

which is one of the simplest recurrence relations involving the contiguous functions. Some of the other recurrence relations are taken up in the exercises.

8.3.2 Integral Representation

To derive an integral representation for the hypergeometric function, we start with the beta-function relation (see Section 2.3)

$$B(n + b, c - b) = \int_0^1 t^{n+b-1}(1 - t)^{c-b-1} dt, \qquad c > b > 0 \quad (8.17)$$

from which we deduce (for $n = 0, 1, 2, \ldots$)

$$\frac{(b)_n}{(c)_n} = \frac{\Gamma(c)}{\Gamma(b)\Gamma(c - b)} \int_0^1 t^{n+b-1}(1 - t)^{c-b-1} dt \qquad (8.18)$$

The substitution of (8.18) into (8.8) yields

$$F(a, b; c; x) = \frac{\Gamma(c)}{\Gamma(b)\Gamma(c - b)} \sum_{n=0}^{\infty} \frac{(a)_n}{n!} x^n \int_0^1 t^{n+b-1}(1 - t)^{c-b-1} dt$$

$$= \frac{\Gamma(c)}{\Gamma(b)\Gamma(c - b)} \int_0^1 t^{b-1}(1 - t)^{c-b-1} \left(\sum_{n=0}^{\infty} \frac{(a)_n}{n!} (xt)^n \right) dt$$

(8.19)

where we have reversed the order of integration and summation. Now, using the relation (from Theorem 8.1)

$$\frac{(a)_n}{n!} = \binom{-a}{n}(-1)^n \qquad (8.20)$$

we recognize the series in (8.19) as a binomial series which has the sum

$$\sum_{n=0}^{\infty} \frac{(a)_n}{n!}(xt)^n = \sum_{n=0}^{\infty} \binom{-a}{n}(-xt)^n = (1 - xt)^{-a} \qquad (8.21)$$

provided $|xt| < 1$. Hence, (8.19) gives us the integral representation

$$F(a, b; c; x) = \frac{\Gamma(c)}{\Gamma(b)\Gamma(c - b)} \int_0^1 t^{b-1}(1 - t)^{c-b-1}(1 - xt)^{-a} dt,$$

$$c > b > 0 \quad (8.22)$$

Although (8.22) was derived under the assumption that $|xt| < 1$, it can be shown that the integral converges for all $|x| \le 1$.* The convergence of (8.22) for $x = 1$ is important in our proof of the following useful theorem.

---

*See E.D. Rainville, *Special Functions*, New York: Chelsea, 1971, pp. 48–49.

Theorem 8.2. For $c \neq 0, -1, -2, \ldots$ and $c - a - b > 0$,

$$F(a, b; c; 1) = \frac{\Gamma(c)\Gamma(c-a-b)}{\Gamma(c-a)\Gamma(c-b)}$$

Proof: We will prove the theorem only with the added restriction $c > b > 0$, although it is valid without this restriction. We simply set $x = 1$ in (8.22) to get

$$F(a, b; c; 1) = \frac{\Gamma(c)}{\Gamma(b)\Gamma(c-b)} \int_0^1 t^{b-1}(1-t)^{c-b-1}(1-t)^{-a}\, dt$$

$$= \frac{\Gamma(c)}{\Gamma(b)\Gamma(c-b)} \int_0^1 t^{b-1}(1-t)^{c-a-b-1}\, dt$$

which, evaluated as a beta integral, yields our result, viz.,

$$F(a, b; c; 1) = \frac{\Gamma(c)\Gamma(b)\Gamma(c-a-b)}{\Gamma(b)\Gamma(c-b)\Gamma(c-a)}$$

$$= \frac{\Gamma(c)\Gamma(c-a-b)}{\Gamma(c-a)\Gamma(c-b)} \qquad \blacksquare$$

8.3.3 The Hypergeometric Equation

The linear second-order DE

$$x(1-x)y'' + [c - (a+b+1)x]y' - aby = 0 \qquad (8.23)$$

is called the *hypergeometric equation* of Gauss. It is so named because the function

$$y_1 = F(a, b; c; x), \qquad c \neq 0, -1, -2, \ldots \qquad (8.24)$$

is a solution. To verify that (8.24) is indeed a solution, we can substitute the series for $F(a, b; c; x)$ directly into (8.23).

Examination of the coefficient of y'' reveals that both $x = 0$ and $x = 1$ are (finite) singular points of the equation. Therefore, to find a second series solution about $x = 0$ would normally require use of the Frobenius method.* Under special restrictions on the parameter c, however, we can produce a second (linearly independent) solution of (8.23) without resorting to this more general method. We simply make the change of dependent variable

$$y = x^{1-c}z \qquad (8.25)$$

---

*For an introductory discussion of the Frobenius method, see L.C. Andrews, *Ordinary Differential Equations with Applications*, Glenview, Ill.: Scott, Foresman, 1982, Chapter 9.

from which we calculate

$$y' = x^{1-c}z' + (1-c)x^{-c}z \qquad (8.26a)$$

$$y'' = x^{1-c}z'' + 2(1-c)x^{-c}z' - c(1-c)x^{-c-1}z \qquad (8.26b)$$

The substitution of (8.25), (8.26a), and (8.26b) into (8.23) leads to (upon algebraic simplification)

$$x(1-x)z'' + [2 - c - (a + b - 2c + 3)x]z'$$
$$- (1 + a - c)(1 + b - c)z = 0 \qquad (8.27)$$

which we recognize as another form of (8.23). Hence, Equation (8.27) has the solution

$$z = F(1 + a - c, 1 + b - c; 2 - c; x), \qquad c \neq 2, 3, 4, \ldots \qquad (8.28)$$

and so we deduce that

$$y_2 = x^{1-c}F(1 + a - c, 1 + b - c; 2 - c; x), \qquad c \neq 2, 3, 4, \ldots \qquad (8.29)$$

is a second solution of (8.23). For $c = 2, 3, 4, \ldots$, the hypergeometric function in (8.29) does not usually exist, while for $c = 1$ the solutions (8.29) and (8.24) are identical. However, if we restrict c to $c \neq 0, \pm 1, \pm 2, \ldots$, then (8.29) is linearly independent of (8.24) and

$$y = C_1 F(a, b; c; x) + C_2 x^{1-c}F(1 + a - c, 1 + b - c; 2 - c; x) \qquad (8.30)$$

is a general solution of Equation (8.23).

To cover the cases when $c = 2, 3, 4, \ldots$, a *hypergeometric function of the second kind* can be introduced (see problem 28). However, beyond its connection as a solution to the hypergeometric equation of Gauss, the hypergeometric function of the second kind has limited usefulness in applications.

Remark: Actually, $y_1 = F(a, b; c; x)$ and $y_2 = x^{1-c}F(1 + a - c, 1 + b - c; 2 - c; x)$ are only two of a total of 24 solutions of Equation (8.23) that can be expressed in terms of the hypergeometric function. For a listing of all 24 solutions, see W.W. Bell, *Special Functions for Scientists and Engineers*, London: Van Nostrand, 1968, pp. 208–209.

EXERCISES 8.3

1. Show that (for $k = 1, 2, 3, \ldots$)

$$\frac{d^k}{dx^k} F(a, b; c; x) = \frac{(a)_k (b)_k}{(c)_k} F(a + k, b + k; c + k; x),$$

2. Show that (for $k = 1, 2, 3, \ldots$)

    (a) $\dfrac{d}{dx}[x^a F(a, b; c; x)] = ax^{a-1} F(a + 1, b; c; x)$,

    (b) $\dfrac{d^k}{dx^k}[x^{a-1+k} F(a, b; c; x)] = (a)_k x^{a-1} F(a + k, b; c; x)$,

    (c) $\dfrac{d^k}{dx^k}[x^{c-1} F(a, b; c; x)] = (c - k)_k x^{c-1-k} F(a, b; c - k; x)$.

In problems 3–6, verify the differentiation formula.

3. $x \dfrac{d}{dx} F(a, b; c; x) + (1 - c) F(a, b; c; x) = aF(a + 1, b; c; x)$
    $+ (1 - c) F(a, b; c - 1; x)$.

4. $x \dfrac{d}{dx} F(a - 1, b; c; x) = (a - 1) F(a, b; c; x)$
    $- (a - 1) F(a - 1, b; c; x)$.

5. $(1 - x) x \dfrac{d}{dx} F(a, b; c; x) = (a + b - c) x F(a, b; c; x)$
    $+ c^{-1}(c - a)(c - b) x F(a, b; c + 1; x)$.

6. $x \dfrac{d}{dx} F(a - 1, b; c; x) = (a - 1) x F(a, b; c; x)$
    $- c^{-1}(a - 1)(c - b) x F(a, b; c + 1; x)$.

In problems 7–13, verify the given contiguous relation by using the results of problems 3–6, or by series representations.

7. $(b - a)(1 - x) F(a, b; c; x) = (c - a) F(a - 1, b; c; x)$
    $- (c - b) F(a, b - 1; c; x)$.

8. $(1 - x) F(a, b; c; x) = F(a - 1, b; c; x)$
    $- c^{-1}(c - b) x F(a, b; c + 1; x)$.

9. $(1 - x) F(a, b; c; x) = F(a, b - 1; c; x)$
    $- c^{-1}(c - a) x F(a, b; c + 1; x)$.

10. $(c - a - b) F(a, b; c; x) + a(1 - x) F(a + 1, b; c; x)$
    $= (c - b) F(a, b - 1; c; x)$.

11. $(c - a - b)F(a, b; c; x) + b(1 - x)F(a, b + 1; c; x)$
    $= (c - a)F(a - 1, b; c; x)$.

12. $(c - b - 1)F(a, b; c; x) + bF(a, b + 1; c; x)$
    $= (c - 1)F(a, b; c - 1; x)$.

13. $[2b - c + (a - b)x]F(a, b; c; x) = b(1 - x)F(a, b + 1; c; x)$
    $- (c - b)F(a, b - 1; c; x)$.

In problems 14 and 15, verify the formula by direct substitution of the series representations.

14. $F(a, b + 1; c; x) - F(a, b; c; x) = \dfrac{ax}{c} F(a + 1, b + 1; c + 1; x)$.

15. $F(a, b; c; x) - F(a, b; c - 1; x)$
    $= -\dfrac{abx}{c(c - 1)} F(a + 1, b + 1; c + 1; x)$.

In problems 16 and 17, use termwise integration to derive the given integral representation.

16. $F(a, b; c; x) = \dfrac{\Gamma(c)}{\Gamma(d)\Gamma(c - d)} \int_0^1 t^{d-1}(1 - t)^{c-d-1} F(a, b; d; xt)\, dt$,
    $c > d > 0$.

17. $F(a, b; c + 1; x) = c \int_0^1 F(a, b; c; xt) t^{c-1}\, dt$, $c > 0$.

18. Show that $(s > 0)$

   (a) $\int_0^\infty e^{-st} F[a, b; 1; x(1 - e^{-t})]\, dt = \dfrac{1}{s} F(a, b; s + 1; x)$,

   (b) $\int_0^\infty e^{-st} F(a, b; 1; 1 - e^{-t})\, dt = \dfrac{\Gamma(s)\Gamma(s + 1 - a - b)}{\Gamma(s + 1 - a)\Gamma(s + 1 - b)}$.

   Hint: Set $x = 1$ in (a).

19. Show that (for $n = 0, 1, 2, \ldots$)

   (a) $F(-n, b; c; 1) = \dfrac{(c - b)_n}{(c)_n}$,

   (b) $F(-n, a + n; c; 1) = (-1)^n \dfrac{(1 + a - c)_n}{(c)_n}$,

   (c) $F(-n, 1 - b - n; c; 1) = \dfrac{(c + b - 1)_{2n}}{(c)_n (c + b - 1)_n}$.

20. Show that

$$F(-\tfrac{1}{2}n, \tfrac{1}{2} - \tfrac{1}{2}n; b + \tfrac{1}{2}; 1) = \dfrac{2^n (b)_n}{(2b)_n}, \qquad n = 0, 1, 2, \ldots$$

21. Using the result of problem 19(a), show that (for $p = 0, 1, 2, \ldots$)

(a) $F(-p, a + n + 1; a + 1; 1) = \begin{cases} 0, & 0 \le n \le p - 1, \\ \dfrac{(-1)^p p!}{(a+1)_p}, & n = p, \end{cases}$

(b) $F(-p, a + n + 2; a + 1; 1)$
$= \begin{cases} 0, & 0 \le n \le p - 2, \\ \dfrac{(-1)^p (n+1)!}{(a+1)_p (n+1-p)!}, & n = p - 1, p. \end{cases}$

22. Given the generating function
$$w(x, t) = (1 - t)^{b-c}(1 - t + xt)^{-b}, \qquad c \ne 0, -1, -2, \ldots$$
show that
$$w(x, t) = \sum_{n=0}^{\infty} \frac{(c)_n}{n!} F(-n, b; c; x) t^n$$
where $F(-n, b; c; x)$ denotes the *hypergeometric polynomials* defined by Equation (8.10).

23. Show that, for $|x| < 1$ and $|x/(1 - x)| < 1$,
$$(1 - x)^{-a} F\left(a, c - b; c; \frac{-x}{1 - x}\right) = F(a, b; c; x)$$

24. By substituting $y = x/(x - 1)$ in problem 23, deduce that

(a) $F(a, c - b; c; x) = (1 - x)^{b-c} F\left(c - a, c - b; c; \dfrac{-x}{1 - x}\right)$,

(b) $F\left(a, c - b; c; \dfrac{-x}{1 - x}\right) = (1 - x)^{c-b} F(c - a, c - b; c; x)$,

(c) $F(a, b; c; x) = (1 - x)^{c-a-b} F(c - a, c - b; c; x)$.

25. Show that
$$(1 - x)^{-a} F\left[\tfrac{1}{2}a, \tfrac{1}{2} + \tfrac{1}{2}a - b; 1 + a - b; \frac{-4x}{(1 - x)^2}\right]$$
$$= F(a, b; 1 + a - b; x)$$

26. Use problems 23–25 to deduce that

(a) $F(a, b; 1 + a - b; -1) = \dfrac{\Gamma(1 + a - b)\Gamma(1 + \tfrac{1}{2}a)}{\Gamma(1 + a)\Gamma(1 + \tfrac{1}{2}a - b)}$,

(b) $F(a, 1 - a; c; \tfrac{1}{2}) = \dfrac{\Gamma(\tfrac{1}{2}c)\Gamma(\tfrac{1}{2}c + \tfrac{1}{2})}{\Gamma(\tfrac{1}{2}a + \tfrac{1}{2}c)\Gamma(\tfrac{1}{2} - \tfrac{1}{2}a + \tfrac{1}{2}c)}$,

(c) $F(2a, 2b; a + b + \tfrac{1}{2}; \tfrac{1}{2}) = \dfrac{\Gamma(a + b + \tfrac{1}{2})\sqrt{\pi}}{\Gamma(a + \tfrac{1}{2})\Gamma(b + \tfrac{1}{2})}$.

27. By assuming a power-series solution of the form

$$y = \sum_{n=0}^{\infty} A_n x^n$$

show that $y = F(a, b; c; x)$ is a solution of the hypergeometric equation

$$x(1 - x)y'' + [c - (a + b + 1)x]y' - aby = 0$$

28. The *hypergeometric function of the second kind* is defined by

$$G(a, b; c; x) = \frac{\Gamma(1 - c)}{\Gamma(a - c + 1)\Gamma(b - c + 1)} F(a, b; c; x)$$
$$+ \frac{\Gamma(c - 1)}{\Gamma(a)\Gamma(b)} x^{1-c} F(1 + a - c, 1 + b - c; 2 - c; x)$$

(a) Show that $G(a, b; c; x)$ is a solution of the hypergeometric equation in problem 27, $c \neq 0, \pm 1, \pm 2, \ldots$.
(b) Show that $G(a, b; c; x) = x^{1-c} G(1 + a - c, 1 + b - c; 2 - c; x)$.

29. Show that the Wronskian of $F(a, b; c; x)$ and $G(a, b; c; x)$ is given by (see problem 28)

$$W(F, G)(x) = - \frac{\Gamma(c)}{\Gamma(a)\Gamma(b)} x^{-c} (1 - x)^{c-a-b-1}$$

30. Derive the generating function relation

$$(1 - xt)^{-a} F\left[\tfrac{1}{2}a, \tfrac{1}{2}a + \tfrac{1}{2}; 1, \frac{t^2(x^2 - 1)}{(1 - xt)^2}\right] = \sum_{n=0}^{\infty} \frac{(a)_n P_n(x)}{n!} t^n$$

where $P_n(x)$ is the nth Legendre polynomial.

8.4 Relation to Other Functions

The hypergeometric function is important in many areas of mathematical analysis and its applications. Partly this is a consequence of the fact that so many elementary and special functions are simply special cases of the hypergeometric function. For example, the specialization

$$F(1, b; b; x) = \sum_{n=0}^{\infty} \frac{(1)_n}{n!} x^n = \sum_{n=0}^{\infty} x^n$$

reveals that

$$F(1, b; b; x) = (1 - x)^{-1} \qquad (8.31)$$

Similarly, it can be established that

$$\arcsin x = xF(\tfrac{1}{2}, \tfrac{1}{2}; \tfrac{3}{2}; x^2) \tag{8.32}$$

and

$$\log(1 - x) = -xF(1, 1; 2; x) \tag{8.33}$$

Example 2: Show that $\arcsin x = xF(\tfrac{1}{2}, \tfrac{1}{2}; \tfrac{3}{2}; x^2)$.

Solution: From the calculus, we recall

$$\arcsin x = \sum_{n=0}^{\infty} \frac{(2n)! \, x^{2n+1}}{2^{2n}(n!)^2 (2n+1)}$$

In order to recognize this series as a hypergeometric series, we need to express the coefficient of $x^{2n+1}/n!$ in terms of Pochhammer symbols. Thus, using the results of problems 13 and 14 in Exercises 8.2, we have

$$(2n)! = 2^{2n}(\tfrac{1}{2})_n n!$$

$$(2n + 1) = \frac{(2n+1)!}{(2n)!} = \frac{(\tfrac{3}{2})_n}{(\tfrac{1}{2})_n}$$

and making these substitutions leads to

$$\arcsin x = x \sum_{n=0}^{\infty} \frac{(\tfrac{1}{2})_n (\tfrac{1}{2})_n}{(\tfrac{3}{2})_n} \frac{x^{2n}}{n!}$$

from which we deduce

$$\arcsin x = xF(\tfrac{1}{2}, \tfrac{1}{2}; \tfrac{3}{2}; x^2)$$

The verification of (8.33) along with several other such relations involving elementary functions is left to the exercises.

A more involved relationship to establish is given by

$$P_n(x) = F\left(-n, n + 1; 1; \frac{1-x}{2}\right) \tag{8.34}$$

where $P_n(x)$ is the nth Legendre polynomial. To prove (8.34), we first observe that

$$(1 - 2xt + t^2)^{-1/2} = \left[(1 - t)^2 - 2t(x - 1)\right]^{-1/2}$$

$$= (1 - t)^{-1} \left[1 - \frac{2t(x-1)}{(1-t)^2}\right]^{-1/2} \tag{8.35}$$

and thus we deduce the relation

$$\sum_{n=0}^{\infty} P_n(x)t^n = (1 - 2xt + t^2)^{-1/2}$$

$$= (1-t)^{-1}\left[1 - \frac{2t(x-1)}{(1-t)^2}\right]^{-1/2}$$

$$= \sum_{k=0}^{\infty} \binom{-\frac{1}{2}}{k} \frac{(-1)^k (2t)^k (x-1)^k}{(1-t)^{2k+1}} \quad (8.36)$$

Our object now is to recognize the right-hand side of (8.36) as a power series in t which has the coefficient $F(-n, n+1; 1; (1-x)/2)$. To obtain powers of t, we further expand $(1-t)^{-2k-1}$ in a binomial series and interchange the order of summation. Hence,

$$\sum_{n=0}^{\infty} P_n(x)t^n = \sum_{m=0}^{\infty}\sum_{k=0}^{\infty} \binom{-\frac{1}{2}}{k}\binom{-2k-1}{m}(-1)^{m+k} 2^k (x-1)^k t^{k+m}$$

$$= \sum_{n=0}^{\infty}\sum_{k=0}^{n} \binom{-\frac{1}{2}}{k}\binom{-2k-1}{n-k}(-1)^n 2^k (x-1)^k t^n \quad (8.37)$$

where the last step is a result of the index change $m = n - k$. Next, from part (3) of Theorem 8.1, we can write

$$\binom{-\frac{1}{2}}{k}\binom{-2k-1}{n-k} = \frac{(-1)^n (\frac{1}{2})_k (2k+1)_{n-k}}{k!(n-k)!} \quad (8.38)$$

but from problems 7 and 13 in Exercises 8.2, we further have

$$(2k+1)_{n-k} = \frac{(n+k)!}{(2k)!} = \frac{n!(n+1)_k}{2^{2k}(\frac{1}{2})_k k!} \quad (8.39)$$

Finally, setting $a = 1$ in problem 11 in Exercises 8.2 leads to

$$(n-k)! = \frac{(-1)^k n!}{(-n)_k} \quad (8.40)$$

so by combining the results of (8.38), (8.39), and (8.40), we find that (8.37) becomes

$$\sum_{n=0}^{\infty} P_n(x)t^n = \sum_{n=0}^{\infty}\left[\sum_{k=0}^{n} \frac{(-n)_k (n+1)_k}{(1)_k k!}\left(\frac{1-x}{2}\right)^k\right]t^n$$

$$= \sum_{n=0}^{\infty} F\left(-n, n+1; 1; \frac{1-x}{2}\right)t^n \quad (8.41)$$

from which (8.34) follows.

8.4.1 Legendre Functions

The relation (8.34) between the nth Legendre polynomial and hypergeometric function provides us with a natural way of introducing the more general function

$$P_\nu(x) = F\left(-\nu, \nu + 1; 1; \frac{1-x}{2}\right) \quad (8.42)$$

where ν is not restricted to integer values. We call $P_\nu(x)$ a *Legendre function of the first kind* of degree ν; it is not a polynomial except in the special case when $\nu = n$ ($n = 0, 1, 2, \ldots$). A *Legendre function of the second kind*, denoted by $Q_\nu(x)$, can also be defined in terms of the hypergeometric function, although we will not discuss it.*

The function $P_\nu(x)$ has many properties in common with the Legendre polynomial $P_n(x)$. For example, by setting $x = 1$ in (8.42), we obtain

$$P_\nu(1) = F(-\nu, \nu + 1; 1; 0) = 1 \quad (8.43)$$

The substitution of $x = 0$ in (8.42) leads to

$$P_\nu(0) = F\left(-\nu, \nu + 1; 1; \tfrac{1}{2}\right) \quad (8.44)$$

and by using the relation [see problem 26(b) in Exercises 8.3]

$$F\left(a, 1 - a; c; \tfrac{1}{2}\right) = \frac{\Gamma(\tfrac{1}{2}c)\Gamma(\tfrac{1}{2}c + \tfrac{1}{2})}{\Gamma(\tfrac{1}{2}a + \tfrac{1}{2}c)\Gamma(\tfrac{1}{2} - \tfrac{1}{2}a + \tfrac{1}{2}c)} \quad (8.45)$$

we deduce that

$$P_\nu(0) = \frac{\sqrt{\pi}}{\Gamma(\tfrac{1}{2} - \tfrac{1}{2}\nu)\Gamma(\tfrac{1}{2}\nu + 1)} \quad (8.46)$$

Recalling the identity

$$\Gamma(x)\Gamma(1 - x) = \frac{\pi}{\sin \pi x} \quad (8.47)$$

we can express (8.46) in the alternative form

$$P_\nu(0) = \frac{\Gamma(\tfrac{1}{2}\nu + \tfrac{1}{2})}{\sqrt{\pi}\,\Gamma(\tfrac{1}{2}\nu + 1)} \cos(\tfrac{1}{2}\nu\pi) \quad (8.48)$$

When ν is a nonnegative integer, we find that (8.48) reduces to the results that we previously derived for the Legendre polynomials (see problem 22).

Various recurrence formulas for $P_\nu(x)$ can be derived by expressing this function in its series representation, or by using properties of the hypergeo-

---

*See T.M. MacRobert, *Spherical Harmonics*, Oxford: Pergamon, 1967, Chapter VI.

metric function. For example, it can be verified that
$$(\nu + 1)P_{\nu+1}(x) - (2\nu + 1)xP_\nu(x) + \nu P_{\nu-1}(x) = 0 \quad (8.49)$$
$$P'_{\nu+1}(x) - xP'_\nu(x) = (\nu + 1)P_\nu(x) \quad (8.50)$$
$$xP'_\nu(x) - P'_{\nu-1}(x) = \nu P_\nu(x) \quad (8.51)$$
and so forth.

The Legendre functions $P_\nu(x)$ are important for theoretical purposes in the general study of spherical harmonics. Their properties are important also from a more practical point of view, since these functions are prominent in solving Laplace's equation in various coordinate systems, such as toroidal coordinates.*

EXERCISES 8.4

In problems 1–8, compare series to deduce the result.

1. $1 = F(0, b; c; x)$.
2. $(1 - x)^{-a} = F(a, b; b; x)$.
3. $\log(1 - x) = -xF(1, 1; 2; x)$.
4. $\log\dfrac{1 + x}{1 - x} = 2xF(\tfrac{1}{2}, 1; \tfrac{3}{2}; x^2)$.
5. $\arctan x = xF(\tfrac{1}{2}, 1; \tfrac{3}{2}; x^2)$.
6. $(1 + x)(1 - x)^{-2a-1} = F(2a, a + 1; a; x)$.
7. $\tfrac{1}{2}(1 + \sqrt{x})^{-2a} + \tfrac{1}{2}(1 - \sqrt{x})^{-2a} = F(a, a + \tfrac{1}{2}; \tfrac{1}{2}; x)$.
8. $\left[\dfrac{1 + \sqrt{1 - x}}{2}\right]^{1-2a} = F(a - \tfrac{1}{2}, a; 2a; x)$.
9. Show that
$$K(x) = \tfrac{1}{2}\pi F(\tfrac{1}{2}, \tfrac{1}{2}; 1; x^2)$$
where $K(x)$ is the *complete elliptic integral of the first kind* defined by
$$K(x) = \int_0^{\pi/2} (1 - x^2\sin^2\phi)^{-1/2}\, d\phi$$

10. Show that
$$E(x) = \tfrac{1}{2}\pi F(-\tfrac{1}{2}, \tfrac{1}{2}; 1; x^2)$$

---
*See Chapter 8 in N.N. Lebedev, *Special Functions and Their Applications*, Dover, 1972.

where $E(x)$ is the *complete elliptic integral of the second kind* defined by

$$E(x) = \int_0^{\pi/2} (1 - x^2 \sin^2\phi)^{1/2}\, d\phi$$

11. Show that the *associated Legendre functions*

$$P_n^m(x) = (1 - x^2)^{m/2} \frac{d^m}{dx^m} P_n(x)$$

satisfy the relation

$$P_n^m(x) = \frac{(n+m)!}{2^m (n-m)! m!} (1-x^2)^{m/2}$$

$$\times F\left(m - n, m + n + 1; m + 1; \frac{1-x}{2}\right)$$

12. Show that the *Chebyshev polynomials of the first kind*

$$T_n(x) = \sum_{k=0}^{[n/2]} \frac{(-1)^k (n-k-1)!}{k!(n-2k)!} (2x)^{n-2k}, \quad n \geq 1$$

satisfy the relation

$$T_n(x) = F\left(-n, n; \tfrac{1}{2}; \frac{1-x}{2}\right)$$

13. Show that the *Chebyshev polynomials of the second kind*

$$U_n(x) = \sum_{k=0}^{[n/2]} \binom{n-k}{k} (-1)^k (2x)^{n-2k}$$

satisfy the relation

$$U_n(x) = (n+1) F\left(-n, n+2; \tfrac{3}{2}; \frac{1-x}{2}\right)$$

14. Show that the *Gegenbauer polynomials*

$$C_n^\lambda(x) = \sum_{k=0}^{[n/2]} \binom{-\lambda}{n-k} \binom{n-k}{k} (-1)^n (2x)^{n-2k}$$

satisfy the relations

(a) $C_{2n}^\lambda(x) = (-1)^n \dfrac{(\lambda)_n}{n!} F(-n, \lambda + n; \tfrac{1}{2}; x^2)$,

(b) $C_{2n+1}^\lambda(x) = (-1)^n \dfrac{(\lambda)_{n+1}}{n!} F(-n, \lambda + n; \tfrac{3}{2}; x^2)$,

(c) $C_n^\lambda(x) = \binom{n + 2\lambda - 1}{n} F\left(-n, 2\lambda + n; \lambda + \tfrac{1}{2}; \dfrac{1-x}{2}\right)$.

The Hypergeometric Function • 291

15. Show that the *Jacobi polynomials*

$$P_n^{(a,b)}(x) = \frac{1}{2^n} \sum_{k=0}^{n} \binom{n+a}{n-k}\binom{n+b}{n-k}(x+1)^{n-k}(x-1)^k$$

satisfy the relations

(a) $P_n^{(a,b)}(x) = (-1)^n \binom{n+b}{b} F\left(-n, n+a+b+1; 1+b; \frac{1+x}{2}\right)$,

(b) $P_n^{(a,b)}(x) = \binom{n+a}{a} F\left(-n, n+a+b+1; 1+a; \frac{1-x}{2}\right)$.

16. Given the *incomplete beta function*

$$B_x(p,q) = \int_0^x t^{p-1}(1-t)^{q-1}\, dt, \qquad p,q > 0$$

show that

(a) $B_x(p,q) = \dfrac{x^p}{p} F(p, 1-q; 1+p; x)$,

(b) $B_1(p,q) = \dfrac{\Gamma(p)\Gamma(q)}{\Gamma(p+q)}$.

17. Show that

$$\sum_{k=0}^{n-1} \binom{a}{k} x^k = (1+x)^a - \binom{a}{n} x^n F(n-a, 1; n+1; -x)$$

18. Verify that

$$P_\nu(x) = \sum_{k=0}^{\infty} \frac{(-\nu)_k (\nu+1)_k}{(k!)^2} \left(\frac{1-x}{2}\right)^k$$

In problems 19–21, use the series representation in problem 18 to deduce the given recurrence formula.

19. $(\nu+1)P_{\nu+1}(x) - (2\nu+1)xP_\nu(x) + \nu P_{\nu-1}(x) = 0$.

20. $P'_{\nu+1}(x) - xP'_\nu(x) = (\nu+1)P_\nu(x)$.

21. $xP'_\nu(x) - P'_{\nu-1}(x) = \nu P_\nu(x)$.

22. Using the relation (8.48), show that (for $k = 0, 1, 2, \ldots$)

(a) $P_{2k+1}(0) = 0$,

(b) $P_{2k}(0) = \dfrac{(-1)^k}{k!} (\tfrac{1}{2})_k$.

23. Show that

$$P'_\nu(0) = \frac{2\Gamma(\tfrac{1}{2}\nu + 1)}{\sqrt{\pi}\,\Gamma(\tfrac{1}{2}\nu + \tfrac{1}{2})} \sin\frac{\nu\pi}{2}$$

24. Show that $P_\nu(x) = P_{-\nu-1}(x)$.

Hint: Recall that $F(a, b; c; x) = F(b, a; c; x)$.

25. By making the substitution $x = 1 - 2z$ in the generalized form of Legendre's equation

$$(1 - x^2)y'' - 2xy' + \nu(\nu + 1)y = 0$$

show that it transforms to Gauss' hypergeometric equation and thus deduce that $y = F\left(-\nu, \nu + 1; 1; \dfrac{1-x}{2}\right)$ is one solution of the generalized Legendre equation.

26. Show that

$$\frac{1}{k!} = \frac{2}{\pi \left(\frac{1}{2}\right)_k} \int_0^{\pi/2} \sin^{2k}\phi \, d\phi, \qquad k = 0, 1, 2, \ldots$$

and then, by expressing $P_\nu(x)$ in its series representation (problem 18), deduce that

$$P_\nu(x) = \frac{2}{\pi} \int_0^{\pi/2} F\left(-\nu, \nu + 1; \tfrac{1}{2}; \frac{1-x}{2}\sin^2\phi\right) d\phi$$

8.5 Summing Series

The hypergeometric function obviously has many areas of application due to its connection with other functions like inverse trigonometric functions, logarithmic functions, and the Legendre polynomials. However, it is also a useful tool in the evaluation or recognition of various series, both finite and infinite.

Example 3: Prove the combinatorial formula

$$\sum_{k=0}^{n} (-1)^k \binom{n}{k}\binom{a+k}{m} = (-1)^n \binom{a}{m-n}, \qquad m = 0, 1, 2, \ldots$$

Solution: From part (3) of Theorem 8.1 and Example 1,

$$\binom{n}{k} = \frac{(-1)^k}{k!}(-n)_k$$

$$\binom{a+k}{m} = \frac{(-1)^m(-a)_m(a+1)_k}{m!(a+1-m)_k}$$

and therefore

$$\sum_{k=0}^{n} (-1)^k \binom{n}{k} \binom{a+k}{m} = \frac{(-1)^m(-a)_m}{m!} \sum_{k=0}^{n} \frac{(-n)_k(a+1)_k}{(a+1-m)_k}$$

$$= \frac{(-1)^m(-a)_m}{m!} F(-n, a+1; a+1-m; 1)$$

Recalling Theorem 8.2,

$$F(-n, a+1; a+1-m; 1) = \frac{\Gamma(a+1-m)\Gamma(n-m)}{\Gamma(a+1+n-m)\Gamma(-m)}$$

$$= \frac{(-1)^n m!}{(m-n)!(a+1-m)_n}$$

where the last step follows from Equation (8.1) and problem 11 in Exercises 2.2. Part (5) of Theorem 8.1 leads to

$$(-a)_m = (-1)^n(-a)_{m-n}(a+1-m)_n$$

and thus by combining results, we obtain

$$\sum_{k=0}^{n} (-1)^k \binom{n}{k} \binom{a+k}{m} = \frac{(-1)^m}{(m-n)!}(-a)_{m-n}$$

$$= (-1)^n \binom{a}{m-n}$$

following another application of part (3) in Theorem 8.1.

The hypergeometric function is useful also in the evaluation of certain integrals, as illustrated in the next example.

Example 4: Show that for $a > -1$,

$$\int_0^\infty x^a \left[L_p^{(a)}(x) \right]^2 e^{-x} dx = \frac{\Gamma(p+a+1)}{p!}, \qquad p = 0, 1, 2, \ldots$$

where $L_p^{(a)}(x)$ is the generalized Laguerre polynomial.

Solution: By writing

$$L_p^{(a)}(x) = \sum_{n=0}^{p} (-1)^n \binom{p+a}{p-n} \frac{x^n}{n!}$$

$$= \binom{p+a}{p} \sum_{n=0}^{p} \frac{(-p)_n}{(a+1)_p} \frac{x^n}{n!}$$

we have

$$\int_0^\infty x^a \left[L_p^{(a)}(x) \right]^2 e^{-x} dx$$

$$= \binom{p+a}{p}^2 \sum_{n=0}^p \frac{(-p)_n}{(a+1)_n n!} \sum_{k=0}^p \frac{(-p)_k}{(a+1)_k k!} \int_0^\infty x^{a+n+k} e^{-x} dx$$

This last integral can be evaluated by using properties of the gamma function and two applications of problem 7 in Exercises 8.2, to get

$$\int_0^\infty x^{a+n+k} e^{-x} dx = \Gamma(a+n+k+1)$$

$$= \Gamma(a+n+1)(a+n+1)_k$$

$$= \Gamma(a+1)(a+1)_n (a+n+1)_k$$

Hence,

$$\int_0^\infty x^a \left[L_p^{(a)}(x) \right]^2 e^{-x} dx$$

$$= \binom{p+a}{p}^2 \Gamma(a+1) \sum_{n=0}^p \frac{(-p)_n}{n!} \sum_{k=0}^p \frac{(-p)_k (a+n+1)_k}{(a+1)_k k!}$$

$$= \binom{p+a}{p}^2 \Gamma(a+1) \sum_{n=0}^p \frac{(-p)_n}{n!} F(-p, a+n+1; a+1; 1)$$

and by using the result of problem 21(a) in Exercises 8.3, we see that

$$F(-p, a+n+1; a+1; 1) = \begin{cases} 0, & 0 \le n \le p-1 \\ \dfrac{(-1)^p p!}{(a+1)_p}, & n = p \end{cases}$$

Finally, the substitution of this last expression for the hypergeometric function leads us to our intended result,

$$\int_0^\infty x^a \left[L_p^{(a)}(x) \right]^2 e^{-x} dx = \binom{p+a}{p}^2 \Gamma(a+1) \frac{(-p)_p (-1)^p p!}{(a+1)_p p!}$$

$$= \binom{p+a}{p}^2 \frac{\Gamma(a+1) p!}{(a+1)_p}$$

$$= \frac{\Gamma(p+a+1)}{p!}$$

EXERCISES 8.5

1. Show that
$$\sum_{k=0}^{n} \frac{(-n)_k (b)_k}{(c)_k k!} = \frac{(c-b)_n}{(c)_n}$$

2. Show that
$$\sum_{k=0}^{n} \frac{(a)_n (b)_{n-k}}{(n-k)! k!} = \frac{(a+b)_n}{n!}$$

3. Show that
$$\sum_{k=1}^{n} \frac{(a)_k (b)_{n-k}}{(n-k)!(k-1)!} = \frac{a(a+b+1)_{n-1}}{(n-1)!}, \qquad n \geq 1$$

In problems 4–20, verify the given identity.

4. $\displaystyle\sum_{k=0}^{n} \binom{n}{k}^2 = \binom{2n}{n}$.

5. $\displaystyle\sum_{k=0}^{n} \binom{2n+1}{k} = 2^{2n}$.

6. $\displaystyle\sum_{k=0}^{n} \binom{k}{m} = \binom{n+1}{m+1}$.

7. $\displaystyle\sum_{k=0}^{n} (-1)^k \binom{m}{k} = (-1)^n \binom{m-1}{n}$, $m > n$.

8. $\displaystyle\sum_{k=1}^{n} k \binom{n}{k}^2 = \frac{(2n-1)!}{[(n-1)!]^2}$.

9. $\displaystyle\sum_{k=0}^{n} \binom{r}{k}\binom{m}{n-k} = \binom{r+m}{n}$.

10. $\displaystyle\sum_{k=0}^{n} \binom{r}{k}\binom{m}{n+k} = \binom{r+m}{r+n}$, $r \geq 0$.

11. $\displaystyle\sum_{k=0}^{n} \binom{2n}{2k}\binom{2k}{k}\binom{2n-2k}{n-k} = \binom{2n}{n}^2$.

12. $\displaystyle\sum_{k=0}^{n} (-1)^k \binom{a+k-1}{k}\binom{a}{n-k} = 0$, $n \geq 1$.

13. $\sum_{k=0}^{n} \binom{n-k}{m}\binom{p+k}{q} = \binom{n+p+1}{m+q+1}$, $p, q = 0, 1, 2, \ldots$.

14. $\sum_{k=0}^{n} \binom{m}{k}\binom{r+k}{n}(-1)^k = (-1)^m \binom{r}{n-m}$, $0 \le m \le n$.

15. $\sum_{k=0}^{n} \binom{2n}{2k} = 2^{2n-1}$, $n > 0$.

16. $\sum_{k=0}^{n} (-1)^k \binom{2n}{k}^2 = (-1)^n \binom{2n}{n}$.

    Hint: Use problem 26 in Exercises 8.3.

17. $\sum_{k=0}^{n-p} \binom{n}{k}\binom{n}{p+k} = \frac{(2n)!}{(n-p)!(n+p)!}$, $0 \le p \le n$.

18. $\sum_{n=0}^{\infty} \frac{1 \times 3 \times 5 \times \cdots \times (2n-1)}{2 \times 4 \times 6 \times \cdots \times (2n)} \left(\frac{1}{2}\right)^n = \sqrt{\frac{2}{3}}$.

19. $\sum_{n=0}^{\infty} \frac{(n!)^2 2^n}{(2n+1)!} = \frac{\pi}{2}$.

    Hint: Use problem 26 in Exercises 8.3.

20. $\sum_{n=0}^{\infty} \frac{[(2n)!]^2}{(2n+1)!(n!)^2} \left(\frac{1}{8}\right)^n = \frac{\pi\sqrt{2}}{4}$.

    Hint: Use problem 26 in Exercises 8.3.

21. Show that $(a > -1)$

$$\int_0^{\infty} x^{a+1} [L_p^{(a)}(x)]^2 e^{-x} \, dx = \frac{\Gamma(a+p+1)}{p!}(2p+a+1)$$

22. Show that* $(m = 0, 1, 2, \ldots)$

$$\int_0^a x^m [L_p^{(m)}(x)]^2 e^{-x} \, dx = 1 - \binom{p+m}{p} e^{-a} \sum_{n=0}^{p} \frac{(-p)_n}{n!}$$

$$\times \sum_{k=0}^{p} \frac{(-p)_k (m+n+1)_k}{(m+1)_k k!} e_{m+n+k}^a$$

where $e_n^x = \sum_{k=0}^{n} \frac{x^k}{k!}$.

---

*The evaluation of this integral is important in determining the total energy contained within the spot size of a Laguerre Gaussian beam. See R.L. Phillips and L.C. Andrews, Spot Size and Divergence for Laguerre Gaussian Beams of Any Order, *Applied Optics*, 22, No. 5, 643–644 (Mar. 1983).

23. Show that the Bessel function $J_p(x)$ satisfies

$$[J_p(x)]^2 = \sum_{n=0}^{\infty} \frac{(-1)^n (2n)!(x/2)^{2n+2p}}{(n!)^2 [\Gamma(n+p+1)]^2}$$

24. Show that the product of zero-order Bessel functions leads to

$$J_0(ax)J_0(bx) = \sum_{n=0}^{\infty} \frac{(-1)^n (ax/2)^{2n}}{(n!)^2} F(-n, -n; 1; b^2/a^2)$$

25. Show that

$$[F(a,b;c;x)]^2 = \sum_{n=0}^{\infty} \frac{(\frac{1}{2})_n (a)_n (b)_n (c-a)_n (c-b)_n}{(c)_n (c)_{2n} (c+n-\frac{1}{2})_n} \cdot \frac{x^{2n}}{n!}$$
$$\times F(2a+2n, 2b+2n; 2c+4n; x)$$

26. Show that

$$\int_0^{\pi/2} [J_1(x \sin\theta)]^2 \csc\theta \, d\theta = \frac{1}{2} - \frac{J_1(2x)}{2x}$$

Hint: Use problem 23.

27. Show that $(n = 0, 1, 2, \ldots)$

$$\frac{1}{\pi} \int_0^{\pi} J_0(2x \sin\phi) \cos 2n\phi \, d\phi = [J_n(x)]^2$$

Hint: Use problem 23.

9

The Confluent Hypergeometric Functions

9.1 Introduction

Whereas Gauss was largely responsible for the systematic study of the hypergeometric function, E.E. Kummer (1810–1893) is the person most associated with developing properties of the related *confluent hypergeometric function*. Kummer published his work on this function in 1836,* and since that time it has been commonly referred to as *Kummer's function*. Like the hypergeometric function, the confluent hypergeometric function is related to a large number of other functions.

Kummer's function satisfies a second-order linear differential equation called the *confluent hypergeometric equation*. A second solution of this DE leads to the definition of the *confluent hypergeometric function of the second kind*, which is also related to many other functions. At the beginning of the twentieth century (1904), Whittaker introduced another pair of confluent hypergeometric functions that now bear his name.[†] The *Whittaker functions* arise as solutions of the confluent hypergeometric equation after a transformation to *Liouville's standard form of the DE*.

---

*E.E. Kummer, Über die Hypergeometrische Reihe $F(a; b; x)$, *J. Reine Angew. Math.*, **15**, 39–83, 127–172 (1836).

[†]E.T. Whittaker, An Expression of Certain Known Functions as Generalized Hypergeometric Series, *Bull. Amer. Math. Soc.*, **10**, 125–134 (1904).

9.2 The Functions $M(a; c; x)$ and $U(a; c; x)$

Perhaps even more important in applications than the hypergeometric function is the related function

$$M(a; c; x) = \sum_{n=0}^{\infty} \frac{(a)_n}{(c)_n} \frac{x^n}{n!}, \quad -\infty < x < \infty \quad (9.1)$$

called the *confluent hypergeometric function*.* It is related to the hypergeometric function according to

$$M(a; c; x) = \lim_{b \to \infty} F(a, b; c; x/b) \quad (9.2)$$

To see this, we note that

$$\lim_{b \to \infty} F(a, b; c; x/b) = \lim_{b \to \infty} \sum_{n=0}^{\infty} \frac{(a)_n (b)_n}{(c)_n} \frac{(x/b)^n}{n!}$$

$$= \sum_{n=0}^{\infty} \frac{(a)_n}{(c)_n} \frac{x^n}{n!} \lim_{b \to \infty} \frac{(b)_n}{b^n}$$

where clearly $(b)_n/b^n = b(b+1)\cdots(b+n-1)/b^n \to 1$ as $b \to \infty$.

Remark: The function $M(a; c; x)$ is also designated by $\Phi(a; c; x)$ or $_1F_1(a; c; x)$, and commas are sometimes used in place of semicolons.

As was the case for the hypergeometric function, the series (9.1) is normally not defined for $c = 0, -1, -2, \ldots$, and if a is a negative integer, the series truncates. By application of the ratio test, it can be shown that the confluent hypergeometric series (9.1) converges for all (finite) x (see problem 1).

9.2.1 Elementary Properties

Because of the similarity of definition to that of $F(a, b; c; x)$, the function $M(a; c; x)$ obviously has many properties analogous to those of the hypergeometric function. For example, it is easy to show that

$$\frac{d}{dx} M(a; c; x) = \frac{a}{c} M(a + 1; c + 1; x) \quad (9.3)$$

---

*The term "confluent" refers to the fact that, due to the transformation (9.2), two singularities in the hypergeometric differential equation (viz., at one and infinity) are merged into one singularity (at infinity) in the confluent hypergeometric differential equation. See Section 9.2.2.

whereas in general
$$\frac{d^k}{dx^k}M(a;c;x) = \frac{(a)_k}{(c)_k}M(a+k;c+k;x), \quad k=1,2,3,\ldots \quad (9.4)$$

The function $M(a;c;x)$ also satisfies recurrence relations involving the contiguous functions $M(a \pm 1; c; x)$ and $M(a; c \pm 1; x)$. From these four contiguous functions, taken two at a time, we find six recurrence relations with coefficients at most linear in x:

$$(c-a-1)M(a;c;x) + aM(a+1;c;x)$$
$$= (c-1)M(a;c-1;x) \quad (9.5)$$

$$cM(a;c;x) - cM(a-1;c;x) = xM(a;c+1;x) \quad (9.6)$$

$$(a-1+x)M(a;c;x) + (c-a)M(a-1;c;x)$$
$$= (c-1)M(a;c-1;x) \quad (9.7)$$

$$c(a+x)M(a;c;x) - acM(a+1;c;x)$$
$$= (c-a)xM(a;c+1;x) \quad (9.8)$$

$$(c-a)M(a-1;c;x) + (2a-c+x)M(a;c;x)$$
$$= aM(a+1;c;x) \quad (9.9)$$

$$c(c-1)M(a;c-1;x) - c(c-1+x)M(a;c;x)$$
$$= (a-c)xM(a;c+1;x) \quad (9.10)$$

The verification of these relations is left to the exercises.

To obtain an integral representation of $M(a;c;x)$, we first recall the identity [Equation (8.18) in Section 8.3.2]

$$\frac{(a)_n}{(c)_n} = \frac{\Gamma(c)}{\Gamma(a)\Gamma(c-a)} \int_0^1 t^{a+n-1}(1-t)^{c-a-1}\,dt, \quad c > a > 0 \quad (9.11)$$

for $n = 0, 1, 2, \ldots$. Thus it follows from (9.1) that

$$M(a;c;x) = \frac{\Gamma(c)}{\Gamma(a)\Gamma(c-a)} \sum_{n=0}^{\infty} \frac{x^n}{n!} \int_0^1 t^{a+n-1}(1-t)^{c-a-1}\,dt$$

$$= \frac{\Gamma(c)}{\Gamma(a)\Gamma(c-a)} \int_0^1 t^{a-1}(1-t)^{c-a-1} \left(\sum_{n=0}^{\infty} \frac{(xt)^n}{n!} \right) dt \quad (9.12)$$

where we have interchanged the order of integration and summation. Recognizing the infinite sum in (9.12) as that of an exponential, we deduce the integral representation

$$M(a;c;x) = \frac{\Gamma(c)}{\Gamma(a)\Gamma(c-a)} \int_0^1 e^{xt} t^{a-1}(1-t)^{c-a-1}\,dt, \quad c > a > 0 \quad (9.13)$$

The integral formula (9.13) can now be used to derive a very important result concerning confluent hypergeometric functions. We simply make the change of variable $t = 1 - u$ to get

$$M(a; c; x) = \frac{\Gamma(c)}{\Gamma(a)\Gamma(c-a)} e^x \int_0^1 e^{-xu} u^{c-a-1}(1-u)^{a-1} du \quad (9.14)$$

which implies

$$M(a; c; x) = e^x M(c - a; c; -x) \quad (9.15)$$

known as *Kummer's transformation*. Even though (9.13) requires that $c > a > 0$, the result (9.15) is valid for all values of the parameters for which the confluent hypergeometric function is defined.

9.2.2 Confluent Hypergeometric Function of the Second Kind

The hypergeometric function $y = F(a, b; c; t)$ is a solution of Gauss's equation

$$t(1-t)\frac{d^2y}{dt^2} + [c - (a + b + 1)t]\frac{dy}{dt} - aby = 0 \quad (9.16)$$

By making the change of variable $t = x/b$, (9.16) becomes

$$x\left(1 - \frac{x}{b}\right)y'' + \left[c - x - \frac{(a+1)}{b}x\right]y' - ay = 0$$

and then allowing $b \to \infty$, we find

$$xy'' + (c - x)y' - ay = 0 \quad (9.17)$$

Now since

$$M(a; c; x) = \lim_{b \to \infty} F\left(a, b; c; \frac{x}{b}\right)$$

it follows that $y_1 = M(a; c; x)$ is a solution of Equation (9.17), which is called the *confluent hypergeometric equation*.

By making the change of variable $y = x^{1-c}z$, we find that (9.17) becomes (after simplification)

$$xz'' + (2 - c - x)z' - (1 + a - c)z = 0 \quad (9.18)$$

Thus, by comparing (9.18) with (9.17), it is clear that

$$z = M(1 + a - c; 2 - c; x), \quad c \neq 2, 3, 4, \ldots$$

is a solution of (9.18), and hence

$$y_2 = x^{1-c} M(1 + a - c; 2 - c; x), \quad c \neq 2, 3, 4, \ldots \quad (9.19)$$

is a second solution of Equation (9.17). Furthermore, if c is not an integer (positive, zero, or negative), then y_2 is linearly independent of $y_1 = M(a;c;x)$, and in this case the general solution of (9.17) is

$$y = C_1 M(a;c;x) + C_2 x^{1-c} M(1+a-c; 2-c; x),$$
$$c \neq 0, \pm 1, \pm 2, \ldots \quad (9.20)$$

where C_1 and C_2 are any constants.

To remove the restriction $c \neq 1, 2, 3, \ldots$ in the general solution (9.20), we introduce the function ($c \neq 0, -1, -2, \ldots$)

$$U(a;c;x) = \frac{\pi}{\sin c\pi}\left[\frac{M(a;c;x)}{\Gamma(1+a-c)\Gamma(c)} - \frac{x^{1-c}M(1+a-c; 2-c; x)}{\Gamma(a)\Gamma(2-c)}\right] \quad (9.21)$$

called the *confluent hypergeometric function of the second kind*. For nonintegral values of c, $U(a;c;x)$ is surely a solution of (9.17), since it is simply a linear combination of two solutions. For $c = 1, 2, 3, \ldots$, we find that (9.21) assumes the indeterminant form $0/0$, and in this case we define (analogous to Bessel functions of the second kind)

$$U(a; n+1; x) = \lim_{c \to n+1} U(a;c;x), \qquad n = 0, 1, 2, \ldots \quad (9.22)$$

which can also be shown to be a solution of (9.17).

To investigate the behavior of $U(a;c;x)$ when a is a nonpositive integer, we set $a = -n$ ($n = 0, 1, 2, \ldots$) in (9.21) to find

$$U(-n;c;x) = \frac{\pi}{\sin \pi c} \frac{M(-n;c;x)}{\Gamma(1-n-c)\Gamma(c)}$$

which, by use of the identity $\Gamma(x)\Gamma(1-x) = \pi/\sin \pi x$, becomes

$$U(-n;c;x) = (-1)^n (c)_n M(-n;c;x), \qquad n = 0, 1, 2, \ldots \quad (9.23)$$

Hence, the functions $U(a;c;x)$ and $M(a;c;x)$ are clearly linearly dependent for $a = 0, -1, -2, \ldots$, and therefore do not constitute a fundamental set of solutions of (9.17) in this case. Nonetheless, for both $a, c \neq 0, -1, -2, \ldots$, it can be shown that $U(a;c;x)$ and $M(a;c;x)$ are linearly independent functions, and in this case the general solution of (9.17) is (see problem 22)

$$y = C_1 M(a;c;x) + C_2 U(a;c;x), \qquad a, c \neq 0, -1, -2, \ldots \quad (9.24)$$

The function $U(a;c;x)$ has many properties like $M(a;c;x)$. Directly from its definition (9.21), we first note that (problem 21)*

$$U(a;c;x) = x^{1-c} U(1+a-c; 2-c; x) \quad (9.25)$$

---

*From (9.25), it follows that $U(a;c;x)$ is defined also for $c = 0, -1, -2, \ldots$.

while the derivative relations are readily found to be (problem 23)

$$\frac{d}{dx}U(a;c;x) = -aU(a+1;c+1;x) \tag{9.26}$$

and

$$\frac{d^k}{dx^k}U(a;c;x) = (-1)^k(a)_k U(a+k;c+k;x), \quad k = 1,2,3,\ldots \tag{9.27}$$

Although more difficult to show, it has the integral representation

$$U(a;c;x) = \frac{1}{\Gamma(a)}\int_0^\infty e^{-xt}t^{a-1}(1+t)^{c-a-1}\,dt, \quad a > 0, \quad x > 0 \tag{9.28}$$

Some additional properties are taken up in the exercises.

9.2.3 Asymptotic Formulas

From the series representation (9.1) of $M(a;c;x)$, it follows immediately that for small values of x,

$$M(a;c;x) \sim 1, \quad x \to 0 \tag{9.29a}$$

or

$$M(a;c;x) \sim 1 + \frac{a}{c}x, \quad x \to 0 \tag{9.29b}$$

by retaining only the first term or two of the series. A similar result can be derived for $U(a;c;x)$, but in this case the functional form of the asymptotic formula will vary somewhat, depending upon the numerical value of c (see problem 27).

To derive an asymptotic formula valid for large values of x, we begin with the integral representation [see (9.14)]

$$M(a;c;x) = \frac{\Gamma(c)}{\Gamma(a)\Gamma(c-a)}e^x\int_0^1 e^{-xu}u^{c-a-1}(1-u)^{a-1}\,du \tag{9.30}$$

This integral can further be expressed as the difference of two integrals by writing

$$\int_0^1 e^{-xu}u^{c-a-1}(1-u)^{a-1}\,du = \int_0^\infty e^{-xu}u^{c-a-1}(1-u)^{a-1}\,du$$
$$- \int_1^\infty e^{-xu}u^{c-a-1}(1-u)^{a-1}\,du$$

Next, we make the substitution $s = xu$ in the first integral on the right and the substitution $t = x(u-1)$ in the second integral on the right. This action yields

$$\int_0^1 e^{-xu} u^{c-a-1}(1-u)^{a-1} du = x^{a-c} \int_0^\infty e^{-s} s^{c-a-1}\left(1 - \frac{s}{x}\right)^{a-1} ds$$

$$- e^{-x} \int_0^\infty e^{-t}\left(1 + \frac{t}{x}\right)^{c-a-1}\left(-\frac{t}{x}\right)^{a-1} dt$$

(9.31)

Hence, for $x \gg s$ and $x \gg t$, we make the approximations

$$\left(1 - \frac{s}{x}\right)^{a-1} \simeq 1, \qquad \left(-\frac{t}{x}\right)^{a-1} \simeq 0$$

and find that (9.31) leads to

$$\int_0^1 e^{-xu} u^{c-a-1}(1-u)^{a-1} du \sim x^{a-c} \int_0^\infty e^{-s} s^{c-a-1} ds$$

$$\sim x^{a-c} \Gamma(c-a)$$

Substituting this result into (9.30), we deduce that

$$M(a; c; x) \sim \frac{\Gamma(c)}{\Gamma(a)} x^{a-c} e^x, \qquad x \to \infty \qquad (9.32)$$

for $a, c \neq 0, -1, -2, \ldots$. If instead of approximating the term $(1 - s/x)^{a-1}$ by unity, we choose to expand it in a binomial series for $x > s$, then we obtain the full asymptotic series (see problem 28)

$$M(a; c; x) \sim \frac{\Gamma(c)}{\Gamma(a)} x^{a-c} e^x \sum_{n=0}^\infty \frac{(1-a)_n (c-a)_n}{n! x^n}, \qquad x \to \infty$$

(9.33)

where again $a, c, \neq 0, -1, -2, \ldots$.

Lastly, if we utilize the integral representation (9.28) for $U(a; c; x)$, it follows in a like manner that (see problem 30)

$$U(a; c; x) \sim x^{-a} \sum_{n=0}^\infty \frac{(-1)^n (a)_n (1+a-c)_n}{n! x^n}, \qquad x \to \infty \quad (9.34)$$

EXERCISES 9.2

1. By applying the ratio test to Equation (9.1), show that the confluent hypergeometric series converges for all x.

2. Show that

   (a) $\dfrac{d}{dx} M(a; c; x) = \dfrac{a}{c} M(a + 1; c + 1; x)$,

   (b) $\dfrac{d^k}{dx^k} M(a; c; x) = \dfrac{(a)_k}{(c)_k} M(a + k; c + k; x)$, $k = 1, 2, 3, \ldots$.

In problems 3–7, verify the differentiation formula.

3. $x \dfrac{d}{dx} M(a; c; x) + aM(a; c; x) = aM(a + 1; c; x)$.

4. $x \dfrac{d}{dx} M(a; c; x) + (c - a - x)M(a; c; x) = (c - a)M(a - 1; c; x)$.

5. $c \dfrac{d}{dx} M(a; c; x) - cM(a; c; x) = (a - c)M(a; c + 1; x)$.

6. $x \dfrac{d}{dx} M(a; c; x) + (c - 1)M(a; c; x) = (c - 1)M(a; c - 1; x)$.

7. $x \dfrac{d}{dx} M(a; c; x) + (c - 1 - x)M(a; c; x)$
   $= (c - 1)M(a - 1; c - 1; x)$.

In problems 8–13, verify the contiguous relation by using the results of problems 3–7, or by using series representations.

8. $(c - a - 1)M(a; c; x) + aM(a + 1; c; x) = (c - 1)M(a; c - 1; x)$.

9. $cM(a; c; x) - cM(a - 1; c; x) = xM(a; c + 1; x)$.

10. $(a - 1 + x)M(a; c; x) + (c - a)M(a - 1; c; x)$
    $= (c - 1)M(a; c - 1; x)$.

11. $c(a + x)M(a; c; x) - acM(a + 1; c; x) = (c - a)xM(a; c + 1; x)$.

12. $(c - a)M(a - 1; c; x) + (2a - c + x)M(a; c; x) = aM(a + 1; c; x)$.

13. $c(c - 1)M(a; c - 1; x) + c(c - 1 + x)M(a; c; x)$
    $= (a - c)xM(a; c + 1; x)$.

14. Show that

    $$M(a + 1; c; x) - M(a; c; x) = \dfrac{x}{c} M(a + 1; c + 1; x)$$

306 • Special Functions for Engineers and Applied Mathematicians

15. Show that

$$M(a; c; x) = \frac{c-a}{c} M(a; c+1; x) + \frac{a}{c} M(a+1; c+1; x)$$

In problems 16–18, derive the integral relation.

16. $M(a; c; x) = \dfrac{\Gamma(c) 2^{1-c}}{\Gamma(a)\Gamma(c-a)} e^{x/2} \int_{-1}^{1} e^{-xt/2} (1+t)^{c-a-1} (1-t)^{a-1} dt,$
$c > a > 0.$

17. $M(a; c; x) = \dfrac{\Gamma(c)}{\Gamma(c-a)} e^x x^{(1-c)/2} \int_{0}^{\infty} e^{-t} t^{\frac{1}{2}(c-1)-a} J_{c-1}(2\sqrt{xt}) \, dt,$
$c > a > 0, \ x > 0.$

18. $\displaystyle\int_0^\infty e^{-st} M(a; c; t) \, dt = \dfrac{1}{s} F\left(a, 1; c; \dfrac{1}{s}\right), \ s > 1.$

19. By substituting its series representation, show directly that $y = M(a; c; x)$ is a solution of

$$xy'' + (c-x)y' - ay = 0$$

20. Show that the confluent hypergeometric function of the second kind (9.21) is also given by

$$U(a; c; x) = \frac{\Gamma(1-c)}{\Gamma(1+a-c)} M(a; c; x)$$
$$+ \frac{\Gamma(c-1)}{\Gamma(a)} x^{1-c} M(1+a-c; 2-c; x)$$

21. Verify the Kummer relation

$$U(a; c; x) = x^{1-c} U(1+a-c; 2-c; x)$$

22. Show that the Wronskian of the confluent hypergeometric functions is given by

$$W(M, U)(x) = -\frac{\Gamma(c)}{\Gamma(a)} x^{-c} e^x$$

Hint: See problem 7 in Exercises 6.5.

23. Show that

(a) $\dfrac{d}{dx} U(a; c; x) = -aU(a+1; c+1; x),$

(b) $\dfrac{d^k}{dx^k} U(a; c; x) = (-1)^k (a)_k U(a+k; c+k; x), \ k = 1, 2, 3, \ldots .$

24. Show that $U(a; c; x)$ has (among others) the contiguous relations
  (a) $U(a; c; x) - aU(a + 1; c; x) = U(a; c - 1; x)$,
  (b) $(c - a)U(a; c; x) + U(a - 1; c; x) = xU(a; c + 1; x)$.

25. From the well-known result of calculus

$$f(x + y) = \sum_{n=0}^{\infty} f^{(n)}(x)\frac{y^n}{n!}, \quad |y| < \rho$$

derive the addition formulas

  (a) $M(a; c; x + y) = \sum_{n=0}^{\infty} \frac{(a)_n}{(c)_n} \frac{y^n}{n!} M(a + n; c + n; x)$,

  (b) $U(a; c; x + y) = \sum_{n=0}^{\infty} \frac{(a)_n}{n!}(-1)^n y^n U(a + n; c + n; x)$.

26. From the result of problem 25, deduce the multiplication formulas

  (a) $M(a; c; xy) = \sum_{n=0}^{\infty} \frac{(a)_n x^n (y - 1)^n}{(c)_n n!} M(a + n; c + n; x)$,

  (b) $U(a; c; xy) = \sum_{n=0}^{\infty} \frac{(a)_n x^n (1 - y)^n}{n!} U(a + n; c + n; x)$.

27. For small arguments, show that

  (a) $U(a; c; x) \sim \dfrac{\Gamma(c - 1)}{\Gamma(a)} x^{1-c}, \quad c > 1, \ x \to 0^+$,

  (b) $U(a; 1; x) \sim -\dfrac{1}{\Gamma(a)}[\log x + \psi(a)], \quad x \to 0^+$.

28. Starting with Equation (9.31), show by expanding $(1 - s/x)^{a-1}$ in a binomial series that

$$M(a; c; x) \sim \frac{\Gamma(c)}{\Gamma(a)} x^{a-c} e^x \sum_{n=0}^{\infty} \frac{(1 - a)_n (c - a)_n}{n! x^n}, \quad x \to \infty$$

29. Using the integral representation (9.28), show that

$$U(a; c; x) \sim x^{-a}, \quad x \to \infty$$

30. Following the technique suggested in problem 28, derive the asymptotic series

$$U(a; c; x) \sim x^{-a} \sum_{n=0}^{\infty} \frac{(-1)^n (a)_n (1 + a - c)_n}{n! x^n}, \quad x \to \infty$$

31. The probability density function for the combined phase of a signal embedded in narrowband Gaussian noise is calculated from*

$$w(\theta) = \int_0^\infty p(r,\theta)\,dr, \qquad 0 \le \theta \le 2\pi$$

where $p(r,\theta)$ is the joint density of the envelope and phase of the signal and noise. If the amplitude of the signal is A and the variance of the noise ψ, it is known that

$$p(r,\theta) = \frac{r}{2\pi\psi}\exp\left[-\frac{r^2 + A^2 - 2Ar\cos\theta}{2\psi}\right],$$

$$0 \le r < \infty, \quad 0 \le \theta \le 2\pi$$

Show that, under these conditions,

$$w(\theta) = \frac{1}{2\pi} + \frac{1}{\pi}\sum_{n=1}^{\infty}\frac{S^{n/2}}{n!}\Gamma\!\left(1+\frac{n}{2}\right)M\!\left(\frac{n}{2};1+n;-S\right)\cos n\theta,$$

$$0 \le \theta \le 2\pi$$

where $S = A^2/2\psi$ represents the signal-to-noise ratio.

Hint: Use the relation $e^{\alpha\cos x} = I_0(\alpha) + \sum_{n=1}^{\infty} I_n(\alpha)\cos nx$.

9.3 Relation to Other Functions

Specializations of either $M(a; c; x)$ or $U(a; c; x)$ lead to most of the other special functions that have been introduced in earlier chapters. For example, it can readily be verified by comparing series or integral representations that

$$e^x = M(a; a; x) \tag{9.35}$$

$$\mathrm{erfc}(x) = \frac{1}{\sqrt{\pi}}e^{-x^2}U(\tfrac{1}{2};\tfrac{1}{2};x^2) \tag{9.36}$$

$$H_{2n}(x) = (-1)^n\frac{(2n)!}{n!}M(-n;\tfrac{1}{2};x^2) \tag{9.37}$$

$$L_n(x) = M(-n;1;x) \tag{9.38}$$

$$\mathrm{Ei}(x) = -e^x U(1;1;-x) \tag{9.39}$$

$$K_p(x) = \sqrt{\pi}\,(2x)^p e^{-x}U(p+\tfrac{1}{2};2p+1;2x) \tag{9.40}$$

among many other such relations (see the exercises).

---

*For a discussion of narrowband Gaussian noise processes, see D. Middleton, *An Introduction to Statistical Communication Theory*, New York: McGraw-Hill, 1960, pp. 335–512.

The validity of (9.35) follows directly from

$$M(a;a;x) = \sum_{n=0}^{\infty} \frac{(a)_n}{(a)_n} \frac{x^n}{n!} = \sum_{n=0}^{\infty} \frac{x^n}{n!} = e^x$$

while (9.36) and (9.37) are proved in Examples 1 and 2 below. Verifying (9.38), (9.39) and (9.40) is left to the exercises.

Example 1: Show that $\text{erfc}(x) = (1/\sqrt{\pi})e^{-x}U(\frac{1}{2};\frac{1}{2};x^2)$.

Solution: By introducing the substitution $t = x\sqrt{1+s}$, we find

$$\text{erfc}(x) = \frac{2}{\sqrt{\pi}} \int_x^{\infty} e^{-t^2} dt$$

$$= \frac{1}{\sqrt{\pi}} xe^{-x^2} \int_0^{\infty} e^{-x^2 s}(1+s)^{-1/2} ds$$

Comparing this last integral with the integral representation (9.28) identifies the parameters $a = 1$ and $c = \frac{3}{2}$, and hence

$$\text{erfc}(x) = \frac{1}{\sqrt{\pi}} xe^{-x^2} U(1;\tfrac{3}{2};x^2)$$

However, by using the identity $U(a;c;x) = x^{1-c}U(1+a-c;2-c;x)$, we can also write

$$\text{erfc}(x) = \frac{1}{\sqrt{\pi}} e^{-x^2} U(\tfrac{1}{2};\tfrac{1}{2};x^2)$$

Example 2: Show that $H_{2n}(x) = (-1)^n \frac{(2n)!}{n!} M(-n;\tfrac{1}{2};x^2)$.

Solution: From the series definition of the Hermite polynomials, we have

$$H_{2n}(x) = \sum_{k=0}^{n} \frac{(-1)^k (2n)!}{k!(2n-2k)!} (2x)^{2n-2k}$$

$$= (-1)^n (2n)! \sum_{j=0}^{n} \frac{(-1)^j (2x)^{2j}}{(n-j)!(2j)!}$$

where the last step has resulted from the index change $j = n - k$. In terms of Pochhammer symbols, we write

$$(2j)! = 2^{2j}(\tfrac{1}{2})_j j!$$

$$(n-j)! = \frac{(-1)^j n!}{(-n)_j}$$

and therefore it follows that

$$H_{2n}(x) = (-1)^n \frac{(2n)!}{n!} \sum_{j=0}^{n} \frac{(-n)_j}{(\tfrac{1}{2})_j} \frac{x^{2j}}{j!} = (-1)^n \frac{(2n)!}{n!} M(-n; \tfrac{1}{2}; x^2)$$

9.3.1 Hermite Functions

The DE

$$y'' - 2xy' + 2\nu y = 0 \qquad (9.41)$$

arises in the solution of Laplace's equation in parabolic coordinates. For $\nu = n$ ($n = 0, 1, 2, \ldots$), this is just the DE satisfied by the Hermite polynomials studied in Chapter 5. Therefore, in the general case where ν is arbitrary we will refer to the solutions of (9.41) as *Hermite functions*.

To find a general solution of (9.41), we start with the change of variable $t = x^2$, which converts the DE to the confluent hypergeometric form

$$t\frac{d^2y}{dt^2} + (\tfrac{1}{2} - t)\frac{dy}{dt} + \frac{\nu}{2}y = 0 \qquad (9.42)$$

Hence, (9.42) is just a special case of (9.17) for which $a = -\nu/2$ and $c = \tfrac{1}{2}$. Recalling Equation (9.20), we see that a general solution of (9.42) is

$$y(t) = C_1 M\left(-\frac{\nu}{2}; \tfrac{1}{2}; t\right) + C_2 t^{1/2} M\left(\frac{1-\nu}{2}; \tfrac{3}{2}; t\right) \qquad (9.43)$$

and so the general solution of (9.41) is

$$y(x) = C_1 M\left(-\frac{\nu}{2}; \tfrac{1}{2}; x^2\right) + C_2 x M\left(\frac{1-\nu}{2}; \tfrac{3}{2}; x^2\right) \qquad (9.44)$$

It is customary to choose the constants C_1 and C_2 to be

$$C_1 = \frac{2^\nu \sqrt{\pi}}{\Gamma\left(\frac{1-\nu}{2}\right)}, \qquad C_2 = -\frac{2^{\nu+1}\sqrt{\pi}}{\Gamma\left(-\frac{\nu}{2}\right)}$$

and then define

$$H_\nu(x) = \frac{2^\nu \sqrt{\pi}}{\Gamma\left(\frac{1-\nu}{2}\right)} M\left(-\frac{\nu}{2}; \tfrac{1}{2}; x^2\right) - \frac{2^{\nu+1}\sqrt{\pi}}{\Gamma\left(-\frac{\nu}{2}\right)} x M\left(\frac{1-\nu}{2}; \tfrac{3}{2}; x^2\right)$$

$$(9.45)$$

which is called the *Hermite function* of degree ν.

Various properties of the Hermite functions can be derived directly from the definition (9.45) in terms of confluent hypergeometric functions. For example, we can immediately deduce that

$$H_\nu(0) = \frac{2^\nu\sqrt{\pi}}{\Gamma\left(\frac{1-\nu}{2}\right)}, \qquad H_\nu'(0) = -\frac{2^{\nu+1}\sqrt{\pi}}{\Gamma\left(-\frac{\nu}{2}\right)} \qquad (9.46)$$

Also, by expressing the confluent hypergeometric functions in (9.45) in series form, a series for $H_\nu(x)$ can be derived for ν not zero or a positive integer (see problem 36).

By comparing (9.45) with the definition of the confluent hypergeometric function of the second kind, it follows that the Hermite function can also be expressed as

$$H_\nu(x) = 2^\nu U\left(-\frac{\nu}{2}; \tfrac{1}{2}; x^2\right) \qquad (9.47)$$

Hence, recalling the result of Example 1, we see that, for example,

$$H_{-1}(x) = \frac{\sqrt{\pi}}{2} e^{x^2} \operatorname{erfc}(x) \qquad (9.48)$$

The basic recurrence formulas for the Hermite polynomials are satisfied as well by the Hermite functions; the proofs are left to the exercises.

9.3.2 Laguerre Functions

The associated Laguerre polynomials are related to the confluent hypergeometric function by (see problem 8)

$$\begin{aligned} L_n^{(a)}(x) &= \frac{(a+1)_n}{n!} M(-n; a+1; x) \\ &= \frac{\Gamma(n+a+1)}{\Gamma(n+1)\Gamma(a+1)} M(-n; a+1; x) \end{aligned} \qquad (9.49)$$

If we choose to replace the index n by the more general index ν (not restricted to nonnegative integer values), we have

$$L_\nu^{(a)}(x) = \frac{\Gamma(\nu+a+1)}{\Gamma(\nu+1)\Gamma(a+1)} M(-\nu; a+1; x) \qquad (9.50)$$

called the *Laguerre function* of degree ν.

For $\nu \neq n$ ($n = 0, 1, 2, \ldots$), it is clear that the Laguerre function is not a polynomial, since the series for $M(-\nu; a+1; x)$ will be infinite in this case. Nonetheless, the basic recurrence formulas for the associated Laguerre polynomials (Section 5.3.3) continue to hold for the more general Laguerre function. In the case where ν is a negative integer, some immediate consequences of the defining relation (9.50) are

$$L_{-n}^{(a)}(x) = 0, \quad n = 1, 2, 3, \ldots, \quad a > -1, \text{ and } a \neq 0, 1, 2, \ldots \tag{9.51}$$

and (for $m = 1, 2, 3, \ldots$)

$$L_{-n}^{(m)}(x) = \begin{cases} 0, & n = 1, 2, 3, \ldots, m \\ \dfrac{(-1)^m (n)_m}{m!}, & n = m+1, m+2, \ldots \end{cases} \tag{9.52}$$

the proofs of which are left to the exercises.

EXERCISES 9.3

In problems 1–15, verify the given relation.

1. $\gamma(a, x) = \dfrac{x^a}{a} M(a; a+1; -x)$.

2. $\Gamma(a, x) = e^{-x} U(1-a; 1-a; x)$.

3. $\text{erf}(x) = \dfrac{2}{\sqrt{\pi}} x M(\tfrac{1}{2}; \tfrac{3}{2}; -x^2)$.

4. $\text{Ei}(x) = -e^x U(1; 1; -x)$.

5. $\text{li}(x) = -x U(1; 1; -\log x)$.

6. $H_{2n+1}(x) = (-1)^n \dfrac{(2n+1)!}{n!} 2x M(-n; \tfrac{3}{2}; x^2)$.

7. $L_n(x) = M(-n; 1; x)$.

8. $L_n^{(a)}(x) = \dfrac{(a+1)_n}{n!} M(-n; a+1; x)$.

9. $J_p(x) = \dfrac{(x/2)^p}{\Gamma(p+1)} e^{-ix} M(p+\tfrac{1}{2}; 2p+1; 2ix)$.

   Hint: Start with the integral representation (6.32) in Section 6.3 and make the change of variable $t = 2s - 1$.

10. $I_p(x) = \dfrac{(x/2)^p}{\Gamma(p+1)} e^{-x} M(p+\tfrac{1}{2}; 2p+1; 2x)$.

11. $K_p(x) = \sqrt{\pi} (2x)^p e^{-x} U(p+\tfrac{1}{2}; 2p+1; 2x)$.

12. $C(x) = \dfrac{x}{2} \left[M(\tfrac{1}{2}; \tfrac{3}{2}; \tfrac{1}{2} i\pi x^2) + M(\tfrac{1}{2}; \tfrac{3}{2}; -\tfrac{1}{2} i\pi x^2) \right]$.

13. $S(x) = \dfrac{x}{2i}\left[M(\tfrac{1}{2};\tfrac{3}{2};\tfrac{1}{2}i\pi x^2) - M(\tfrac{1}{2};\tfrac{3}{2};-\tfrac{1}{2}i\pi x^2)\right].$

14. $\text{Ci}(x) = -\tfrac{1}{2}[e^{-ix}U(1;1;ix) + e^{ix}U(1;1;-ix)].$

15. $\text{Si}(x) = \dfrac{\pi}{2} + \dfrac{1}{2i}[e^{-ix}U(1;1;ix) - e^{ix}U(1;1;-ix)].$

In problems 16–23, verify the special cases.

16. $M(1;2;-x) = \dfrac{1}{x}(1 - e^{-x}).$

17. $M(2;1;-x) = (1 - x)e^{-x}.$

18. $M(1;3;-x) = \dfrac{2}{x^2}(x + e^{-x} - 1).$

19. $M(3;2;-x) = (1 - x/2)e^{-x}.$

20. $M(\tfrac{1}{2};1;-x) = e^{-x/2}I_0(x/2).$

21. $M(\tfrac{1}{2};2;-x) = e^{-x/2}[I_0(x/2) + I_1(x/2)].$

22. $M(\tfrac{3}{2};2;-x) = e^{-x/2}[I_0(x/2) - I_1(x/2)].$

23. $M(\tfrac{3}{2};3;-x) = \dfrac{4}{x}e^{-x/2}I_1(x/2).$

24. Show that $(x > 0)$

  (a) $\lim\limits_{a\to\infty} M(a;c;-x/a) = \Gamma(c)x^{(1-c)/2}J_{c-1}(2\sqrt{x}),$

  (b) $\lim\limits_{a\to\infty} M(a;c;x/a) = \Gamma(c)x^{(1-c)/2}I_{c-1}(2\sqrt{x}).$

In problems 25 and 26, use properties of the confluent hypergeometric function to sum the series.

25. $\displaystyle\sum_{n=0}^{\infty} \dfrac{n!}{(2n+1)!} x^{2n+1} = \sqrt{\pi}\, e^{x^2/4}\text{erf}(x/2).$

26. $\displaystyle\sum_{k=0}^{\infty} \dfrac{(n+k)!}{n!(k!)^2}(-1)^k x^k = e^x L_n(x),\ n = 0,1,2,\ldots\,.$

In problems 27–29, verify the integral relation.

27. $\displaystyle\int_0^{\infty} x^{2n+1}\cos bx\, e^{-ax^2/2}\,dx = \dfrac{2^n n!}{a^{n+1}} M\!\left(n+1;\tfrac{1}{2};-\dfrac{b^2}{2a}\right),\ a > 0.$

28. $\displaystyle\int_0^{\infty} x^{2n}\sin bx\, e^{-ax^2/2}\,dx = \dfrac{2^n n!}{a^{n+1}} bM\!\left(n+1;\tfrac{3}{2};-\dfrac{b^2}{2a}\right),\ a > 0.$

29. $\displaystyle\int_0^{\infty} J_0(ax)x^{2\mu-1}e^{-b^2x^2}\,dx = \dfrac{\Gamma(\mu)}{2b^{2\mu}} M\!\left(\mu;1;-\dfrac{a^2}{4b^2}\right),\ a \ge 0,\ b > 0,$
$\mu \ne 0,-1,-2,\ldots\,.$

In problems 30 and 31, use the result of problem 28 in Exercises 9.2 to derive the asymptotic formula.

30. $\mathrm{erf}(x) \sim 1$, $x \to \infty$.

31. $I_p(x) \sim \dfrac{e^x}{\sqrt{2\pi x}} \sum_{n=0}^{\infty} \dfrac{(\frac{1}{2}-p)_n (\frac{1}{2}+p)_n}{n!(2x)^n}$, $x \to \infty$.

In problems 32–35, use the result of problem 30 in Exercises 9.2 to derive the asymptotic formula.

32. $\mathrm{erfc}(x) \sim \dfrac{e^{-x}}{x\sqrt{\pi}} \sum_{n=0}^{\infty} \dfrac{(-1)^n (\frac{1}{2})_n}{x^n}$, $x \to \infty$.

33. $\mathrm{Ei}(x) \sim \dfrac{e^x}{x} \sum_{n=0}^{\infty} \dfrac{n!}{x^n}$, $x \to \infty$.

34. $\mathrm{li}(x) \sim \dfrac{x}{\log x} \sum_{n=0}^{\infty} \dfrac{n!}{(\log x)^n}$, $x \to \infty$.

35. $K_p(x) \sim \sqrt{\dfrac{\pi}{2x}}\, e^{-x} \sum_{n=0}^{\infty} \dfrac{(-1)^n (\frac{1}{2}+p)_n (\frac{1}{2}-p)_n}{n!(2x)^n}$, $x \to \infty$.

36. By expressing the confluent hypergeometric functions in (9.45) by their series representations, deduce that

$$H_\nu(x) = \dfrac{1}{2\Gamma(-\nu)} \sum_{k=0}^{\infty} (-1)^k \Gamma\!\left(\dfrac{k-\nu}{2}\right) \dfrac{(2x)^k}{k!},$$

$\nu \neq n$ $(n = 0, 1, 2, \ldots)$

In problems 37–39, use the result of problem 36 to deduce the given relation.

37. $H'_\nu(x) = 2\nu H_{\nu-1}(x)$.

38. $2\nu H'_{\nu-1}(x) = 2xH'_\nu(x) - 2\nu H_\nu(x)$.

39. $H_{\nu+1}(x) - 2xH_\nu(x) + 2\nu H_{\nu-1}(x) = 0$.

40. For $\nu < 0$, show that

$$H_\nu(x) = \dfrac{1}{\Gamma(-\nu)} \int_0^\infty e^{-t^2 - 2xt} t^{-\nu-1}\, dt$$

41. Using the result of problem 40, deduce the asymptotic series (for $\nu < 0$)

$$H_\nu(x) \sim (2x)^\nu \sum_{n=0}^{\infty} \dfrac{(-1)^n (-\nu)_{2n}}{n!(2x)^{2n}}, \quad x \to \infty$$

42. Show that
$$H_{-1/2}(x) = \sqrt{\frac{x}{2\pi}}\, e^{x^2/2} K_{1/4}(x^2/2)$$

43. Derive the series representation $(a > -1)$
$$L_\nu^{(a)}(x) = \frac{\Gamma(\nu + a + 1)}{\Gamma(-\nu)\Gamma(\nu + 1)} \sum_{k=0}^{\infty} \frac{\Gamma(k - \nu)}{\Gamma(k + a + 1)} \frac{x^k}{k!}$$

In problems 44–46, use the result of problem 43 to deduce the given relation.

44. $L_\nu^{(a)\prime}(x) = -L_{\nu-1}^{(a+1)}(x)$.

45. $xL_\nu^{(a)\prime\prime}(x) - \nu L_\nu^{(a)\prime}(x) + (\nu + a)L_{\nu-1}^{(a)}(x) = 0$.

46. $(\nu + 1)L_{\nu+1}^{(a)}(x) + (x - 1 - 2\nu - a)L_\nu^{(a)}(x) + (\nu + a)L_{\nu-1}^{(a)}(x) = 0$.

47. Show that
$$L_{-n}^{(a)}(x) = 0, \quad n = 1, 2, 3, \ldots, \quad a > -1 \text{ and } a \neq 0, 1, 2, \ldots$$

48. Show that (for $m = 1, 2, 3, \ldots$)
$$L_{-n}^{(m)}(x) = \begin{cases} 0, & n = 1, 2, 3, \ldots, m \\ \dfrac{(-1)^m (n)_m}{m!}, & n = m + 1, m + 2, \ldots \end{cases}$$

9.4 Whittaker Functions

For purposes of developing certain theories concerning DEs, it is sometimes helpful to transform the equation to what is called the *Liouville standard form*.* To derive this form, we first write the DE in *normal form*

$$y'' + A(x)y' + B(x)y = 0 \tag{9.53}$$

Next, we set $y = u(x)v(x)$, for which
$$y' = uv' + u'v$$
$$y'' = uv'' + 2u'v' + u''v$$

and when these expressions are substituted into (9.53), we get

$$vu'' + (2v' + Av)u' + (v'' + Av' + Bv)u = 0 \tag{9.54}$$

---
*Sometimes called the *normal form*, not to be confused with (9.53).

By selecting $2v' + Av = 0$, the coefficient of u' can be made to vanish. A function v which gives this result is

$$v(x) = \exp\left(-\frac{1}{2}\int A(x)\,dx\right) \tag{9.55}$$

and thus (9.54) reduces to the *Liouville standard form*

$$u'' + Q(x)u = 0 \tag{9.56}$$

where

$$Q(x) = B(x) - \tfrac{1}{4}[A(x)]^2 - \tfrac{1}{2}A'(x) \tag{9.57}$$

If we rewrite the confluent hypergeometric DE

$$xy'' + (c - x)y' - ay = 0$$

in normal form, i.e.,

$$y'' + \left(\frac{c}{x} - 1\right)y' - \frac{a}{x}y = 0 \tag{9.58}$$

we can then identify the functions

$$A(x) = \frac{c}{x} - 1, \qquad B(x) = -\frac{a}{x} \tag{9.59}$$

Hence, the Liouville standard form of the confluent hypergeometric DE is

$$u'' + \left[-\frac{1}{4} + \frac{c - 2a}{2x} + \frac{2c - c^2}{4x^2}\right]u = 0 \tag{9.60}$$

From (9.55), we calculate

$$v(x) = \exp\left[-\frac{1}{2}\int\left(\frac{c}{x} - 1\right)dx\right] = e^{x/2}x^{-c/2} \tag{9.61}$$

and since $y = u(x)v(x)$ (or $u = y/v$), it follows that one solution of (9.60) is given by

$$u_1 = e^{-x/2}x^{c/2}M(a;c;x), \qquad c \neq 0, -1, -2, \ldots \tag{9.62}$$

It is customary to introduce new parameters m and k by means of the transformations

$$\begin{aligned}\frac{c}{2} &= m + \tfrac{1}{2} & (c = 2m + 1) \\ \frac{c}{2} - a &= k & (a = \tfrac{1}{2} + m - k)\end{aligned} \tag{9.63}$$

so that in terms of these parameters, Equation (9.60) becomes

$$u'' + \left[-\frac{1}{4} + \frac{k}{x} + \frac{\frac{1}{4} - m^2}{x^2} \right] u = 0 \qquad (9.64)$$

with solution $u_1 = M_{k,m}(x)$, where

$$M_{k,m}(x) = e^{-x/2} x^{m+\frac{1}{2}} M(\tfrac{1}{2} + m - k; 2m + 1; x),$$
$$2m \neq -1, -2, -3, \ldots \quad (9.65)$$

We call $M_{k,m}(x)$ a *Whittaker function of the first kind*.

We have previously shown (Section 9.2.3) that when $c \neq 2, 3, 4, \ldots$, the function

$$y_2 = x^{1-c} M(1 + a - c; 2 - c; x)$$

is a second linearly independent solution of (9.58). Using the parameters m and k, and the relation $u = y/v$, it follows that when $2m$ is not an integer, the function

$$u_2 = e^{-x/2} x^{-m+\frac{1}{2}} M(\tfrac{1}{2} - m - k; -2m + 1; x) \qquad (9.66)$$

is a second linearly independent solution of (9.64). However, comparison of (9.66) with (9.65) identifies $u_2 = M_{k,-m}(x)$, and therefore a general solution of (9.64) is

$$u = C_1 M_{k,m}(x) + C_2 M_{k,-m}(x), \qquad 2m \neq 0, \pm 1, \pm 2, \ldots \quad (9.67)$$

The solutions $M_{k, \pm m}(x)$ of (9.64) are not always the most convenient ones to use in forming a general solution, because of the restriction that $2m$ cannot be an integer. Therefore, in certain situations we find it preferable to introduce the *Whittaker function of the second kind*

$$W_{k,m}(x) = e^{-x/2} x^{m+\frac{1}{2}} U(\tfrac{1}{2} + m - k; 2m + 1; x) \qquad (9.68)$$

It can be shown that $W_{k,m}(x)$ is a solution of (9.64) that is linearly independent of $M_{k,m}(x)$, even when $2m = 0, 1, 2, \ldots$. That this is so follows from the linear independence of the confluent hypergeometric functions of the first and second kinds. In terms of $W_{k,m}(x)$, the general solution of (9.64) reads

$$u = C_1 M_{k,m}(x) + C_2 W_{k,m}(x), \qquad 2m \neq -1, -2, -3, \ldots \quad (9.69)$$

The Whittaker functions clearly have many properties which follow directly from those of Kummer's functions, some of which are discussed in the exercises. In most applications the choice of using Kummer's functions or Whittaker's functions is mostly a matter of convenience. Both sets of functions commonly occur in reference material, although the functions of Whittaker are somewhat less prominent.

EXERCISES 9.4

1. Given Bessel's equation
$$x^2 y'' + xy' + (x^2 - p^2) y = 0$$
   (a) find the Liouville standard form.
   (b) For $p = \frac{1}{2}$, use (a) to deduce that the general solution of Bessel's DE for this special case can be expressed as
$$y = C_1 x^{-1/2} \cos x + C_2 x^{-1/2} \sin x$$

2. It can be shown that oscillatory solutions of $u'' + Q(x)u = 0$ exist only if $Q(x) > 0$. Use this criterion to deduce that
$$my'' + cy' + ky = 0$$
has oscillatory solutions only if $c^2 - 4mk < 0$.

3. Use the criterion stated in problem 2 to deduce that Bessel's modified equation
$$x^2 y'' + xy' - (x^2 + p^2) y = 0$$
has no oscillatory solutions.

In problems 4–8, verify the given relation.

4. $\operatorname{erf}(x) = \dfrac{2}{\sqrt{\pi x}} e^{-x^2/2} M_{-\frac{1}{4},\frac{1}{4}}(x^2)$.

5. $\operatorname{erfc}(x) = \dfrac{1}{\sqrt{\pi x}} e^{-x^2/2} W_{-\frac{1}{4},\frac{1}{4}}(x^2)$.

6. $\gamma(a, x) = \Gamma(a) - x^{(a-1)/2} e^{-x/2} W_{\frac{1}{2}(a-1),\frac{1}{2}a}(x)$.

7. $M_{0,m}(2x) = \Gamma(m+1) 2^{2m+\frac{1}{2}} \sqrt{x}\, I_m(x)$.

8. $W_{0,m}(2x) = \sqrt{\dfrac{2x}{\pi}}\, K_m(x)$.

9. Show that
   (a) $W_{k,m}(x) = W_{k,-m}(x)$,
   (b) $W_{-k,m}(-x) = W_{-k,-m}(-x)$.

10. Show that $(n = 1, 2, 3, \ldots)$

   (a) $\dfrac{d^n}{dx^n}[e^{x/2}x^{m-\frac{1}{2}}M_{k,m}(x)]$
   $= (-1)^n(-2m)_n x^{m-\frac{1}{2}n-\frac{1}{2}}e^{x/2}M_{k-\frac{1}{2}n,\,m-\frac{1}{2}n}(x),$

   (b) $\dfrac{d^n}{dx^n}[e^{x/2}x^{m-\frac{1}{2}}W_{k,m}(x)]$
   $= (-1)^n(\tfrac{1}{2}-m-k)_n x^{m-\frac{1}{2}n-\frac{1}{2}}e^{x/2}W_{k-\frac{1}{2}n,\,m-\frac{1}{2}n}(x).$

In problems 11–13, derive the asymptotic formula.

11. $M_{k,m}(x) \sim x^{m+\frac{1}{2}},\ x \to 0^+.$

12. $M_{k,m}(x) \sim \dfrac{\Gamma(2m+1)}{\Gamma(\frac{1}{2}+m-k)} x^{-k} e^{-x/2},\ x \to \infty.$

13. $W_{k,m}(x) \sim x^k e^{-x/2},\ x \to \infty.$

14. Show that the *parabolic cylinder function* defined by
$$D_n(x) = 2^{\frac{1}{2}n+\frac{1}{4}} x^{-1/2} W_{\frac{1}{2}n+\frac{1}{4},\,-\frac{1}{4}}(x^2/2),\ n = 0, \pm 1, \pm 2, \ldots$$
satisfies the DE
$$y'' + \left(n + \frac{1}{2} - \frac{x^2}{4}\right) y = 0$$

15. Verify that (see problem 14)
$$\int_{-\infty}^{\infty} [D_0(x)]^2\, dx = \sqrt{2\pi}$$

16. Verify that (see problem 14)

   (a) $D_n(x) = 2^{-n/2} e^{-x^2/4} H_n(x/\sqrt{2}).$

   (b) $D_{-1}(x) = \sqrt{\dfrac{\pi}{2}}\, e^{x^2/4}\mathrm{erfc}(x/\sqrt{2})$

17. Show that (for $\nu + \tfrac{1}{2} + m > 0,\ k - \nu > 0$)
$$\int_0^{\infty} e^{-bt/2} t^{\nu-1} M_{k,m}(bt)\, dt$$
$$= \dfrac{\Gamma(k-\nu)\Gamma(\frac{1}{2}+m+\nu)\Gamma(2m+1)}{\Gamma(\frac{1}{2}+m+k)\Gamma(\frac{1}{2}+m-\nu)} b^{-\nu}$$

18. Evaluate the integrals

   (a) $\displaystyle\int_0^{\infty} [M_{k,m}(x)]^2\, dx,$

   (b) $\displaystyle\int_0^{\infty} x^{-1}[M_{k,m}(x)]^2\, dx.$

19. Show that the Wronskian of Whittaker's functions is

$$W\{M_{k,m}, W_{k,m}\}(x) = -\frac{\Gamma(2m+1)}{\Gamma(m-k+\tfrac{1}{2})},$$

$$2m \neq -1, -2, -3, \ldots$$

20. Using the integral representation for $U(a; c; x)$, show that

(a) $W_{k,m}(x) = \dfrac{x^k e^{-x/2}}{\Gamma(m-k+\tfrac{1}{2})} \displaystyle\int_0^{\infty} e^{-t} t^{m-k-\tfrac{1}{2}} (1+t/x)^{m+k-\tfrac{1}{2}}\, dt$,

$m - k + \tfrac{1}{2} > 0$.

(b) From (a), deduce the asymptotic series

$$W_{k,m}(x) \sim e^{-x/2} x^k \sum_{n=0}^{\infty} \frac{(-1)^n (m-k+\tfrac{1}{2})_n (\tfrac{1}{2}-m-k)_n}{n! x^n},$$

$$x \to \infty$$

21. Given the set of polynomials*

$$G_n^m(x) = \sum_{k=0}^{n} \binom{m+n-k}{m} \frac{x^k}{k!}$$

show that

(a) $G_n^m(x) = \dfrac{1}{n!} x^{m+n+1} U(m+1; m+n+2; x)$,

(b) $G_n^m(x) = \dfrac{1}{n!} x^a e^{x/2} W_{b, a+\tfrac{1}{2}}(x)$, $a = \dfrac{m+n}{2}$, $b = \dfrac{n-m}{2}$.

---

*The polynomials $G_n^m(x)$ arise in the problem of finding the probability density function for the output of a cross correlator. For example, see L.E. Miller and J.S. Lee, The Probability Density Function for the Output of an Analog Cross Correlator, *IEEE Trans. Inform. Theory*, **IT-20**, 433–440 (July 1974), and L.C. Andrews and C.S. Brice, "The PDF and CDF for the Sum of N Filtered Outputs of an Analog Cross Correlator with Bandpass Inputs, *IEEE Trans. Inform. Theory*, **IT-29**, 299–306 (March 1983).

10

Generalized Hypergeometric Functions

10.1 Introduction

The special properties associated with the hypergeometric and confluent hypergeometric functions have spurred a number of investigations into developing functions even more generalized than these. Some of this work was done in the nineteenth century by Clausen, Appell, and Lauricella (among others), but much of it has occurred during the last seventy years. Even the most recent names are too numerous to mention, but those of MacRobert and Meijer are among the most famous.

The importance of working with generalized functions of any kind stems from the fact that the majority of special functions are simply special cases of them, and thus each recurrence formula or identity developed for the generalized function becomes a master formula from which a large number of relations for other functions can be deduced. New relations for some of the special functions have been discovered in just this way. Also, the use of generalized functions often facilitates the analysis by permitting complex expressions to be represented more simply in terms of some generalized function. Operations such as differentiation and integration can sometimes be performed more readily on the resulting generalized functions than on the original complex expression, even though the two are equivalent. Finally, in many situations we resort to expressing our results in terms of these generalized functions because there are no simpler functions that we can call upon.

Our treatment of generalized hypergeometric functions is brief. For a deeper discussion the interested reader should consult one of the many publications devoted entirely to functions of this nature.*

10.2 The Set of Functions $_pF_q$

In general, we say that a series $\Sigma u_n(x)$ is a *hypergeometric-type series* if the ratio $u_{n+1}(x)/u_n(x)$ is a rational function of n. A general series of this type is

$$_pF_q(a_1,\ldots,a_p;c_1,\ldots,c_q;x) = \sum_{n=0}^{\infty} \frac{(a_1)_n \cdots (a_p)_n}{(c_1)_n \cdots (c_q)_n} \frac{x^n}{n!} \quad (10.1)$$

where p and q are nonnegative integers and no c_k ($k = 1, 2, \ldots, q$) is zero or a negative integer. The function defined by (10.1), which we denote simply $_pF_q$, is called a *generalized hypergeometric function*. Clearly, (10.1) includes the special cases $_2F_1$ and $_1F_1$, which are the *hypergeometric* and *confluent hypergeometric functions*, respectively.

Applying the ratio test to (10.1) leads to

$$\lim_{n \to \infty} \left| \frac{u_{n+1}(x)}{u_n(x)} \right| = |x| \lim_{n \to \infty} \left| \frac{(a_1 + n) \cdots (a_p + n)}{(c_1 + n) \cdots (c_q + n)(1 + n)} \right| \quad (10.2)$$

and hence, provided the series does not terminate, we see that if:

1. $p < q + 1$, the series *converges* for all (finite) x.
2. $p = q + 1$, the series *converges* for $|x| < 1$ and diverges for $|x| > 1$.
3. $p > q + 1$, the series *diverges* for all x except $x = 0$.

The series (10.1) is therefore meaningful when $p > q + 1$ only if it truncates [see (10.9) below].

Because of its generality, the function $_pF_q$ includes a great variety of functions as special cases. Some of these special cases are given by the

---

*For example, see A.M. Mathai and R.K. Saxena, *Generalized Hypergeometric Functions with Applications in Statistics and Physical Sciences*, Lecture Notes in Mathematics, New York: Springer, 1973.

following:*

$$e^x = \sum_{n=0}^{\infty} \frac{x^n}{n!} = {}_0F_0(-;-;x) \tag{10.3}$$

$$(1-x)^{-a} = \sum_{n=0}^{\infty} (a)_n \frac{x^n}{n!} = {}_1F_0(a;-;x) \tag{10.4}$$

$$\cos x = \sum_{n=0}^{\infty} \frac{(-1)^n}{(\frac{1}{2})_n} \frac{(x^2/4)^n}{n!} = {}_0F_1\left(-;\tfrac{1}{2}; -\frac{x^2}{4}\right) \tag{10.5}$$

$$J_0(x) = \sum_{n=0}^{\infty} \frac{(-1)^n}{(1)_n} \frac{(x^2/4)^n}{n!} = {}_0F_1\left(-;1; -\frac{x^2}{4}\right) \tag{10.6}$$

$$F(a,b;c;x) = \sum_{n=0}^{\infty} \frac{(a)_n (b)_n}{(c)_n} \frac{x^n}{n!} = {}_2F_1(a,b;c;x) \tag{10.7}$$

$$M(a;c;x) = \sum_{n=0}^{\infty} \frac{(a)_n}{(c)_n} \frac{x^n}{n!} = {}_1F_1(a;c;x) \tag{10.8}$$

An important terminating series for the case $p > q + 1$ is the Hermite polynomial

$$H_n(x) = (2x)^n {}_2F_0\left(-\frac{n}{2}, \frac{1-n}{2}; -; -\frac{1}{x^2}\right) \tag{10.9}$$

The series (10.1) has been studied extensively over the years, and many of the important properties associated with various ${}_pF_q$ have been developed. For this reason it is often advantageous to express a given series in the form of (10.1), since it provides a standard form by which to classify the series. In this standard form one may then be able to identify the function that the series defines; if not, at least some general theory concerning the specific function ${}_pF_q$ is probably available.

If the ratio of successive terms is a rational function of n, a given series can always be expressed in terms of Pochhammer symbols such as Equation (10.1). There are several ways in which this can be accomplished, but perhaps the easiest way in general is to examine the ratio

$$\frac{u_{n+1}(x)}{u_n(x)} = \frac{(a_1+n)\cdots(a_p+n)x}{(c_1+n)\cdots(c_q+n)(1+n)} \tag{10.10}$$

Here we can quickly identify the parameters $a_1,\ldots,a_p, c_1,\ldots,c_q$, and the argument x. If the factor $1+n$ is not in the denominator, it can be

---

*The absence of a parameter in ${}_pF_q$ is emphasized by a dash.

introduced by multiplying both numerator and denominator by it. Let us illustrate the technique with an example.

Example 1: Determine the function $_pF_q$ defined by the series

$$f(x) = \sum_{n=0}^{\infty} \frac{(-1)^n (2n)!(x/2)^{2n}}{(n!)^2 [\Gamma(n+p+1)]^2}$$

Solution: Using properties of factorials and the gamma function, we find

$$\frac{u_{n+1}(x)}{u_n(x)} = -\frac{(2n+2)!(x/2)^{2n+2}}{[(n+1)!]^2 [\Gamma(n+p+2)]^2} \cdot \frac{(n!)^2 [\Gamma(n+p+1)]^2}{(2n)!(x/2)^{2n}}$$

$$= \frac{(\tfrac{1}{2}+n)(-x^2)}{(p+1+n)^2 (1+n)}$$

and thus deduce that

$$f(x) = {}_1F_2(\tfrac{1}{2}; p+1, p+1; -x^2)$$

EXERCISES 10.2

1. Show that

$$\frac{d}{dx}\left[{}_pF_q(a_1,\ldots,a_p; c_1,\ldots,c_q; x)\right]$$

$$= \frac{\prod_{k=1}^{p} a_k}{\prod_{j=1}^{q} c_j} {}_pF_q(a_1+1,\ldots,a_p+1; c_1+1,\ldots,c_q+1; x)$$

2. For $a+b \neq 1$, show that

$${}_0F_1(-; a; x)\, {}_0F_1(-; b; x)$$

$$= {}_2F_3\left(\frac{a+b-1}{2}, \frac{a+b}{2}; a+b-1, a, b; 4x\right)$$

3. Use the result of problem 2 to deduce that

$$[J_0(x)]^2 = {}_1F_2(\tfrac{1}{2}; 1, 1; -x^2)$$

4. Verify *Ramanujan's theorem*

$$_1F_1(a;b;x)\,_1F_1(a;b;-x) = {}_2F_3\left(a, b-a; b, \frac{b}{2}, \frac{b+1}{2}; \frac{x^2}{4}\right)$$

5. Show that (for $n = 0, 1, 2, \ldots$)

$$_2F_0(-n, a; -; x) = (a)_n(-1)^n x^n \,_1F_1\left(-n; 1-a-n; -\frac{1}{x}\right)$$

6. Use the result of problem 5 to show that ($n = 0, 1, 2, \ldots$)

(a) $L_n^{(a)}(x) = \dfrac{(-1)^n}{n!} x^n \,_2F_0\left(-n, -n-a; -\dfrac{1}{x}\right),$

(b) $H_n(x) = (2x)^n \,_2F_0\left(-\dfrac{n}{2}, \dfrac{1-n}{2}; -; -\dfrac{1}{x^2}\right).$

7. Verify *Kummer's second formula* $[2a \neq -(2n+1), n = 0, 1, 2, \ldots]$

$$e^{-x}\,_1F_1(a; 2a; 2x) = {}_0F_1(-; a+\tfrac{1}{2}; x^2/4)$$

8. Use the result of problem 7 to show that ($x > 0$)

$$_0F_1(-; a+\tfrac{1}{2}; x^2/4) = (x/2)^{\frac{1}{2}-a}\Gamma(a+\tfrac{1}{2})I_{a-\frac{1}{2}}(x)$$

9. Show that (for $n = 0, 1, 2, \ldots$)

$$_3F_2(-n, a, b; c, 1-c+a+b-n; 1) = \frac{(c-a)_n(c-b)_n}{(c)_n(c-a-b)_n}$$

Hint: Expand the relation

$$F(c-a, c-b; c; x) = (1-x)^{a+b-c} F(a, b; c; x)$$

in series form and compare like coefficients.

10. Show that

$$\int_0^x {}_0F_1[-; 1; -\tfrac{1}{4}t(x-t)]\,dt = x\,_0F_1\left(-; \tfrac{3}{2}; -\frac{x^2}{16}\right) = 2\sin\frac{x}{2}$$

11. Show that

$$\int_0^x t^{1/2}(x-t)^{-1/2}\left[1 - t^2(x-t)^2\right]^{-1/2} dt = \frac{\pi}{2} x \,_2F_1\left(\tfrac{1}{4}, \tfrac{3}{4}; 1; \frac{x^4}{16}\right)$$

12. Show that ($s > 1$)

$$\int_0^\infty e^{-st} t^\nu \,_pF_q(a_1, \ldots, a_p; c_1, \ldots, c_q; t)\,dt$$

$$= \frac{\Gamma(\nu+1)}{s^{\nu+1}} \,_{p+1}F_q\left(\nu+1, a_1, \ldots, a_p; c_1, \ldots, c_q; \frac{1}{s}\right)$$

In problems 13–16, express the series as a function $_pF_q$.

13. $\sum_{n=0}^{\infty} \dfrac{(2n)!(2n+1)!}{2^{4n}(n!)^4} x^n.$

14. $x + \sum_{n=1}^{\infty} \dfrac{1 \times 3 \times \cdots \times (2n-1) x^{2n+1}}{(2n+1)(2n+3) \cdots (4n+1)}.$

15. $\sum_{k=0}^{n} \binom{2n}{2k} k! x^k.$

16. $\sum_{k=0}^{[n/2]} \dfrac{(-1)^k n! (2x)^{n-2k}}{(\tfrac{1}{2})_k (n-2k)!}.$

17. Bessel polynomials are defined by*

$$b_n(x) = {}_2F_0\left(-n, 1+n; -; -\dfrac{x}{2}\right)$$

Show that

(a) $K_{n+\frac{1}{2}}(x) = \dfrac{\pi}{2x} e^{-x} b_n(1/x),$

(b) $G_n^n(x) = \dfrac{x^n}{n!} b_n(2/x),$ where the polynomials $G_n^m(x)$ are defined in problem 21, Exercises 9.4.

18. For the polynomials defined by†

$$Z_n(x) = {}_2F_2(-n, 1+n; 1, 1; x)$$

show that

(a) $nZ_n'(x) - nZ_n(x) = -nZ_{n-1}(x) - xZ_{n-1}'(x),$

(b) $(1-t)^{-1} {}_1F_1\left(\tfrac{1}{2}; 1; \dfrac{-4xt}{(1-t)^2}\right) = \sum_{n=0}^{\infty} Z_n(x) t^n,$

(c) $(1-t)^{-1} \exp\left[\dfrac{-2xt}{(1-t)^2}\right] I_0\left[\dfrac{-2xt}{(1-t)^2}\right] = \sum_{n=0}^{\infty} Z_n(x) t^n.$

In problems 19–22, verify the formulas for products of Bessel functions.

19. $J_p(x) J_\nu(x) = \dfrac{(x/2)^{p+\nu}}{\Gamma(p+1)\Gamma(\nu+1)} {}_2F_3\left(\dfrac{p+\nu+1}{2}, \dfrac{p+\nu+2}{2}; p+1, \nu+1, p+\nu+1; -x^2\right).$

20. $[J_p(x)]^2 = \dfrac{(x/2)^{2p+1}}{[\Gamma(p+1)]^2} {}_1F_2(p+\tfrac{1}{2}; p+1, 2p+1; -x^2).$

---

*Bessel polynomials were first studied by H.L. Krall and O. Frink, A New Class of Orthogonal Polynomials: The Bessel Polynomials, Trans. Amer. Math. Soc., 65, 100–115 (1949). See also E. Grosswald, The Bessel Polynomials, Lecture Notes in Mathematics, New York: Springer, 1978.

†These polynomials were introduced by H. Bateman, Two Systems of Polynomials for the Solution of Laplace's Integral Equation, Duke Math. J., 2, 569–577 (1936).

21. $J_p(x)J_{p+1}(x) = \dfrac{(x/2)^{2p+1}}{\Gamma(p+1)\Gamma(p+2)} \, {}_1F_2(p + \tfrac{3}{2}; p + 2, 2p + 2; -x^2).$

22. $J_p(x)I_p(x) = \dfrac{(x/2)^{2p}}{[\Gamma(p+1)]^2} \, {}_0F_3\!\left(-; \dfrac{p+1}{2}, \dfrac{p+2}{2}, p+1; -\dfrac{x^4}{64}\right).$

10.3 Other Generalizations

In the first half of the twentieth century new theories concerning generalized functions began to flourish. Most of this work followed Barnes's use of the gamma function in 1907 to develop a new theory of the hypergeometric function ${}_2F_1$. In the 1930s, both the *E-function* of MacRobert and the *G-function* of Meijer were introduced in an attempt to give meaning to the symbol ${}_pF_q$ for the case $p > q + 1$. The *E*-function is actually a special case of the *G*-function, and for that reason is less prominent in the literature.

10.3.1 The Meijer G-Function

In 1936, C.S. Meijer introduced the *G*-function*

$$G^{m,n}_{p,q}\!\left(x \, \Big| \, \begin{matrix} a_1,\ldots,a_p \\ c_1,\ldots,c_q \end{matrix}\right) = \sum_{k=1}^{m} \frac{\prod'_{j=1}^{m} \Gamma(c_j - c_k) \prod_{j=1}^{n} \Gamma(1 + c_k - a_j) x^{c_k}}{\prod_{j=m+1}^{q} \Gamma(1 + c_k - c_j) \prod_{j=n+1}^{p} \Gamma(a_j - c_k)}$$

$$\times {}_pF_{q-1}\!\left[1 + c_k - a_1,\ldots, 1 + c_k - a_p; 1 + c_k - c_1,\ldots, *, \ldots,\right.$$

$$\left. 1 + c_k - c_q; (-1)^{p-m-n} x\right] \quad (10.11)$$

where $1 \leq m \leq q$, $0 \leq n \leq p \leq q - 1$, no two of the c_k's ($k = 1, 2, \ldots, m$) differ by zero or an integer, and $a_j - c_k \neq 1, 2, 3, \ldots$ for $j = 1, 2, \ldots, n$ and $k = 1, 2, \ldots, m$. If $p = q$, we restrict $|x| < 1$. For notational convenience, we often write

$$G^{m,n}_{p,q}\!\left(x \, \Big| \, \begin{matrix} a_p \\ c_q \end{matrix}\right) \equiv G^{m,n}_{p,q}\!\left(x \, \Big| \, \begin{matrix} a_1,\ldots,a_p \\ c_1,\ldots,c_q \end{matrix}\right) \quad (10.12)$$

or, if confusion is not likely, we simply write $G^{m,n}_{p,q}(x)$.

---

*C.S. Meijer, Einige Integraldarstellungen aus der Theorie der Besselschen und der Whittaker Funktionen, *K. Akad. Wet. Amst. Proc.*, **39**, 394–403, 519–527 (1936). The prime in the product symbol $\prod'$ denotes the omission of the term when $j = k$. Also, in the parameter set of ${}_pF_{q-1}$ the parameter corresponding to $1 + c_k - c_k$ (indicated by *) is to be omitted. Lastly, an empty product is interpreted as unity.

Some relations between the G-function and other functions are given by the following:

$$G_{01}^{10}(x|a) = x^a e^{-x} \tag{10.13}$$

$$G_{11}^{11}\left(x\left|\begin{matrix}1-a\\0\end{matrix}\right.\right) = \Gamma(a)(1+x)^{-a} \tag{10.14}$$

$$G_{02}^{10}(x|a,b) = x^{\frac{1}{2}(a+b)} J_{a-b}(2\sqrt{x}) \tag{10.15}$$

$$G_{12}^{11}\left(x\left|\begin{matrix}\frac{1}{2}\\p,-p\end{matrix}\right.\right) = \sqrt{\pi}\, e^{-x/2} I_p(x/2) \tag{10.16}$$

$$G_{02}^{20}(x|a,b) = 2x^{\frac{1}{2}(a+b)} K_{a-b}(2\sqrt{x}) \tag{10.17}$$

$$G_{12}^{11}\left(x\left|\begin{matrix}1-a\\0,1-c\end{matrix}\right.\right) = \frac{\Gamma(a)}{\Gamma(c)}\,{}_1F_1(a;c;-x) \tag{10.18}$$

$$G_{22}^{12}\left(x\left|\begin{matrix}1-a,1-b\\0,1-c\end{matrix}\right.\right) = \frac{\Gamma(a)\Gamma(b)}{\Gamma(c)}\,{}_2F_1(a,b;c;-x) \tag{10.19}$$

Meijer redefined the G-function in 1941* in terms of a Barnes contour integral in the complex plane that ultimately led to an interpretation of the symbol ${}_pF_q$ when $p > q + 1$. In particular, as a consequence of his more general definition, we have the important property

$$G_{p,q}^{m,n}\left(\frac{1}{x}\left|\begin{matrix}a_p\\c_q\end{matrix}\right.\right) = G_{q,p}^{n,m}\left(x\left|\begin{matrix}1-c_q\\1-a_p\end{matrix}\right.\right) \tag{10.20}$$

which allows us to transform from a G-function for which $p > q$ to one for which $p < q$ (and vice versa).

The basic properties of the G-function are far too numerous for us to discuss in any detail. Also, the proofs of many of these properties (and any real understanding of this function) require knowledge of complex-variable theory. Hence, for our purposes, we will be content to merely list a few of the simplest properties, and them without justification.

If one of the parameters in the numerator set coincides with one of the parameters in the denominator set, the order of the G-function may decrease. For example, if $a_j = c_k$ for some $j = 1, 2, \ldots, n$ and some $k = m+1, m+2, \ldots, q$, then

$$G_{p,q}^{m,n}\left(x\left|\begin{matrix}a_p\\c_q\end{matrix}\right.\right) = G_{p-1,q-1}^{m,n-1}\left(x\left|\begin{matrix}a_1,\ldots,a_{j-1},a_{j+1},\ldots,a_p\\c_1,\ldots,c_{k-1},c_{k+1},\ldots,c_q\end{matrix}\right.\right) \tag{10.21}$$

An analogous relationship exists if $a_j = c_k$ for some $j = n+1, n+2, \ldots, p$ and some $k = 1, 2, \ldots, n$. In this case it is m, and not n, that decreases by one unit in addition to p and q decreasing by one unit.

---

*C.S. Meijer, Neue Integraldarstellungen für WHITTAKERsche Funktionen, *Proc. Ned. Akad. v. Wetensch.*, Amsterdam, **44**, 81–92 (1941).

Multiplication of the Meijer G-function by powers of x leads to the simple relation

$$x^r G_{p,q}^{m,n}\left(x \bigg| \begin{matrix} a_p \\ c_q \end{matrix}\right) = G_{p,q}^{m,n}\left(x \bigg| \begin{matrix} a_p + r \\ c_q + r \end{matrix}\right) \qquad (10.22)$$

where the implication is that each numerator and denominator parameter is increased by the power r. Differentiation of this function is also easily performed, although there are several varieties of formulas. A particularly simple differentiation formula is given by

$$\frac{d}{dx}\left[x^{-c_1} G_{p,q}^{m,n}\left(x \bigg| \begin{matrix} a_p \\ c_q \end{matrix}\right)\right] = -x^{-1-c_1} G_{p,q}^{m,n}\left(x \bigg| \begin{matrix} a_1,\ldots,a_p \\ c_1+1, c_2, \ldots, c_q \end{matrix}\right)$$
$$(10.23)$$

As an illustration of the use of the last two properties, consider the special example

$$\frac{d}{dx} G_{02}^{20}(x|a, 0) = -x^{-1} G_{02}^{20}(x|a, 1)$$
$$= -G_{02}^{20}(x|a-1, 0) \qquad (10.24)$$

To give a check on this result, and also to emphasize the efficiency of the G-function notation, we note from (10.17) above that

$$G_{02}^{20}(x|a, 0) = 2x^{a/2} K_a(2x^{1/2})$$

Thus, (10.24) is equivalent to the formula

$$\frac{d}{dx}\left[2x^{a/2} K_a(2x^{1/2})\right] = -2x^{\frac{1}{2}(a-1)} K_{a-1}(2x^{1/2}) \qquad (10.25)$$

Of course, we can derive this result directly through application of the product formula and chain rule, which yields

$$\frac{d}{dx}\left[2x^{a/2} K_a(2x^{1/2})\right] = 2x^{a/2} K_a'(2x^{1/2}) x^{-1/2} + ax^{-1+a/2} K_a(2x^{1/2})$$
$$= 2x^{a/2}\left[-x^{-1/2} K_{a-1}(2x^{1/2})\right] \qquad (10.26)$$

where the last step is obtained through application of the identity

$$K_p'(x) = -K_{p-1}(x) - \frac{p}{x} K_p(x) \qquad (10.27)$$

Further simplification of (10.26) yields the same result as (10.25).

This last example gives some hint of the power and economy in using the G-function as a tool of analysis. The difficulty that often exists in working with this function is the recognition of the particular $G_{p,q}^{m,n}(x)$ as one of the elementary or special functions. However, there are countless instances in which the G-function of interest is not related to any known function.

One of the major areas of application where the G-function has proven to be effective is in probability theory. For example, the probability density function associated with the product of n random variables of the same distribution has been found in terms of G-functions.* While certain special cases of the G-functions associated with such products can be expressed in terms of simpler functions, the general case most likely cannot. In such instances, the G-functions must be dealt with directly for computational purposes.

10.3.2 The MacRobert E-Function

In the late 1930s, T.M. MacRobert also made an attempt to give meaning to the symbol ${}_pF_q$ when $p > q + 1$.† For the values $p \leq q + 1$, he introduced the function (called the *E-function*)

$$E(a_1,\ldots,a_p:c_1,\ldots,c_q:x) = \frac{\prod_{k=1}^{p}\Gamma(a_k)}{\prod_{j=1}^{q}\Gamma(c_j)} \, {}_pF_q\!\left(a_1,\ldots,a_p;c_1,\ldots,c_q;-\frac{1}{x}\right)$$

(10.28)

where $x \neq 0$ if $p < q$ and $|x| > 1$ if $p = q + 1$, while for the values $p \geq q + 1$,‡

$$E(a_1,\ldots,a_p:c_1,\ldots,c_q:x)$$

$$= \sum_{n=1}^{p} \frac{\prod_{k=1}^{p}{}'\Gamma(a_k - a_n)}{\prod_{j=1}^{q}\Gamma(c_j - c_n)} \Gamma(a_n) x^{a_n}$$

$$\times {}_{q+1}F_{p-1}\!\left[a_n, a_n - c_1 + 1,\ldots, a_n - c_q + 1; \right.$$
$$\left. a_n - a_1,\ldots, *,\ldots, a_n - a_p + 1; (-1)^{p+q}x\right]$$

(10.29)

where $|x| < 1$ if $p = q + 1$.

---

*M.D. Springer and W.E. Thompson, The Distribution of Products of Beta, Gamma and Gaussian Random Variables, *SIAM J. Appl. Math.*, **18**, No. 4 (June 1970).

†T.M. MacRobert, Induction Proofs of the Relations between Certain Asymptotic Expansions and Corresponding Generalized Hypergeometric Series, *Proc. Roy. Soc. Edinburgh*, **58**, 1–13 (1937–1938).

‡See the footnote at the beginning of Section 10.3.1.

MacRobert's E-function never gained wide acceptance in the literature, mostly because it was found to be a special case of the Meijer G-function, i.e.,

$$E(a_1,\ldots,a_p:c_1,\ldots,c_q:x) = G_{q+1,p}^{p,1}\left(x\,\Big|\,\begin{matrix}1,c_1,\ldots,c_q\\a_1,\ldots,a_p\end{matrix}\right) \qquad (10.30)$$

Hence, all properties of the E-function are simple consequences of properties of the G-function.

EXERCISES 10.3

1. From the definition (10.11), show that

$$_pF_q(a_1,\ldots,a_p;c_1,\ldots,c_q;x)$$

$$= \frac{\prod_{j=1}^{q}\Gamma(c_j)}{\prod_{j=1}^{p}\Gamma(a_j)} G_{p,q+1}^{1,p}\left(-x\,\Big|\,\begin{matrix}1-a_1,\ldots,1-a_p\\0,1-c_1,\ldots,1-c_q\end{matrix}\right)$$

In problems 2–15, use the result of problem 1 and properties of the G-function to deduce the given relation.

2. $G_{01}^{10}(x|0) = e^{-x}$.

3. $G_{01}^{10}(x|a) = x^a e^{-x}$, $x \geq 0$.

4. $G_{11}^{10}\left(x\,\Big|\,\begin{matrix}a+b+1\\a\end{matrix}\right) = \dfrac{x^a(1-x)^b}{\Gamma(b+1)}$, $0 < x < 1$.

5. $G_{11}^{11}\left(x\,\Big|\,\begin{matrix}1-a\\0\end{matrix}\right) = \Gamma(a)(1+x)^{-a}$, $|x| < 1$.

6. $G_{02}^{10}\left(\dfrac{x^2}{4}\,\Big|\,\tfrac{1}{2},0\right) = \dfrac{1}{\sqrt{\pi}}\sin x$, $x \geq 0$.

7. $G_{02}^{10}\left(\dfrac{x^2}{4}\,\Big|\,0,\tfrac{1}{2}\right) = \dfrac{1}{\sqrt{\pi}}\cos x$.

8. $G_{02}^{10}\left(x^2\,\Big|\,\tfrac{1}{2}p,-\tfrac{1}{2}p\right) = J_p(2x)$, $x \geq 0$.

9. $G_{02}^{10}(x^2|a,b) = x^{a+b}J_{a-b}(2x)$, $x \geq 0$.

10. $G_{22}^{10}\left(x\,\Big|\,\begin{matrix}1,1\\1,0\end{matrix}\right) = \log(1+x)$, $|x| < 1$.

11. $G_{12}^{11}\left(x\left|\begin{matrix}1-a\\0,1-c\end{matrix}\right.\right) = \dfrac{\Gamma(a)}{\Gamma(c)}\,_1F_1(a;c;-x).$

12. $G_{22}^{12}\left(x\left|\begin{matrix}1-a,1-b\\0,1-c\end{matrix}\right.\right) = \dfrac{\Gamma(a)\Gamma(b)}{\Gamma(c)}\,_2F_1(a,b;c;-x),\ |x|<1.$

13. $G_{12}^{11}\left(x\left|\begin{matrix}\tfrac{1}{2}\\p,-p\end{matrix}\right.\right) = \sqrt{\pi}\,e^{-x/2}I_p(x/2),\ x\geq 0.$

14. $G_{12}^{11}\left(x\left|\begin{matrix}a\\b,c\end{matrix}\right.\right) = \dfrac{\Gamma(1-a+b)}{\Gamma(1-c+b)}z^b\,_1F_1(1-a+b;1+b-c;-x).$

15. $G_{13}^{11}\left(x\left|\begin{matrix}\tfrac{1}{2}\\p,0,-p\end{matrix}\right.\right) = \sqrt{\pi}\left[J_p(\sqrt{x})\right]^2,\ x\geq 0.$

In problems 16–21, verify the relation.

16. $J_p'(x) = G_{13}^{11}\left(\dfrac{x^2}{4}\left|\begin{matrix}-\tfrac{1}{2}\\ \tfrac{1}{2}(p-1),-\tfrac{1}{2}(p+1),\tfrac{1}{2}\end{matrix}\right.\right).$

17. $K_p(2\sqrt{x}) = \tfrac{1}{2}x^{-p/2}G_{02}^{20}(x|p,0).$

    Hint: Use problem 8 in Exercises 10.2.

18. $K_p(x) = \sqrt{\pi}\,e^x G_{12}^{20}\left(2x\left|\begin{matrix}\tfrac{1}{2}\\p,-p\end{matrix}\right.\right).$

19. $L_n^{(a)}(x) = \dfrac{(-1)^n}{n!}e^x G_{12}^{20}\left(x\left|\begin{matrix}-n-a\\0,-a\end{matrix}\right.\right).$

20. $M_{k,m}(x) = \dfrac{\Gamma(2m+1)}{\Gamma(m+k+\tfrac{1}{2})}e^{x/2}G_{12}^{11}\left(x\left|\begin{matrix}1-k\\ \tfrac{1}{2}+m,\tfrac{1}{2}-m\end{matrix}\right.\right).$

21. $W_{k,m}(x) = \dfrac{e^{-x/2}}{\Gamma(\tfrac{1}{2}-k+m)\Gamma(\tfrac{1}{2}-k-m)}G_{12}^{21}\left(x\left|\begin{matrix}1+k\\ \tfrac{1}{2}+m,\tfrac{1}{2}-m\end{matrix}\right.\right).$

22. Use the results of problems 6 and 8 to deduce that $(x>0)$

$$J_{1/2}(x) = \sqrt{\dfrac{2}{\pi x}}\,\sin x$$

23. Verify that $(p\leq q+1)$

$$\dfrac{d}{dx}E(a_1,\ldots,a_p:c_1,\ldots,c_q:x)$$
$$= x^{-2}E(a_1+1,\ldots,a_p+1:c_1+1,\ldots,c_q+1:x)$$

Bibliography

THE AVAILABLE LITERATURE ON SPECIAL FUNCTIONS IS VAST, both in textbooks and in research papers. Rather than attempt to list any substantial part of it, we have opted to give a short list of some of the classical books as well as some more recent references. However, each of these references in turn supplies numerous additional references, including many of the early research papers on special functions.

M. Abramowitz, and I.A. Stegun (Eds.), *Handbook of Mathematical Functions*, New York: Dover, 1965.

G. Arfken, *Mathematical Methods for Physicists*, 2nd ed., New York: Academic, 1970.

E. Artin, *The Gamma Function*, New York: Holt, Rinehart and Winston, 1964.

R.A. Askey (Ed.), *Theory and Application of Special Functions*, New York: Academic, 1975.

W.W. Bell, *Special Functions for Scientists and Engineers*, London: Van Nostrand, 1968.

B.C. Carlson, *Special Functions of Applied Mathematics*, New York: Academic, 1977.

A. Erdelyi et al., *Higher Transcendental Functions*, 3 vols., Bateman Manuscript Project, New York: McGraw-Hill, 1953, 1955.

I.S. Gradshteyn and I.M. Ryzhik, *Table of Integrals, Series, and Products*, New York: Academic, 1980.

H. Hochstadt, *The Functions of Mathematical Physics*, New York: Wiley, 1971.

D. Jackson, *Fourier Series and Orthogonal Polynomials*, Carus Math. Monogr. 6, Menasha, Wis.: Math. Assoc. Amer., 1941.

E. Jahnke et al., *Tables of Higher Functions*, 6th ed., New York: McGraw-Hill, 1960.

N.N. Lebedev, *Special Functions and Their Applications* (R.A. Silverman, Transl. and Ed.) New York: Dover, 1972.

Y.L. Luke, *The Special Functions and Their Approximations*, 2 vols., New York: Academic, 1969.

F.W.J. Olver, *Asymptotics and Special Functions*, New York: Academic, 1974.

W. Magnus and F. Oberhettinger, *Formulas and Theorems for the Special Functions of Mathematical Physics*, New York: Springer, 1969.

E.D. Rainville, *Special Functions*, New York: Chelsea, 1960.

L.J. Slater, *Confluent Hypergeometric Functions*, London: Cambridge U.P., 1966.

I.N. Sneddon, *Special Functions of Mathematical Physics and Chemistry*, 2nd ed., Edinburgh: Oliver and Boyd, 1961.

G.N. Watson, *A Treatise on the Theory of Bessel Functions*, 2nd ed., London: Cambridge U.P., 1952.

Appendix: A List of Special-Function Formulas

FOR EASY REFERENCE, the following is a selected list of formulas for many of the special functions discussed in the text.

Gamma Function

1. $\Gamma(x) = \int_0^\infty e^{-t} t^{x-1} \, dt, \; x > 0$

2. $\Gamma(x) = 2 \int_0^\infty e^{-t^2} t^{2x-1} \, dt, \; x > 0$

3. $\Gamma(x) = \int_0^1 \left(\log \frac{1}{t}\right)^{x-1} dt, \; x > 0$

4. $\Gamma(x) = \lim_{n \to \infty} \dfrac{n! \, n^x}{x(x+1)(x+2) \cdots (x+n)}$

5. $\dfrac{1}{\Gamma(x)} = x e^{\gamma x} \prod_{n=1}^\infty \left(1 + \dfrac{x}{n}\right) e^{-x/n}$

6. $\Gamma(x+1) = x \Gamma(x)$

7. $\Gamma(x) \Gamma(1-x) = \dfrac{\pi}{\sin \pi x}$

8. $\Gamma(\tfrac{1}{2} + x)\Gamma(\tfrac{1}{2} - x) = \dfrac{\pi}{\cos \pi x}$

9. $\sqrt{\pi}\,\Gamma(2x) = 2^{2x-1}\Gamma(x)\Gamma(x + \tfrac{1}{2})$

10. $\Gamma(1) = 1$

11. $\Gamma(\tfrac{1}{2}) = \sqrt{\pi}$

12. $\Gamma(n + 1) = n!, \quad n = 0, 1, 2, \ldots$

13. $\Gamma(n + \tfrac{1}{2}) = \dfrac{(2n)!}{2^{2n}n!}\sqrt{\pi}, \quad n = 0, 1, 2, \ldots$

14. $\displaystyle\int_0^{\pi/2} \cos^{2x-1}\theta \sin^{2y-1}\theta\, d\theta = \dfrac{\Gamma(x)\Gamma(y)}{2\Gamma(x + y)}, \quad x > 0,\ y > 0$

15. $\Gamma(x + 1) \sim \sqrt{2\pi x}\, x^x e^{-x}\left(1 + \dfrac{1}{12x} + \cdots\right), \quad x \to \infty$

16. $n! \sim \sqrt{2\pi n}\, n^n e^{-n}, \quad n \gg 1$

Beta Function

17. $B(x, y) = \displaystyle\int_0^1 t^{x-1}(1 - t)^{y-1}\, dt, \quad x > 0,\ y > 0$

18. $B(x, y) = \displaystyle\int_0^\infty \dfrac{t^{x-1}}{(1 + t)^{x+y}}\, dt, \quad x > 0,\ y > 0$

19. $B(x, y) = 2\displaystyle\int_0^{\pi/2} \cos^{2x-1}\theta \sin^{2y-1}\theta\, d\theta, \quad x > 0,\ y > 0$

20. $B(x, y) = B(y, x)$

21. $B(x, y) = \dfrac{\Gamma(x)\Gamma(y)}{\Gamma(x + y)}, \quad x > 0,\ y > 0$

Incomplete Gamma Function

22. $\gamma(a, x) = \displaystyle\int_0^x e^{-t} t^{a-1}\, dt, \quad a > 0$

23. $\Gamma(a, x) = \displaystyle\int_x^\infty e^{-t} t^{a-1}\, dt, \quad a > 0$

24. $\gamma(a, x) + \Gamma(a, x) = \Gamma(a)$

25. $\gamma(a, x) = x^a \displaystyle\sum_{n=0}^\infty \dfrac{(-1)^n x^n}{n!(n + a)}$

26. $\Gamma(a, x) = \Gamma(a) - x^a \displaystyle\sum_{n=0}^\infty \dfrac{(-1)^n x^n}{n!(n + a)}$

27. $\gamma(a + 1, x) = a\gamma(a, x) - x^a e^{-x}$

28. $\Gamma(a+1, x) = a\Gamma(a, x) + x^a e^{-x}$

29. $\gamma(n+1, x) = n![1 - e^{-x} e_n(x)], \quad n = 0, 1, 2, \ldots$

30. $\Gamma(n+1, x) = n! e^{-x} e_n(x), \quad n = 0, 1, 2, \ldots$

31. $\Gamma(a, x) \sim \Gamma(a) x^{a-1} e^{-x} \sum_{n=0}^{\infty} \frac{x^{-n}}{\Gamma(a-n)}, \quad a > 0, \; x \to \infty$

Digamma and Polygamma Functions

32. $\psi(x) = \dfrac{d}{dx} \log \Gamma(x) = \dfrac{\Gamma'(x)}{\Gamma(x)}$

33. $\psi(x) = -\gamma + \sum_{n=0}^{\infty} \left(\dfrac{1}{n+1} - \dfrac{1}{n+x} \right)$

34. $\psi(x+1) = \psi(x) + \dfrac{1}{x}$

35. $\psi(1-x) - \psi(x) = \pi \cot \pi x$

36. $\psi(x) + \psi(x + \tfrac{1}{2}) - 2 \log 2 = 2\psi(2x)$

37. $\psi(1) = -\gamma$

38. $\psi(n+1) = -\gamma + \sum_{k=1}^{n} \dfrac{1}{k}, \quad n = 1, 2, 3, \ldots$

39. $\psi(x+1) = -\gamma + \sum_{n=1}^{\infty} (-1)^{n+1} \zeta(n+1) x^n, \quad -1 < x < 1$

40. $\psi(x+1) \sim \log x + \dfrac{1}{2x} - \dfrac{1}{2} \sum_{n=1}^{\infty} \dfrac{B_{2n}}{n} x^{-2n}, \quad x \to \infty$

41. $\psi^{(m)}(x) = \dfrac{d^{m+1}}{dx^{m+1}} \log \Gamma(x), \quad m = 1, 2, 3, \ldots$

42. $\psi^{(m)}(x) = (-1)^{m+1} m! \sum_{n=0}^{\infty} \dfrac{1}{(n+x)^{m+1}}$

43. $\psi^{(m)}(1) = (-1)^{m+1} m! \zeta(m+1)$

44. $\psi^{(m)}(x+1) = (-1)^{m+1} \sum_{n=0}^{\infty} (-1)^n \dfrac{(m+n)!}{n!} \zeta(m+n+1) x^n,$
$-1 < x < 1$

Error Functions

45. $\operatorname{erf}(x) = \dfrac{2}{\sqrt{\pi}} \int_0^x e^{-t^2} dt$

338 • Special Functions for Engineers and Applied Mathematicians

46. $\text{erf}(x) = \dfrac{2}{\sqrt{\pi}} \sum_{n=0}^{\infty} \dfrac{(-1)^n x^{2n+1}}{n!(2n+1)}, \quad -\infty < x < \infty$

47. $\text{erf}(0) = 0$

48. $\text{erf}(\infty) = 1$

49. $\text{erfc}(x) = \dfrac{2}{\sqrt{\pi}} \int_x^{\infty} e^{-t^2}\, dt$

50. $\text{erfc}(x) = 1 - \text{erf}(x)$

51. $\text{erfc}(x) \sim \dfrac{e^{-x^2}}{\sqrt{\pi}\, x}\left[1 + \sum_{n=1}^{\infty} (-1)^n \dfrac{1 \times 3 \times \cdots \times (2n-1)}{(2x^2)^n}\right],$
   $x \to \infty$

Exponential Integrals

52. $\text{Ei}(x) = \displaystyle\int_{-\infty}^{x} \dfrac{e^t}{t}\, dt, \quad x \neq 0$

53. $E_1(x) = -\text{Ei}(-x) = \displaystyle\int_x^{\infty} \dfrac{e^{-t}}{t}\, dt, \quad x > 0$

54. $E_1(x) = \Gamma(0, x)$

55. $E_1(x) = -\gamma - \log x - \displaystyle\sum_{n=1}^{\infty} \dfrac{(-1)^n x^n}{n!\, n}, \quad x > 0$

56. $\text{Ei}(x) \sim \dfrac{e^x}{x} \displaystyle\sum_{n=0}^{\infty} \dfrac{n!}{x^n}, \quad x \to \infty$

57. $E_1(x) \sim \dfrac{e^{-x}}{x} \displaystyle\sum_{n=0}^{\infty} \dfrac{(-1)^n n!}{x^n}, \quad x \to \infty$

Elliptic Integrals

58. $F(m, \phi) = \displaystyle\int_0^{\phi} \dfrac{d\theta}{\sqrt{1 - m^2 \sin^2\theta}}, \quad 0 < m < 1$

59. $E(m, \phi) = \displaystyle\int_0^{\phi} \sqrt{1 - m^2 \sin^2\theta}\, d\theta, \quad 0 < m < 1$

60. $\Pi(m, \phi, a) = \displaystyle\int_0^{\phi} \dfrac{d\theta}{\sqrt{1 - m^2 \sin^2\theta}\,(1 + a^2 \sin^2\theta)}, \quad 0 < m < 1,$
   $a \neq m, 0$

61. $K(m) = \displaystyle\int_0^{\pi/2} \dfrac{d\theta}{\sqrt{1 - m^2 \sin^2\theta}}, \quad 0 < m < 1$

62. $E(m) = \int_0^{\pi/2} \sqrt{1 - m^2\sin^2\theta}\, d\theta,\ 0 < m < 1$

63. $\Pi(m, a) = \int_0^{\pi/2} \dfrac{d\theta}{\sqrt{1 - m^2\sin^2\theta}\,(1 + a^2\sin^2\theta)},\ 0 < m < 1,$
    $a \ne m, 0$

64. $K(m) = \dfrac{\pi}{2} \sum_{n=0}^{\infty} \binom{-\frac{1}{2}}{n}^2 m^{2n}$

65. $E(m) = \dfrac{\pi}{2} \sum_{n=0}^{\infty} \binom{\frac{1}{2}}{n}\binom{-\frac{1}{2}}{n} m^{2n}$

Legendre Polynomials

66. $P_n(x) = \sum_{k=0}^{[n/2]} \dfrac{(-1)^k (2n - 2k)! x^{n-2k}}{2^n k!(n - k)!(n - 2k)!}$

67. $P_n(x) = \dfrac{1}{2^n n!} \dfrac{d^n}{dx^n}[(x^2 - 1)^n]$

68. $(1 - 2xt + t^2)^{-1/2} = \sum_{n=0}^{\infty} P_n(x) t^n$

69. $P_n(1) = 1;\ P_n(-1) = (-1)^n$

70. $P_n'(1) = \tfrac{1}{2}n(n + 1);\ P_n'(-1) = (-1)^{n-1}\tfrac{1}{2}n(n + 1)$

71. $P_{2n}(0) = \dfrac{(-1)^n (2n)!}{2^{2n}(n!)^2};\ P_{2n+1}(0) = 0$

72. $(n + 1)P_{n+1}(x) - (2n + 1)xP_n(x) + nP_{n-1}(x) = 0$

73. $P_{n+1}'(x) - 2xP_n'(x) + P_{n-1}'(x) - P_n(x) = 0$

74. $P_{n+1}'(x) - xP_n'(x) - (n + 1)P_n(x) = 0$

75. $xP_n'(x) - P_{n-1}'(x) - nP_n(x) = 0$

76. $P_{n+1}'(x) - P_{n-1}'(x) = (2n + 1)P_n(x)$

77. $(1 - x^2)P_n'(x) = nP_{n-1}(x) - nxP_n(x)$

78. $(1 - x^2)P_n''(x) - 2xP_n'(x) + n(n + 1)P_n(x) = 0$

79. $P_n(x) = \dfrac{1}{\pi} \int_0^{\pi} [x + (x^2 - 1)^{1/2}\cos\phi]^n\, d\phi$

80. $|P_n(x)| \le 1,\ |x| \le 1$

81. $|P_n(x)| < \left[\dfrac{\pi}{2n(1 - x^2)}\right]^{1/2},\ |x| < 1,\ n = 1, 2, 3, \ldots$

82. $\int_{-1}^{1} P_n(x)P_k(x)\,dx = 0, \ k \neq n$

83. $\int_{-1}^{1} [P_n(x)]^2\,dx = \dfrac{2}{2n+1}$

84. $\displaystyle\sum_{k=0}^{n} (2k+1)P_k(t)P_k(x) = \dfrac{n+1}{t-x}[P_{n+1}(t)P_n(x) - P_n(t)P_{n+1}(x)]$

Associated Legendre Functions

85. $P_n^m(x) = (1-x^2)^{m/2}\dfrac{d^m}{dx^m}[P_n(x)], \ m = 1,2,3,\ldots$

86. $P_n^0(x) = P_n(x)$

87. $P_n^{-m}(x) = (-1)^m \dfrac{(n-m)!}{(n+m)!} P_n^m(x)$

88. $(n-m+1)P_{n+1}^m(x) - (2n+1)xP_n^m(x) + (n+m)P_{n-1}^m(x) = 0$

89. $(1-x^2)P_n^{m\prime\prime}(x) - 2xP_n^{m\prime}(x) + \left[n(n+1) - \dfrac{m^2}{1-x^2}\right]P_n^m(x) = 0$

90. $\int_{-1}^{1} P_n^m(x)P_k^m(x)\,dx = 0, \ k \neq n$

91. $\int_{-1}^{1} [P_n^m(x)]^2\,dx = \dfrac{2(n+m)!}{(2n+1)(n-m)!}$

Hermite Polynomials

92. $H_n(x) = \displaystyle\sum_{k=0}^{[n/2]} \dfrac{(-1)^k n!}{k!(n-2k)!}(2x)^{n-2k}$

93. $H_n(x) = (-1)^n e^{x^2} \dfrac{d^n}{dx^n}(e^{-x^2})$

94. $\exp(2xt - t^2) = \displaystyle\sum_{n=0}^{\infty} H_n(x)\dfrac{t^n}{n!}$

95. $H_{2n}(0) = (-1)^n \dfrac{(2n)!}{n!}; \ H_{2n+1}(0) = 0$

96. $H_{n+1}(x) - 2xH_n(x) + 2nH_{n-1}(x) = 0$

97. $H_n'(x) = 2nH_{n-1}(x)$

98. $H_n''(x) - 2xH_n'(x) + 2nH_n(x) = 0$

99. $\int_{-\infty}^{\infty} e^{-x^2} H_n(x)H_k(x)\,dx = 0, \ k \neq n$

100. $\int_{-\infty}^{\infty} e^{-x^2}[H_n(x)]^2\, dx = 2^n n! \sqrt{\pi}$

Laguerre Polynomials

101. $L_n(x) = \sum_{k=0}^{n} \dfrac{(-1)^k n! x^k}{(k!)^2 (n-k)!}$

102. $L_n(x) = \dfrac{e^x}{n!} \dfrac{d^n}{dx^n}(x^n e^{-x})$

103. $(1-t)^{-1} \exp\left[-\dfrac{xt}{1-t}\right] = \sum_{n=0}^{\infty} L_n(x) t^n$

104. $L_n(0) = 1$

105. $(n+1)L_{n+1}(x) + (x - 1 - 2n)L_n(x) + n L_{n-1}(x) = 0$

106. $L_n'(x) - L_{n-1}'(x) + L_{n-1}(x) = 0$

107. $x L_n'(x) = n L_n(x) - n L_{n-1}(x)$

108. $x L_n''(x) + (1-x) L_n'(x) + n L_n(x) = 0$

109. $\int_0^{\infty} e^{-x} L_n(x) L_k(x)\, dx = 0,\ k \neq n$

110. $\int_0^{\infty} e^{-x}[L_n(x)]^2\, dx = 1$

Associated Laguerre Polynomials

111. $L_n^{(m)}(x) = (-1)^m \dfrac{d^m}{dx^m}[L_{n+m}(x)],\ m = 1, 2, 3, \ldots$

112. $L_n^{(m)}(x) = \sum_{k=0}^{n} \dfrac{(-1)^k (m+n)! x^k}{(n-k)!(m+k)! k!}$

113. $L_n^{(m)}(x) = \dfrac{e^x}{n!} x^{-m} \dfrac{d^n}{dx^n}(x^{n+m} e^{-x})$

114. $(1-t)^{-m-1} \exp\left[-\dfrac{xt}{1-t}\right] = \sum_{n=0}^{\infty} L_n^{(m)}(x) t^n$

115. $L_n^{(m)}(0) = \dfrac{(n+m)!}{n! m!}$

116. $(n+1)L_{n+1}^{(m)}(x) + (x - 1 - 2n - m)L_n^{(m)}(x) + (n+m)L_{n-1}^{(m)}(x) = 0$

117. $L_{n-1}^{(m)}(x) + L_n^{(m-1)}(x) - L_n^{(m)}(x) = 0$

118. $L_n^{(m)\prime}(x) = -L_{n-1}^{(m+1)}(x)$

119. $xL_n^{(m)\prime\prime}(x) + (m + 1 - x)L_n^{(m)\prime}(x) + nL_n^{(m)}(x) = 0$

120. $\int_0^\infty e^{-x}x^m L_n^{(m)}(x)L_k^{(m)}(x)\,dx = 0,\ k \neq n$

121. $\int_0^\infty e^{-x}x^m [L_n^{(m)}(x)]^2\,dx = \dfrac{\Gamma(n + m + 1)}{n!}$

Gegenbauer Polynomials

122. $C_n^\lambda(x) = (-1)^n \sum\limits_{k=0}^{[n/2]} \binom{-\lambda}{n-k}\binom{n-k}{k}(2x)^{n-2k}$

123. $(1 - 2xt + t^2)^{-\lambda} = \sum\limits_{n=0}^\infty C_n^\lambda(x)t^n$

124. $C_n^\lambda(1) = (-1)^n \binom{-2\lambda}{n};\ C_n^\lambda(-1) = \binom{-2\lambda}{n}$

125. $C_{2n}^\lambda(0) = \binom{-\lambda}{n};\ C_{2n+1}^\lambda(0) = 0$

126. $(n + 1)C_{n+1}^\lambda(x) - 2(\lambda + n)xC_n^\lambda(x) + (2\lambda + n - 1)C_{n-1}^\lambda(x) = 0$

127. $C_n^{\lambda\prime}(x) = 2\lambda C_{n+1}^{\lambda+1}(x)$

128. $(1 - x^2)C_n^{\lambda\prime\prime}(x) - (2\lambda + 1)xC_n^{\lambda\prime}(x) + n(n + 2\lambda)C_n^\lambda(x) = 0$

129. $\int_{-1}^1 (1 - x^2)^{\lambda - \frac{1}{2}} C_n^\lambda(x)C_k^\lambda(x)\,dx = 0,\ k \neq n$

130. $\int_{-1}^1 (1 - x^2)^{\lambda - \frac{1}{2}} [C_n^\lambda(x)]^2\,dx = \dfrac{2^{1-2\lambda}\pi}{(n+\lambda)}\dfrac{\Gamma(n + 2\lambda)}{[\Gamma(\lambda)]^2 n!}$

Chebyshev Polynomials

131. $T_n(x) = \dfrac{n}{2}\sum\limits_{k=0}^{[n/2]} \dfrac{(-1)^k(n-k-1)!}{k!(n-2k)!}(2x)^{n-2k},\ n = 1, 2, 3, \ldots$

132. $U_n(x) = \sum\limits_{k=0}^{[n/2]} \binom{n-k}{k}(-1)^k(2x)^{n-2k}$

133. $\dfrac{1 - xt}{1 - 2xt + t^2} + \sum\limits_{n=0}^\infty T_n(x)t^n$

134. $(1 - 2xt + t^2)^{-1} = \sum\limits_{n=0}^\infty U_n(x)t^n$

135. $T_n(1) = 1$; $U_n(1) = n + 1$
136. $T_{2n}(0) = (-1)^n$; $T_{2n+1}(0) = 0$
137. $U_{2n}(0) = (-1)^n$; $U_{2n+1}(0) = 0$
138. $T_{n+1}(x) - 2xT_n(x) + T_{n-1}(x) = 0$
139. $U_{n+1}(x) - 2xU_n(x) + U_{n-1}(x) = 0$
140. $T_n(x) = U_n(x) - xU_{n-1}(x)$
141. $(1 - x^2)U_n(x) = xT_n(x) - T_{n+1}(x)$
142. $(1 - x^2)T_n''(x) - xT_n'(x) + n^2 T_n(x) = 0$
143. $(1 - x^2)U_n''(x) - 3xU_n'(x) + n(n+2)U_n(x) = 0$
144. $\int_{-1}^{1} (1 - x^2)^{-1/2} T_n(x) T_k(x)\, dx = 0,\ k \neq n$
145. $\int_{-1}^{1} (1 - x^2)^{1/2} U_n(x) U_k(x)\, dx = 0,\ k \neq n$
146. $\int_{-1}^{1} (1 - x^2)^{-1/2} [T_n(x)]^2\, dx = \begin{cases} \pi, & n = 0 \\ \dfrac{\pi}{2}, & n \geq 1 \end{cases}$
147. $\int_{-1}^{1} (1 - x^2)^{1/2} [U_n(x)]^2\, dx = \dfrac{\pi}{2}$

Bessel Functions of the First Kind

148. $J_p(x) = \sum_{k=0}^{\infty} \dfrac{(-1)^k (x/2)^{2k+p}}{k!\, \Gamma(k + p + 1)}$
149. $\exp\left[\tfrac{1}{2} x\left(t - \dfrac{1}{t}\right)\right] = \sum_{n=-\infty}^{\infty} J_n(x) t^n,\ t \neq 0$
150. $J_{-n}(x) = (-1)^n J_n(x),\ n = 0, 1, 2, \ldots$
151. $J_0(0) = 1$; $J_p(0) = 0,\ p > 0$
152. $\dfrac{d}{dx}[x^p J_p(x)] = x^p J_{p-1}(x)$
153. $\dfrac{d}{dx}[x^{-p} J_p(x)] = -x^{-p} J_{p+1}(x)$
154. $\left(\dfrac{d}{x\, dx}\right)^m [x^p J_p(x)] = x^{p-m} J_{p-m}(x)$
155. $\left(\dfrac{d}{x\, dx}\right)^m [x^{-p} J_p(x)] = (-1)^m x^{-p-m} J_{p+m}(x)$

156. $J_p'(x) + \dfrac{p}{x}J_p(x) = J_{p-1}(x)$

157. $J_p'(x) - \dfrac{p}{x}J_p(x) = -J_{p+1}(x)$

158. $J_{p-1}(x) + J_{p+1}(x) = \dfrac{2p}{x}J_p(x)$

159. $J_{p-1}(x) - J_{p+1}(x) = 2J_p'(x)$

160. $x^2 J_p''(x) + x J_p'(x) + (x^2 - p^2)J_p(x) = 0$

161. $J_n(x) = \dfrac{1}{\pi}\int_0^\pi \cos(n\phi - x\sin\phi)\,d\phi, \quad n = 0, 1, 2, \ldots$

162. $J_p(x) = \dfrac{(x/2)^p}{\sqrt{\pi}\,\Gamma(p + \frac{1}{2})}\int_{-1}^1 e^{ixt}(1 - t^2)^{p-\frac{1}{2}}\,dt, \quad p > -\frac{1}{2},\ x > 0$

163. $\int x^p J_{p-1}(x)\,dx = x^p J_p(x) + C$

164. $\int x^{-p} J_{p+1}(x)\,dx = -x^{-p} J_p(x) + C$

165. $\int_0^b x J_p(k_m x) J_p(k_n x)\,dx = 0,\ m \neq n;\ J_p(k_n b) = 0,\ n = 1, 2, 3, \ldots$

166. $\int_0^b x[J_p(k_n x)]^2\,dx = \tfrac{1}{2}b^2[J_{p+1}(k_n b)]^2;\ J_p(k_n b) = 0,\ n = 1, 2, 3, \ldots$

167. $J_p(x) \sim \dfrac{(x/2)^p}{\Gamma(p + 1)},\quad p \neq -1, -2, -3, \ldots,\ x \to 0^+$

168. $J_p(x) \sim \sqrt{\dfrac{2}{\pi x}}\cos[x - (p + \tfrac{1}{2})\pi/2],\ x \to \infty$

Bessel Functions of the Second Kind

169. $Y_p(x) = \dfrac{(\cos p\pi)J_p(x) - J_{-p}(x)}{\sin p\pi}$

170. $Y_0(x) = \dfrac{2}{\pi}J_0(x)\left[\log\left(\dfrac{x}{2}\right) + \gamma\right]$
$\quad - \dfrac{2}{\pi}\sum_{k=1}^\infty \dfrac{(-1)^k(x/2)^{2k}}{(k!)^2}\left(1 + \dfrac{1}{2} + \cdots + \dfrac{1}{k}\right)$

171. $Y_n(x) = \dfrac{2}{\pi}J_n(x)\log\left(\dfrac{x}{2}\right) - \sum_{k=0}^{n-1}\dfrac{(n - k - 1)!}{k!}(x/2)^{2k-n}$
$\quad - \dfrac{1}{\pi}\sum_{k=0}^\infty \dfrac{(-1)^k(x/2)^{2k+n}}{k!(k + n)!}[\psi(k + n + 1) + \psi(k + 1)],$
$n = 1, 2, 3, \ldots$

172. $Y_{-n}(x) = (-1)^n Y_n(x)$, $n = 0, 1, 2, \ldots$

173. $\dfrac{d}{dx}[x^p Y_p(x)] = x^p Y_{p-1}(x)$

174. $\dfrac{d}{dx}[x^{-p} Y_p(x)] = -x^{-p} Y_{p+1}(x)$

175. $Y_p'(x) + \dfrac{p}{x} Y_p(x) = Y_{p-1}(x)$

176. $Y_p'(x) - \dfrac{p}{x} Y_p(x) = -Y_{p+1}(x)$

177. $Y_{p-1}(x) + Y_{p+1}(x) = \dfrac{2p}{x} Y_p(x)$

178. $Y_{p-1}(x) - Y_{p+1}(x) = 2Y_p'(x)$

179. $x^2 Y_p''(x) + x Y_p'(x) + (x^2 - p^2) Y_p(x) = 0$

180. $Y_0(x) \sim \dfrac{2}{\pi} \log x$, $x \to 0^+$

181. $Y_p(x) \sim -\dfrac{\Gamma(p)}{\pi} \left(\dfrac{2}{x}\right)^p$, $p > 0$, $x \to 0^+$

182. $Y_p(x) \sim \sqrt{\dfrac{2}{\pi x}} \sin\left[x - \dfrac{(p + \frac{1}{2})\pi}{2}\right]$, $x \to \infty$

Modified Bessel Functions of the First Kind

183. $I_p(x) = \sum\limits_{k=0}^{\infty} \dfrac{(x/2)^{2k+p}}{k!\,\Gamma(k + p + 1)}$

184. $\exp\left[\tfrac{1}{2} x\left(t + \dfrac{1}{t}\right)\right] = \sum\limits_{n=-\infty}^{\infty} I_n(x) t^n$, $t \neq 0$

185. $I_{-n}(x) = I_n(x)$, $n = 0, 1, 2, \ldots$

186. $I_0(0) = 1$; $I_p(0) = 0$, $p > 0$

187. $\dfrac{d}{dx}[x^p I_p(x)] = x^p I_{p-1}(x)$

188. $\dfrac{d}{dx}[x^{-p} I_p(x)] = x^{-p} I_{p+1}(x)$

189. $I_p'(x) + \dfrac{p}{x} I_p(x) = I_{p-1}(x)$

190. $I_p'(x) - \dfrac{p}{x} I_p(x) = I_{p+1}(x)$

191. $I_{p-1}(x) + I_{p+1}(x) = 2 I_p'(x)$

192. $I_{p-1}(x) - I_{p+1}(x) = \dfrac{2p}{x} I_p(x)$

193. $x^2 I_p''(x) + x I_p'(x) - (x^2 + p^2) I_p(x) = 0$

194. $I_0(x) = \dfrac{1}{\pi} \displaystyle\int_0^\pi e^{\pm x\cos\theta}\, d\theta$

195. $I_p(x) = \dfrac{(x/2)^p}{\sqrt{\pi}\,\Gamma(p+\frac{1}{2})} \displaystyle\int_{-1}^{1} e^{-xt}(1-t^2)^{p-\frac{1}{2}}\, dt,\quad p > -\frac{1}{2},\ x > 0$

196. $I_p(x) \sim \dfrac{(x/2)^p}{\Gamma(p+1)},\quad p \neq -1,-2,-3,\ldots,\ x \to 0^+$

197. $I_p(x) \sim \dfrac{e^x}{\sqrt{2\pi x}},\quad x \to \infty$

Modified Bessel Functions of the Second Kind

198. $K_p(x) = \dfrac{\pi}{2}\, \dfrac{I_{-p}(x) - I_p(x)}{\sin p\pi}$

199. $K_0(x) = -I_0(x)\left[\log\left(\dfrac{x}{2}\right) + \gamma\right]$
$\quad + \displaystyle\sum_{k=1}^{\infty} \dfrac{(x/2)^{2k}}{(k!)^2}\left(1 + \dfrac{1}{2} + \cdots + \dfrac{1}{k}\right)$

200. $K_n(x) = (-1)^{n-1} I_n(x) \log\left(\dfrac{x}{2}\right)$
$\quad + \dfrac{1}{2}\displaystyle\sum_{k=0}^{n-1} \dfrac{(-1)^k (n-k-1)!}{k!}\left(\dfrac{x}{2}\right)^{2k-n}$
$\quad + \dfrac{(-1)^n}{2}\displaystyle\sum_{k=0}^{\infty}\dfrac{(x/2)^{2k+n}}{k!(k+n)!}[\psi(k+n+1) + \psi(k+1)],$
$n = 1, 2, 3, \ldots$

201. $K_{-p}(x) = K_p(x)$

202. $\dfrac{d}{dx}[x^p K_p(x)] = -x^p K_{p-1}(x)$

203. $\dfrac{d}{dx}[x^{-p} K_p(x)] = -x^{-p} K_{p+1}(x)$

204. $K_p'(x) + \dfrac{p}{x} K_p(x) = -K_{p-1}(x)$

205. $K_p'(x) - \dfrac{p}{x} K_p(x) = -K_{p+1}(x)$

206. $K_{p-1}(x) + K_{p+1}(x) = -2 K_p'(x)$

207. $K_{p-1}(x) - K_{p+1}(x) = -\dfrac{2p}{x} K_p(x)$

208. $x^2 K_p''(x) + x K_p'(x) - (x^2 + p^2) K_p(x) = 0$

209. $K_p(x) = \dfrac{\sqrt{\pi}\,(x/2)^p}{\Gamma(p+\frac{1}{2})} \int_1^\infty e^{-xt}(t^2-1)^{p-\frac{1}{2}}\,dt,\ p > -\frac{1}{2},\ x > 0$

210. $K_0(x) \sim -\log x,\ x \to 0^+$

211. $K_p(x) \sim \dfrac{\Gamma(p)}{2}\left(\dfrac{2}{x}\right)^p,\ p > 0,\ x \to 0^+$

212. $K_p(x) \sim \sqrt{\dfrac{\pi}{2x}}\,e^{-x},\ x \to \infty$

Hypergeometric Function

2.13. $F(a,b;c;x) = \sum_{n=0}^{\infty} \dfrac{(a)_n (b)_n}{(c)_n}\dfrac{x^n}{n!},\ |x| < 1$

214. $F(a,b;c;x) = F(b,a;c;x)$

215. $\dfrac{d^k}{dx^k} F(a,b;c;x) = \dfrac{(a)_k (b)_k}{(c)_k} F(a+k, b+k; c+k; x),\ k = 1,2,3,\ldots$

216. $F(a,b;c;x) = \dfrac{\Gamma(c)}{\Gamma(b)\Gamma(c-b)} \int_0^1 t^{b-1}(1-t)^{c-b-1}(1-xt)^{-a}\,dt,\ c > b > 0$

217. $F(a,b;c;1) = \dfrac{\Gamma(c)\Gamma(c-a-b)}{\Gamma(c-a)\Gamma(c-b)}$

Confluent Hypergeometric Function

218. $M(a;c;x) = \sum_{n=0}^{\infty} \dfrac{(a)_n}{(c)_n}\dfrac{x^n}{n!},\ -\infty < x < \infty$

219. $M(a;c;x) = e^x M(c-a;c;-x)$

220. $\dfrac{d^k}{dx^k} M(a;c;x) = \dfrac{(a)_k}{(c)_k} M(a+k; c+k; x),\ k = 1,2,3,\ldots$

221. $M(a;c;x) = \dfrac{\Gamma(c)}{\Gamma(a)\Gamma(c-a)} \int_0^1 e^{xt} t^{a-1}(1-t)^{c-a-1}\,dt,\ c > a > 0$

222. $M(a;c;x) \sim 1,\ x \to 0$

223. $M(a;c;x) \sim \dfrac{\Gamma(c)}{\Gamma(a)} x^{a-c} e^x \sum_{n=0}^{\infty} \dfrac{(1-a)_n (c-a)_n}{n!\,x^n},\ x \to \infty$

Selected Answers to Exercises

Chapter 1

EXERCISES 1.2

3. 1   **7.** 2121/999   **9.** converges absolutely

12. converges conditionally

EXERCISES 1.3

3. converges only for $x = 0$ **6.** converges uniformly

18. $\sum_{n=1}^{\infty} (-1)^{n+1} \dfrac{2^{2n-1} x^{2n}}{(2n)!}$ **20.** $1 - \frac{3}{2}x^2 + \frac{21}{24}x^4 - \frac{61}{240}x^6 + \cdots$

EXERCISES 1.5

3. $\dfrac{\pi}{2} - \dfrac{4}{\pi} \sum_{n=1}^{\infty} \dfrac{\cos(2n+1)x}{(2n+1)^2}$ **11.** (a) $\frac{3}{4}$ (b) $\frac{3}{4}$ (c) $\frac{3}{4}$

Chapter 2

EXERCISES 2.2

3. 60 **7.** $\frac{10}{9}$ **31.** $\pi\sqrt{2}/4$

35. $\dfrac{\Gamma(x/p)}{ps^{x/p}}$ **46.** $\dfrac{[\Gamma(\frac{1}{4})]^2}{2\sqrt{\pi}}$

EXERCISES 2.3

1. $2\pi/\sqrt{3}$ **11.** $\pi/8$ **14.** π **17.** $\pi a^6/16$

Chapter 3

EXERCISES 3.3

15. $\text{Si}(b) - \text{Si}(a)$

EXERCISES 3.4

11. $12E(1/\sqrt{3})$

Chapter 4

EXERCISES 4.2

15. (a) $n = 1, 3, 5, \ldots$ (b) $n = 0, 2, 4, \ldots$

EXERCISES 4.4

27. $x^3 = \frac{3}{5}P_1(x) + \frac{2}{5}P_3(x)$

36. $f(x) = \displaystyle\sum_{n=0}^{\infty} \dfrac{(-1)^{n+1}(4n+1)(2n-2)!}{2^{2n}(n+1)!(n-1)!} P_{2n}(x)$

EXERCISES 4.5

1. none of these **5.** smooth

EXERCISES 4.6

3. $y = C_1 P_3(x) + C_2 Q_3(x)$

Chapter 6

EXERCISES 6.4

4. $f(x) = 2 \sum_{n=0}^{\infty} \dfrac{J_0(k_n x)}{k_n J_1(k_n)}$

EXERCISES 6.5

3. $y = C_1 J_{1/4}(x) + C_2 J_{-1/4}(x)$

EXERCISES 6.6

3. $y = C_1 J_k(x^2) + C_2 Y_k(x^2)$ 5. $y = x^{-n}[C_1 J_n(x) + C_2 Y_n(x)]$
8. $y = x^{1/2}[C_1 J_{1/2}(x) + C_2 J_{-1/2}(x)]$
11. $y = x^{1/2}[C_1 J_0(2x^{1/2}) + C_2 Y_0(2x^{1/2})]$
13. (b) $y(t) = C_1 J_0(2\sqrt{a/m}\, t) + C_2 Y_0(2\sqrt{a/m}\, t)$
    (c) $y(x) = C_1 J_0(2\sqrt{a/m}\, e^{mx}) + C_2 Y_0(2\sqrt{a/m}\, e^{mx})$

EXERCISES 6.7

2. $y = C_1 I_1(x) + C_2 K_1(x)$
25. $y = x^{1/2}[C_1 I_{1/3}(\tfrac{2}{3}x^{3/2}) + C_2 K_{1/3}(\tfrac{2}{3}x^{3/2})]$
27. $y = x^2[C_1 I_{2/3}(x^3) + C_2 K_{2/3}(x^3)]$

Chapter 7

EXERCISES 7.2

1. (a) $u(r,\phi) = 1$ (b) $u(r,\phi) = r\cos\phi$
   (c) $u(r,\phi) = \tfrac{2}{3}r^2 P_2(\cos\phi) + \tfrac{1}{3}$ (d) $u(r,\phi) = \tfrac{4}{3}r^2 P_2(\cos\phi) - \tfrac{1}{3}$
8. $u(r,\phi) = \tfrac{2}{3}T_0[1 - (r/c)^2 P_2(\cos\phi)]$

EXERCISES 7.3

8. $u(r,z) = \dfrac{I_0(3r)}{I_0(3)} \sin 3z$

Index

Abel, N. H., 92
Abel's formula, 226
Abel's theorem, 19
Abramowitz, M., 207
Absolute convergence, 5, 39, 40
Addition theorems for Bessel functions, 201
   confluent hypergeometric functions, 307
   Hermite polynomials, 176
   Laguerre polynomials, 181
Alternating harmonic series, 8, 18
   series test, 6
Airy's equation, 230
Andrews, L. C., 155, 215, 228, 280, 296, 320
Anger function, 247
Appel's symbol, 273
Arfken, G., 248
Associated Laguerre polynomials, 179–182, 293–294, 296, 312, 325, 332
   addition formula for, 181
   definition of, 179
   differential equation for, 181
   generating function for, 180
   orthogonality of, 183
   recurrence relations for, 180
   relation to confluent hypergeometric function, 312
   Rodrigues's formula for, 180
   series representation of, 180
Associated Legendre functions, 160–165, 259–261, 290
   definition of, 162
   differential equation for, 160, 259–260
   generating function for, 165
   orthogonality of, 164
   recurrence relations for, 163
   relation to hypergeometric function, 290
   Rodrigues's formula for, 162
Asymptotic series, 26–31, 72, 78–80, 82, 90, 95, 103–104, 107, 248–251, 303–304
   about infinity, 27–31
   zero, 27
   definition of, 27–28
   for Bessel functions, 248–251, 314
   confluent hypergeometric functions, 303–304
   digamma function, 78–80
   error functions, 95, 314
   exponential integrals, 104, 107, 314
   Fresnel integrals, 103
   gamma function, 82
   incomplete gamma function, 72
   logarithmic integral, 107, 314
   polygamma functions, 90

Bateman, H., 326
Bell, W. W., 156, 281
Bernfeld, M., 204
Bernoulli, J., 79, 195

Bernoulli numbers, 79
  relation to zeta function, 88
Bessel, F. W., 195
Bessel functions, 44, 73, 184, 195–251, 265–271, 297, 312–315, 323, 326–329, 331–332
  Anger, 247
  asymptotic formulas for, 248–251, 314
  Hankel, 224–225, 234–236, 242
  integral, 246
  Kelvin, 247
  modified, 232–241, 268–271, 308, 312, 314, 318, 326, 328, 332
  of the first kind, 44, 73, 184, 196–222, 224, 226–232, 236, 248–251
    addition theorem for, 201
    argument zero, 197
    differential equation for, 200, 216, 221, 228, 265, 318
    generating function for, 196–197
    graph of, 198
    half-integral order, 203
    infinite series of, 215–220, 267
    integral representations of, 205–207, 211–212
    integrals of, 207–214
    Jacobi-Anger expansion for, 201
    Lommel's formula for, 203
    negative order, 197–198
    orthogonality of, 215
    recurrence relations for, 198–200
    relation to confluent hypergeometric function, 312
    series representation of, 197–198
    Weber's integral formula for, 210
  of the second kind, 220–228
    definition of, 221
    graph of, 225
    recurrence formulas for, 225–226
    series representation of, 221–224
  of the third kind, 224–225
  spherical, 241–248, 260–261
  Struve, 246
  Weber, 247
  Wronskians of, 226–227, 237–238, 245–246
Bessel polynomials, 326
Bessel series, 215–220, 267
  convergence of, 218
Beta function, 66–71, 206, 279, 291
  definition of, 66–67
  incomplete, 70–71, 291
  relation to gamma function, 67
  symmetry property of, 67

Binomial coefficients, 10–13
Binomial series, 13, 22, 26, 118, 209, 279
Boundary-value problems, 252–271
  circular domains, 264–271
  cylindrical domains, 264–271
  of the first kind, 254
  spherical domains, 253–264

Campbell, R., 52
Cauchy-Euler equation, 255, 262
Cauchy product, 10, 23
Chebyshev polynomials, 128, 187–189, 193, 290
  argument negative one, 193–194
    unity, 193–194
    zero, 193–194
  differential equations for, 188
  generating functions for, 128, 193
  orthogonality of, 188
  relation to hypergeometric function, 290
  series representation of, 187
Christoffel-Darboux lemma, 151
Comparison tests, 6
Confluent hypergeometric function, 298–320, 323, 325, 328, 332
  addition formulas for, 303–304
  as limit of hypergeometric function, 299
  asymptotic formulas for, 303–304
  contiguous function relations for, 300, 305
  convergence of series for, 299
  derivatives of, 299–300, 305
  differential equation for, 301, 306, 316
    Liouville standard form of, 317
  integral representation of, 300–301, 306
  Kummer's transformation for, 301
  multiplication formulas for, 307
  of the second kind, 301–303, 307–315
    asymptotic formulas for, 304, 307
    contiguous relations for, 307
    derivatives of, 303, 306
    integral representations of, 303
    Kummer relation for, 306
    relation to other functions, 308–315
  relation to other functions, 308–315
  series representation of, 299
  Whittaker functions, 315–320
  Wronskian of, 306
Contiguous function relations, 278, 282–283, 300, 305, 307
  confluent hypergeometric functions, 300, 307
  hypergeometric function, 278, 282–283

Convergence, 2, 4–7, 15, 34–35, 39–41
  absolute, 5, 39–40
  conditional, 5, 39–40
  pointwise, 15, 34, 40–43, 148
  tests of, 4–7, 16, 39–41
  uniform, 15–19, 35, 40–43
Cook, C. E., 204
Cosine integral, 106, 108, 313
Cylinder function (*see* Bessel functions)

Debnath, L., 61
Differential equation,
  Airy, 230
  associated Laguerre, 181
  associated Legendre, 160, 259–260
  Bessel, 200, 216, 221, 228, 265, 318
    Liouville standard form of, 318
    second solution of, 221
  Cauchy-Euler, 259, 262
  Chebyshev, 188
  confluent hypergeometric, 301, 306, 316
    Liouville standard form of, 317
    second solution of, 301–302
  Gegenbauer, 186
  heat, 96, 101, 253, 263
  Helmholtz, 241, 260
  Hermite, 170, 173, 310
  hypergeometric, 280, 285
    second solution of, 281, 285
  Jacobi, 191
  Laguerre, 178
  Laplace, 253, 257, 268
  Legendre, 126, 129, 135, 137, 158–159, 255
  modified Bessel, 233, 268, 318
  related to Bessel's equation, 228–232, 240
  spherical Bessel, 241, 260–261
Digamma function, 74–91, 222–224
  argument positive integer, 76
    zero, 75
  asymptotic series for, 78–80
  definition of, 74
  graph of, 75
  integral representations of, 77–78
  recurrence formula for, 76
  series representation of, 74, 84
Dirichlet problem, 254
Double infinite series, 9, 259
Duplication formula, 58

E-function, 330–331
Elliptic functions, 112–115

Elliptic integrals, 108–115, 289–290
  complete, 109, 289–290
    limiting values of, 110
    relations to hypergeometric function, 289–290
    series representation of, 110
  definition of, 109
Erdelyi, A., 278
Error function, 93–102, 308–309, 312–314, 318–319
  argument infinite, 93, 314
    negative, 93
    zero, 93
  complementary, 93, 95, 308–309, 311, 314, 319
    asymptotic series for, 95, 314
    relation to confluent hypergeometric function, 309, 318
  definition of, 93
  derivative of, 99
  graph of, 94
  integral of, 101
  relation to confluent hypergeometric function, 312, 318
    Fresnel integrals, 99
  series representation of, 93
Euler, L., 50, 92, 195, 272
Euler formulas, 24, 65
Euler-Mascheroni constant, 59, 74, 88–89, 223
Euler product for gamma function, 65
Even function, 35–36
Exponential integral, 103–104, 106–108, 308, 312, 314
  asymptotic formula for, 104, 107
  definition of, 103
  graph of, 104
  relation to confluent hypergeometric function, 312
    incomplete gamma function, 103
  series representation of, 104

Fourier coefficients, 32, 35, 142, 171, 179, 218
Fourier trigonometric series, 32–37, 49, 269
  convergence of, 34–35
Fox, L., 189
Fractional-order derivatives, 61–62
Fresnel integrals, 97–99, 102–103, 312–313
  asymptotic series for, 103
  definition of, 97
  graph of, 98

relation to confluent hypergeometric
    function, 312–313
  error function, 99
  series representation of, 102
Frink, O., 326
Frullani integral representation, 77

G-function (see Meijer G-function)
Gamma function, 50–67, 71–73, 81–82, 99,
    103, 312, 318
  argument infinite, 55
    negative, 51, 56, 59, 65
    negative integer, 51
    odd half-integer, 59
    one-half, 58
    positive integer, 52
    zero, 54, 56
  asymptotic series for, 82
  definition of, 51, 53, 59
  derivatives of, 54
  duplication formula for, 58
  graph of, 55
  incomplete, 71–73, 99, 103, 312, 318
  infinite product for, 59, 65
  integral representations of, 53–58, 63
  minimum value of, 55
  recurrence formula for, 52
  relation to beta function, 67
  Stirling's formula for, 81–82
Gauss, C., 53, 55, 92, 272, 278, 298
Gegenbauer polynomials, 185–187, 190–193,
    290
  differential equation for, 186, 192
  generating function for, 185
  orthogonality of, 186
  recurrence relations for, 185–186
  relation to Chebyshev polynomials, 187
    Hermite polynomials, 186–187
    hypergeometric function, 290
    Jacobi polynomials, 191
    Legendre polynomials, 186
  series representation of, 185
Generalized hypergeometric functions,
    321–332
  convergence of series for, 322
  derivative of, 324
  MacRobert E-function, 330–331
  Meijer G-function, 327–330
  relations to other functions, 323–327
Generating functions for
  associated Laguerre polynomials, 180
  associated Legendre functions, 165
  Bessel functions, 196, 239

Chebyshev polynomials, 128, 193
Gegenbauer polynomials, 185
Hermite polynomials, 167, 174
hypergeometric polynomials, 284
Jacobi polynomials, 189
Laguerre polynomials, 176, 180
Legendre polynomials, 117–119, 135, 285
Geometric series, 3–4, 13, 15, 26, 276
Grosswald, E., 326

Hankel functions, 224–225, 234–236, 242
Harmonic oscillator, 172–173
Harmonic series, 7
Heat equation, 96, 101, 253, 263
Helmholtz equation, 241, 260
Hermite functions, 310–311, 314–315
  recurrence relations for, 314
  series representation of, 314
Hermite polynomials, 167–176, 183, 186,
    308–310, 312, 319
  addition formula for, 176
  argument zero, 173
  differential equation for, 170, 173, 310
  generating function for, 167, 174
  infinite series of, 170–172
  integral representations of, 168, 174
  orthogonality of, 170
  recurrence relations for, 169–170
  relation to confluent hypergeometric
    function, 308–309, 312
      Gegenbauer polynomials, 186
      Laguerre polynomials, 183
      Legendre polynomials, 168
  Rodrigues's formula for, 168
  series representation of, 167
  table of, 168
Hermite series, 170–172
  convergence of, 171
Hypergeometric functions, 272–297, 299,
    323, 325, 328, 332
  argument negative one, 284
    one-half, 284
    unity, 280, 283–284
  contiguous function relations for, 278,
    282–283
  convergence of series for, 276
  derivatives of, 277–278, 282
  differential equation for, 280, 285
  integral representations of, 279, 283
  of the second kind, 285
  relation to other functions, 285–292
  symmetry property of, 277

Wronskian of, 285
(*see also* generalized hypergeometric functions)
Hypergeometric polynomials, 277, 284
  generating function for, 284
Hypergeometric series, 276, 322
  generalized, 322

Improper integrals, 38–44
  convergence of, 38–43
  differentiation of, 42
  divergence of, 38–43
  integration of, 42
  partial integral of, 41
  Weierstrass M-test for, 41
Incomplete gamma function, 71–73, 99, 103, 312, 318
  complementary, 71
    asymptotic series for, 72
  relation to confluent hypergeometric function, 312, 318
  series representation of, 71
Infinite product, 45–49
  associated series, 46
  convergence of, 45, 47
  divergence of, 45
  for cosine function, 59, 65
    gamma function, 59, 65
    sine function, 48, 49, 60, 88
  partial product of, 45
Infinite sequence, 2
Infinite series, 2–37
  alternating, 5
    harmonic, 8, 18
    series test, 6
  asymptotic, 26–31, 72, 78–80, 95, 103–104, 107, 248–251, 303–304
  binomial, 22, 26, 118, 209, 279
  confluent hypergeometric, 299
  convergence of, 2, 4–7, 19, 34
  differentiation of, 18
  divergence of, 3, 4–7
  double, 9, 259
  Fourier, 32–37, 49, 269
  geometric, 3–4, 13, 15, 26, 276
  harmonic, 7
  hypergeometric, 276, 322
  integration of, 17
  of constants, 1–4
    functions, 15–26, 32–37
  operations with, 7–10, 23–25
  partial sum of, 2, 15, 149, 154
  p-series, 7

  positive, 5
  power, 19–26
  product of, 9–10
  uniqueness of, 22
  Weierstrass M-test for, 16
Integral Bessel function, 246
Integral test for series, 7

Jackson, D., 153
Jacobi-Anger expansion, 202
Jacobian elliptic functions, 112–115
Jacobi, C. G. J., 92
Jacobi polynomials, 189–191, 194, 291
  differential equation for, 191
  generating function for, 189
  orthogonality of, 191
  recurrence formula for, 191
  relation to associated Laguerre polynomials, 190
    Gegenbauer polynomials, 191
    hypergeometric function, 291
    Legendre polynomials, 189
  series representation of, 189
Jordan inequality, 135

Kemble, E. C., 173
Kellogg, O. D., 117
Kelvin functions, 247
Krall, H. L., 326
Kummer, E. E., 298
Kummer's function (*see* confluent hypergeometric functions)
Kummer's relation, 306
Kummer's second formula, 325
Kummer's transformation, 301

Laguerre functions, 311–312, 315
  recurrence relations for, 315
  series representation of, 315
Laguerre polynomials, 176–184, 308, 312
  argument zero, 182
  associated, 179–182, 293–294, 296, 312, 325, 332
  differential equation for, 178
  generating function for, 176
  infinite series of, 179
  orthogonality of, 178
  recurrence relations for, 177–178
  relation to confluent hypergeometric function, 312
    Hermite polynomials, 183
  Rodrigues's formula for, 176

series representation of, 176
  table of, 177
Laguerre series, 178–179
  convergence of, 179
Landen, J., 92
Laplacian, 253, 262–263
Laplace integral formula, 132
Laplace's equation, 253–254, 257, 268
Lebedev, N. N., 29, 98, 106, 171, 237, 289
Lee, J. S., 320
Legendre, A., 50, 53, 92, 117
Legendre duplication formula, 58
Legendre functions, 155–165, 288–289, 292
  associated, 160–165, 259–261, 290
  of the first kind, 288–289, 292
  of the second kind, 155–160, 288
    graph of, 157
  recurrence relations for, 289
  Wronskian of, 158
Legendre polynomials, 44, 116–165, 186, 204, 214, 255, 285–287, 313
  argument minus one, 122
    negative, 120
    unity, 122
    zero, 123
  bounds on, 133–134
  differential equation for, 126, 129, 135, 137, 158–159, 255
  finite series of, 139–141
  generating function for, 117–119, 135, 285
  graph of, 121–122
  infinite series of, 142–143, 147, 256
    convergence of, 147–154
  integral representation of, 132
  orthogonality of, 137, 145
  recurrence relations for, 124–126
  relation to Gegenbauer polynomials, 186
    Hermite polynomials, 168
    hypergeometric function, 286
    Jacobi polynomials, 189
  Rodrigues's formula for, 132
  series representation of, 119–120, 131, 135
  tables of, 120–121
  trigonometric argument, 121
Legendre series, 137–154, 256
  convergence of, 147–154
Leibniz formula, 160, 177
Liouville standard form, 315–318
  for Bessel's equation, 318
  confluent hypergeometric equation, 317
Logarithmic derivative function (*see* digamma function)

Logarithmic integral, 105, 107, 312, 314
  asymptotic formula for, 107, 314
  relation to confluent hypergeometric function, 312
Lommel's formula, 203

Macdonald's function, 233
MacRobert E-function, 330–331
MacRobert, T. M., 288, 327, 330
Mathai, A. M., 322
Meijer, C. S., 327
Meijer G-function, 327–330
  definition of, 327
  relation to other functions, 328, 331–332
Modified Bessel function, 232–241, 248–251, 268, 308, 312–315, 318
  asymptotic formulas for, 248–251, 314
  differential equation for, 233, 268, 318
  generating function for, 239
  graphs of, 235
  half-integral order, 238
  integral representations of, 236–237
  of the first kind, 233
    second kind, 233
  recurrence relations for, 236
  relation to confluent hypergeometric function, 312, 318
  series representation of, 233–234
  spherical, 245–246
  Wronskian of, 237–238
Multiplication formulas, 307

Neuman formula, 158
Normal form, 315

Odd function, 35–36
Olver, R. W. J., 28
Orthogonal polynomials, 137, 166, 170, 178, 186, 188, 191

P-series, 7
Parabolic cylinder function, 319
Parker, I. B., 189
Partial integral, 41
  product, 45
  sum, 2, 15, 149, 154
Phillips, R. L., 296
Piecewise continuous function, 34, 148–149
  smooth function, 149
Pochhammer symbol, 273–276
Poincaré, J. H., 28
Pointwise convergence, 15, 34, 40–43, 148
Poisson, S. D., 206

Polygamma functions, 83–85, 90–91
  argument unity, 83
  asymptotic series for, 90
  definition of, 83
  series representation of, 83–84
  (*see also* digamma function)
Power formula for series, 25
Powers, D., 97
Product (*see* infinite product)
Psi function (*see* digamma function)

Rainville, E. D., 47, 51, 276, 279
Ramanujan's theorem, 325
Ratio test, 6
Riemann, G., 51
Riemann lemma, 149
Riemann zeta function, 83–91
  definition of, 85
  graph of, 86
  infinite product representation of, 86
  integral representation of, 86, 89
  relation to Bernoulli numbers, 88
  special values of, 89

Saxena, R. K., 322
Semiconvergent series (*see* asymptotic series)
Separation of variables, 254, 257–258, 260, 265, 268
Sequence, 2
Series (*see* infinite series)
Sine integral, 106, 108, 313
Smooth function, 149
Spherical Bessel functions, 241–251, 260–261
  asymptotic formulas for, 249, 251
  definition of, 242
  differential equation for, 241, 260
  graphs of, 244
  Hankel, 242
  modified, 245–246
  recurrence relations for, 242
  relation to trigonometric functions, 243
  series representation of, 243
  Wronskian of, 245
Springer, M. D., 330

Stegun, I., 207
Stirling, J., 81
Stirling's formula, 81–82
Stirling's series, 82
Struve function, 246
Superposition principle, 255

Taylor series, 21
Tchebysheff (*see* Chebyshev)
Tests of convergence for
  improper integrals, 39–41
  infinite series, 4–7, 16

Ultraspherical polynomials (*see* Gegenbauer polynomials)
Uniform convergence, 15–19, 35, 41–43

Wallis's formula, 48
Watson, G. N., 86, 157
Wave equation, 263, 264
Weber function, 247–248
Weber's integral formula, 210
Webster, A. G., 260
Weierstrass infinite product, 59, 74
Weierstrass, K., 50, 59, 61
Weierstrass M-test for
  improper integrals, 41
  infinite series, 16
Whittaker, E. T., 86, 157, 298
Whittaker functions, 315–320, 332
  asymptotic formulas for, 319
  definitions of, 317
  relations to other functions, 318–320
  Wronskian of, 320
Wronskian of
  Bessel functions, 226–227
  confluent hypergeometric functions, 306
  hypergeometric functions, 285
  Legendre functions, 158
  modified Bessel functions, 237–238
  spherical Bessel functions, 245–246
  Whittaker functions, 320

Zeta function (*see* Riemann zeta function)